# The Development of MULTIPLICATIVE REASONING in the Learning of Mathematics

edited by
GUERSHON HAREL
and
JERE CONFREY

State University of New York Press

Published by
State University of New York Press, Albany

For information, address the State University of New York Press,
State University Plaza, Albany, NY 12246

Production by Bernadine Dawes
Marketing by Bernadette LaManna

Library of Congress Cataloging-in-Publication Data

The Development of multiplicative reasoning in the learning of
    mathematics / edited by Guershon Harel and Jere Confrey.
        p.      cm. — (SUNY series, reform in mathematics education)
    Includes index.
    ISBN 0-7914-1763-8 (acid-free) : $73.50. — ISBN 0-7914-1764-6
(pbk. : acid-free) : $24.95
        1. Mathematics—Study and teaching (Elementary)
    2. Multiplication.    3. Division.    I. Harel, Guershon, 1952–    .
    II. Confrey, Jere.    III. Series.
    QA135.5.D49   1994
    372.7'2044—dc20                                              93-19443
                                                                      CIP

1 2 3 4 5 6 7 8 9 10

# CONTENTS

# INTRODUCTION

## Jere Confrey
## Guershon Harel

### HISTORICAL PRECEDENTS

The interest in compiling a book of recent research on the development of multiplicative concepts evolved from a variety of efforts. Research on multiplication and division revived in the early 1980s with the work of Fischbein, Deri, Nello, and Marino (1985) who hypothesized that "Each fundamental operation of arithmetic generally remains linked to an implicit, unconscious, and primitive intuitive model" and argued that these models impose constraints on students' predictions of the operation needed when solving multiplication and division with different decimal numbers. They conjectured that the primitive intuitive model for multiplication was repeated addition and for division was based in either partitioning (sharing) or repeated subtraction. This work built on and was further complemented by a line of research by Bell, Swan, and Taylor (1981) and subsequently by Bell, Fischbein, and Greer (1984), Luke (1988), and Graeber and Tirosh (1988)— all of whom investigated how these models would lead to the assumption that multiplication makes bigger and division makes smaller. Recent work has examined further the impact of numeric form and value on students' and teachers' selection of operation and is reported in this book (Harel, Behr, Post, and Lesh, Chapter 10). An issue of *The Journal of Mathematical Behavior* (vol. 7, no. 3, December 1988) with guest editor B. Greer was devoted to the topic.

In 1983, in the *Acquisition of Mathematics Concepts and Processes* (R. Lesh and M. Landau, editors), the research activities of two groups set the stage for a second wave of research in this arena. Vergnaud introduced to an American audience the idea that a conceptual field is a "set of problems and situations for the treatment of which concepts, procedures and representations of different but narrowly interconnected types are necessary" (p. 127). He discussed the treatment of a multiplicative conceptual field as an example and identified the broad strands of this conceptual field to include: multiplication, division, fractions, ratio, rational number, linear and $n$-linear functions, dimensional analysis, and vector spaces.

In the same volume, results from the Rational Number Project were reported by Behr, Lesh, Post, and Silver. By reviewing and analyzing the seminal work of such scholars as Dienes, Karplus, Kieran, and Hart, they offered a synthesis of the field and identified six "subconstructs" of rational number: part to whole comparison, decimal, ratio, indicated division (quotient), operator, and measure of continuous or discrete quantities. These two papers established a common theme. It was no longer sufficient to analyze the cognitive development of these ideas in isolation. There was a recognition that the ideas were interwoven into a field of related concepts, whose acquisition would not be linear or piece by piece. As with a spider's web, contact with one strand would reverberate across the entire space.

In 1986, a series of conferences were held reviewing significant bodies of research under National Science Foundation's Research Agenda Project for Mathematics Education (Sowder, 1988). One of these groups produced *Number Concepts and Operations in the Middle Grades* ( J. Hiebert and M. Behr, editors). The majority of these papers treated topics from the multiplicative conceptual field. This collection provided a case for rethinking the development of the number concept through the grades. As the editors put it:

> Mastery of many of the number concepts and number relationships in the middle grades appears to require a reconceptualization of number, a significant change from the primary grades in the way number is conceived. . . . Given the fundamental changes in the nature of number . . . , it is not surprising that significant cognitive reorientations are needed to construct and comprehend such changes. This means that it is likely that there are not smooth continuous paths from early addition and subtraction to multiplication and division, nor from whole numbers to rational numbers. Multiplication is not simply repeated addition, and rational numbers are not simply ordered pairs of whole numbers. The new concepts are not the sums of previous ones. Com-

petence with middle school number concepts requires a break with simpler concepts of the past, and a reconceptualization of number itself. (Hiebert and Behr, 1988, p. 8)

In addition to work by the researchers just cited, the book included such work as that conducted by Nesher, which introduced an instructional dimension into the discussion of models of multiplication. Nesher made distinctions between "mapping rule" multiplication (three books per shelf and four shelves), multiplicative compare problems (three times as many this as that), and Cartesian multiplication (four shirts, three pants, how many outfits?). She connected these distinctions to distinctions made by Schwartz (1988) between different types of quantity and multiplication of these various types of quantity by each other: intensive by extensive, scalar by extensive, and extensive by extensive. Nesher reported that students' success in solving multiplication problems seemed heavily dependent on the instructional setting and linguistic cues.

A follow-up conference to the Research Agenda Project meeting was held in San Diego, and a subgroup was formed to pursue further the ideas of the "multiplicative conceptual field." With support from the National Center for Research in Mathematical Sciences Education and the National Science Foundation, this group met four times over the period of 1989–1990, chaired by the editors of this book. The chapters in this volume are, for the most part, a result of those meetings and discussions. A few were solicited separately when their authors were unable to attend the meetings. The editors wish to acknowledge the contributions of Kenneth Carlstrom to publication of this book. His careful reviews and comments contributed significantly to improvement of the manuscript. Support for these meetings and manuscript preparation came from NSF grant MDR 9053590.

## SHARED COMMITMENTS

The contributors to this monograph share four basic commitments:

1. *The topics included under the rubric "multiplicative conceptual field" (MCF), possess an interconnectedness and complexity that presents a unique research challenge.* This

sense of connectedness is repeatedly expressed throughout the volume: "[R]esearchers should study conceptual fields and not isolated situations or isolated concepts" (Vergnaud, p. 46); "The domain represents a critical juncture at which many types of mathematical knowledge are called into play, and a point beyond which a student's understanding in the mathematical sciences will be greatly hampered if the conceptual coordination of all the contributing domains is not attained." (Lamon, p. 90);

2. *The ideas of the MCF develop over considerable time periods, and new topics require the reconsideration of old topics as one develops a mature understanding of the field.* A developmental approach makes it essential for researchers of elementary, middle and secondary levels to discuss, debate and share their insights. "[O]ne of the goals of current research is to identify important mathematical processes, themes, or connections by which thinking becomes progressively more sophisticated from early childhood through early adulthood" (Lamon, p. 90). "This chapter is concerned with the extension of meaning of multiplication and division from their early conceptualizations" (Greer, p. 61). "Such guidance, necessarily, would have to start from points that are accessible to the child; and to establish these starting points it seems indispensable to gain some insight into the child's conceptual structures and methods, no matter how wayward or ineffective they might seem" (Steffe, p. 5).

3. *Among the authors there is a willingness to think deeply about and to rethink and revise the epistemological content of the area.* "We are proposing an alternative experiential basis for the construction of number in a primitive cognitive scheme we label *splitting*" (Confrey, p. 291). "In undertaking historical analysis, we are not advocating the naive view that individual learning should follow historical development, but rather looking for ways in which a 'rational reconstruction' . . . of the historical genesis of mathematical concepts may complement our work with students, helping us see students' work from a different perspective" (Smith and Confrey, p. 332). "We will now examine the several forms of competent, but informal, reasoning that have been commonly observed in missing-value problem situations, and then contrast these with the more formal equation-building approach typically taught in schools" (Kaput and West, p. 245). "In our recent effort to better understand the multiplicative concep-

tual field, and in particular the transition phase from the additive structure to the multiplicative structure, we probed into [the question of incongruity between the meaning of] multiplication and division in the whole number domain [and that in] the rational number domain" (Harel, Behr, Post, and Lesh, p. 363). "This perspective calls for deep, careful, and detailed analysis of mathematical constructs both to exhibit their mathematical structure and to hypothesize about the cognitive structures necessary for understanding them. Such analysis would lead to a theory about mathematical knowing and learning that could guide cognitive research" (Behr, Harel, Post, and Lesh, p. 124).

4. *To learn about a multiplicative conceptual field, one must examine its relationship to the situations in which multiplicative reasoning occurs and not view its ideas as isolated abstractions.* "The premise behind this investigation is that when people reason mathematically about situations, they are reasoning about *things* and *relationships*. The 'things' reasoned about are not objects of direct experience and they are not abstract mathematical entities. They are objects derived from experience—objects that have been constituted conceptually to have qualities that we call *mathematical*" (Thompson, p. 180).

## THE CONTENTS

The need for a new publication on the topic was due to the progress that has been made in this field. As one reads through this volume, it becomes apparent that certain issues concerning units, ratios, rates, and recursion are emerging as fundamental ones. These are not treated uniformly in the book, but they continually resurface as explanatory concepts for the research reported. A second reason for the production of a new volume is that the topics included in MCF seem increasingly critical in school mathematics:

1. Mullis, Dossey, Owens, and Phillips (1990) report results that suggest that these topics are still poorly learned. At grade 8 only 49 percent of the students answered the following question correctly: "The weight of an object on the Moon is 1/6 the weight of that object on the Earth. An object that weighs 30 pounds on Earth would weigh how many pounds on the Moon? Answer __5__ " (p. 476).

2. These topics create the critical juncture in middle school, separating those students who persist and those who drop out.

3. There is evidence that these topics are poorly understood by elementary teachers, and hence, effective methods for approaching them will have an impact not only on students but on teachers (Harel, Behr, Post, and Lesh, Chapter 10 in this volume; Simon and Blume, 1992).

4. These topics are critical in technological environments, where decimal notation typically replaces ratio and root displays and where issues of scaling require a deep understanding of the rational numbers and operations.

5. These topics provide a rich and fertile source of problems as mathematics educators recognize the need for increased use of situations, contexts, and problem solving.

Shulman (1978) argued for the importance of critical research sites. A critical research site is one in which the production of new insights will lead to dramatic reconceptualization in both that arena and others. All too often researchers choose topics whose primary criterion for selection is their amenability to available research methods, using the field's current concepts and forms of methodology, an approach which slows down progress in the field.

In recognizing the complexity of the topics in MCF, the work reported in this volume represents a rather substantial departure from traditional mathematics education research. The research projects represent an attempt to consider seriously the difficulties and challenges students and teachers face when approaching the topics of multiplication, division, ratio and proportion and to find new ways to think and talk about those difficulties and challenges. In this volume, to create some of those new descriptions, the contributors have moved beyond the presentation of particular research results to articulate and examine their conceptual and methodological commitments, placing their results in a larger framework.

## Part I. Theoretical Approaches

In the first part, the authors seek to establish a conceptual paradigm for their work and provide examples of what is produced from such a paradigm. Leslie Steffe locates his work in radical constructivism. He contrasts Piaget's claim that the operations of the child may not mirror adult operations in any simple way, with "The current notion of school mathematics [that] is based almost exclusively on formal mathematical procedures and concepts that, of their nature, are very remote from the conceptual world of the children who are to learn them" (Steffe, p. 5). Steffe calls for very close attention to be paid to the current structures of the child and argues for the teaching experiment as the most viable way to allow sustained interactions in which a child's initial organizations and operations can be inferred and the development of new schemes can be observed.

In contrast, Vergnaud begins from a different starting point. Rejecting the trends in psychology to explain all cognition from non-discipline-based positions (information processing, and so on) he argues for subject-matter specificity of research on multiplication and its related concepts. In doing so, he seeks a conceptual analysis of the domain. In this contribution, Vergnaud clearly articulates what he means by the *domain* to be analyzed. He does not mean exclusively the mathematical concepts of "multiplication and division; linear and bilinear (and *n*-linear) functions; ratio, rate, fraction, and rational numbers; dimensional analysis; and linear mapping and linear combinations of magnitudes" (Vergnaud, p. 46), which he identifies as the mathematical bundle of topics included in MCF. He proposes to investigate these related ideas of mathematics in such a way as to include the "conceptual operations needed to progressively master this field" (p. 42), "the situations and problems that offer a sound experiential reference" (buying and sharing sweets, speed, concentration, density, similarity, probability), a "bulk" (he uses this in preference to set, which connotes too strongly well-defined borders) of concepts for analysis, and language and symbols for communicating and thinking. Vergnaud, like Steffe, locates his work in Piagetian and Vygotskian traditions and emphasizes the developmental, nonlinear acquisition of these ideas. He is explicit in his educational intentions: prediction of comparative difficulty of problems and the design of instructional situations.

Interestingly enough, both researchers end by identifying a scheme as a primary explanatory construct in their work. For Vergnaud, "Schemes are the most essential part of a theory of conceptual fields, as they generate actions" (p. 55). He defines the role of a scheme as "the invariant organization of action for a certain class of situations" (p. 53). Schemes require "concepts in action" to provide the categories for obtaining information and "theorems in action" to allow one to derive rules and expectations.

Steffe defines a scheme, in accordance with von Glasersfeld's definition, as having three parts: "First, there is an experiential situation (i.e., a "trigger" situation, as perceived by the child, with which an activity and its result have been associated); second, there is the child's specific activity or procedure; and third, there is a result (again, as perceived or conceived by the child)" (p. 7). Steffe quotes Piaget, who wrote, "All action that is repeatable or generalized through application to new objects engenders . . . a 'scheme'" (quoted on p. 6). Steffe's goal is to "seek to learn the child's schemes and how the child might modify those schemes in the context of solving my situations" (p. 7).

Given that both researchers end up emphasizing the importance of "schemes", one might expect a measure of convergence in their claims about multiplication, division, and related concepts. However, differences are striking, perhaps a matter of degree and emphasis, but nonetheless clearly evident in the two chapters. In the Steffe work, the child is ever present. His definition of scheme emphasizes the child's perspective, and he devotes considerable attention to articulating how a child might be viewing a task as a continuation of earlier actions. Grounded on his seminal work on counting (Steffe, von Glasersfeld, Richards, Cobb, 1983), his approach starts from counting and number sequences, from which he articulates a theory of unit types to explain how a child comes to understand multiplication. Multiplication requires the coordination of two counts along with the construction of a set of objects as a unit. In the end of the chapter, he addresses the formal mathematical properties of commutativity and Krutetskii's notion of "curtailment", but he uses children's schemes to explain the properties, not the reverse.

Vergnaud's analyses are more immediately informed by the formal mathematical principles that he sees as applying across problems and representational forms as hidden mathematical structure, at different levels of abstraction. He states that the goal of teaching is to assist students in seeking and utilizing the "invariance across situations" thus promoting generalization and transfer. He explicitly rejects the idea that conceptual analysis can be conducted a priori and offers a broad outline of a developmental path through MCF. His descriptions of effective approaches come from such conceptual analysis, informed by student work and teaching situations. Vergnaud's analysis has led researchers to articulate even more differentiated maps of the field. In contrast to Steffe, however, Vergnaud does not discuss explicitly the child's voice and ways of talking. The children's methods are cast into the more formal presentation of "theorems in action" of the child.

Greer, in "Extending the Meaning of Multiplication and Division," takes the position that mathematical ideas have their birth in limited frameworks, such as the integers, and their meaning is gradually extended to allow for the inclusion of the real numbers. This claim allows him to make a fundamental epistemological claim, "that epistemological and psychological questions of comparable interest and importance are raised at every stage of reconceptualization through the long development of the conceptual field," (p. 73) and "the extension of concepts, relations, functions, and so forth from one domain to a more general domain is a characteristic mode of development within mathematics and an appropriate subject for study" (p. 74). He takes the position that categories of word problems, such as rate, multiplicative comparison, rectangular array, and product of measures, should be orthogonally considered in relation to the types of numbers used in them. Three categories are proposed: counted integers, integers or fractions derived from division, and decimals as measures of quantities.

He reviews the literature on the misconception "multiplication makes bigger and division makes smaller" (MMBDMS). Logically, he argues, the numbers in a problem that can be modeled by a single operation carry no information as to the appropriate operation. Nevertheless, they drastically influence the difficulty of the choice-of-operations task. When the multiplier is nonintegral, the problem increases in difficulty. In division, factors other than numeric values are cited as playing significant roles. Greer describes the tendency of students to alter their predictions as "nonconservation of operations" (Greer, 1987).

Again we witness a clear commitment to the situation:

"understanding of the application of the operations in modeling situations is weak" (p. 69). Greer uses the term *schema* to describe that which can be developed robustly when the student exhibits an "appreciation of the invariance of the operation over the numbers involved, which is the keystone of extension of meaning for multiplication and division" (p. 70). Citing the importance of historical or ontogenic terms, Greer takes the position that such extensions take time. He sees fractions, in particular the unit fraction, as providing the linking role, both as a divider and as a multiplier (as do Vergnaud (this volume), and Confrey (this volume) in her discussion of earlier work by Greer).

While Greer's chapter can be viewed as an elegant encapsulation of a line of studies to which he was a major contributor, in it he also leaves the reader with a set of provocative new issues to consider: (1) fractions may be useful to form a bridge to decimal multiplication and division, particularly the unit fraction, and (2) situational problems can be presented in configurations that progressively raise certain challenges to allow students to gradually overcome obstacles.

Taken together, the first three chapters of the book give the reader a variety of choices about how to conduct research in the multiplicative conceptual field. Each builds from a Piagetian framework, two using most explicitly the idea of schemes (Vergnaud, Steffe) and the third working with a conservation argument. In Steffe, the examination is built from the ground up, starting from the actions of the child and building a descriptive model informed by the discipline of mathematics, but concerned primarily with adequate explanation of the child's view of a task and methods of proceeding. In Vergnaud, an analysis of the conceptual field seeks to weave together mathematical explanation with educational purpose using the logical apparatus of "theorems in action" as the guiding construct. And Greer seeks an explanatory construct for the puzzling and robust tendency of students to alter their prediction of operation in light of the type of decimal presented. He offers "nonconservation of operation" as a psychological construct to be used in evaluating the maturity of a mathematical mind in terms of one's successful extension of understanding from the limited but intuitive class of natural numbers.

## Part II. The Role of the Unit

In the second part of the book, the contributors continue a theme present in the introductory theoretical part concerning the role of the unit. The concept of a unit, an entity that is treated as a whole, is key throughout the book. Why does this particular construct become so essential in the multiplicative literature, when its importance in the addition and subtraction literature is limited? Dienes and Golding (1966), writing about "sets of sets" anticipate the need for a distinction between the entities operated on in addition and those in multiplication:

> It is important to realize that in this operation we have gone beyond the idea of addition. It is true that the same answer can be obtained to the problem by an addition of the three terms as by multiplication by three. Just because the answer is the same does not mean that the operation is the same. Multiplication involves a new kind of variable, namely the multiplier, which counts *sets*. The multiplier is a property of sets of sets. The multiplicand is a property of sets. So the two factors do not refer to the same universe. . . . Every number refers to sets in addition, whereas in multiplication some refer to sets of sets and others refer to sets. This is a very great difference and the exercise children will have had in dealing with sets and in dealing with sets of sets and even with sets of sets of sets, will serve them in good stead in coming to grips with the problems of multiplication at this stage. (p. 34)

In this volume, that distinction is strengthened through widespread discussions of the role of the unit. To create an understanding of a set of sets, one must first be able to treat a set, a collection, as a unit. This fundamental process of treating a collection as a whole is named by a variety of the authors in the volume and elsewhere as *constructing a composite unit* (Steffe, this volume), *unitizing* (Lamon, this volume), and in the context of repeated multiplication as *re-unitizing* (Confrey, 1988), and *reinitializing* (Confrey, this volume). Behr, Harel, Post, and Lesh (this volume) create a way of symbolizing this action through the use of parentheses. Kaput and West (this volume) use rectangular cells in their software to represent this same action.

The emphasis in the book is that a new type of unit is

constructed in multiplication and division. If one returns to the original development of a unit in the act of a counting, a basic difference in the addition and multiplication literature is apparent. As one counts, the unit is made, and simultaneously, the unit makes counting possible. That is, the creation of number sequences is achieved by a repeated action, and from this action one creates both the numbers, 1, 2, 3, 4, 5 and the operation, a count. This interrelationship between an operation and the construction of number is an important insight expressed in a number of the chapters in the book (Steffe, Lamon, Behr et al., Kaput and West, Confrey, Smith and Confrey). In repeated multiplication, one sees a change, however. There is a repeat action, what Confrey calls *splitting*, but the initial whole is continuously revised into a new whole. In an earlier paper (Confrey, 1988), Confrey labels this *reunitizing* but later renames it *reinitializing* so as not to confuse it with the role of the multiplicative unit. This creates a recursive view of the multiplicative process in which the action of multiplication itself (with a primitive in splitting) is taken as a unit.

In the first chapter of this part, "Ratio and Proportion: Cognitive Foundations in Unitizing and Norming," Lamon recognizes the complexity of the domain of ratio and proportion and proceeds to offer two elegant constructs for organizing it. *Unitizing* is "the ability to construct a reference unit or a unit whole" (p. 92), and *norming* is "to reinterpret a situation in terms of that unit." She reports on the variety of strategies students used in solving four problems on ratio and proportion. In each case, she provides careful description of how the students approached the problems and reasoned out their answers. In her analysis and discussion, she argues that the students used a building up strategy to solve "missing value" problems, and in doing so, she suggests that they were treating the ratio itself as a unit by which they reconceptualized the problem. In her problems, in which the students are asked to compare the amounts of food distributed over a set of children and subsequently "aliens," she describes how the students used a rate (so many $x$ per so many $y$) to create and build up comparisons. Her presentation of student methods demonstrates a number of her distinctions.

In the second chapter of this section, Behr et al. create

notational languages to represent the mathematics that is or might be used by children, and they demonstrate the applicability of these languages in additive and multiplicative conceptual fields. By analyzing these fields from the perspective of the unit, they point to some of their common structures. Based on this commonality they argue that using a units approach to elementary multiplicative and divisional relations (which allows for nonunitary units, i.e., composite units) would greatly enhance students' entry and successful acquisition of rational number proficiency.

They see a need for this notational language because they have found standard mathematical symbolism inadequate to represent children's mathematics in terms of the various unit types, such as those demonstrated by Lamon. Their notational language consists of two systems. The first is iconic and the second is linguistic, with the word *unit* as the basic element, from which units of units can be formed. Their use of two systems provides a stable interpretative structure and assists one in viewing the world from a "units of quantity" perspective. These systems, they emphasize, are not designed for use with children in instructional activities. They aid researchers to communicate about children's conceptions of specific additive and multiplicative situations, to hypothesize the cognitive structures that develop, or need to be developed, in acquiring an understanding of the concepts discussed, and to suggest kinds of learning events that children ought to experience so that they have an opportunity to develop these structures.

## Part III. Ratio and Rate

The third part of the book concerns the topics of ratio and rate. In this part, the authors, Thompson and Kaput and West, seek to create a bridge between the role in multiplicative structures of rate, as a description across a set of particular instances, and ratio, as a multiplicative relation between two specific quantities.

Thompson's chapter on children's concepts of rate and ratio must be understood in the context of his theoretical perspective on quantitative reasoning, a perspective that brings forth the mathematics of rate, but that does not apply the ideas learned in abstract and largely symbolic forms. In his framework, consistent with that of Steffe, Confrey, Smith and Confrey, and Kaput and West, actions are internalized into mental images of situations. The automaticity and freedom inherent in these images varies, and at the most sophisticated level, the transformation of the objects are carried out at an operational level. This progression from action to operational transformation is a key activity in Thompson's argument about the construction of quantities, which, for him, are "schematic;" that is, "a quantity . . . is composed of an object, a quality of the object, an appropriate unit or dimension, and a process by which to assign a numerical value to the quality" (p. 184). Quantification is the act of assigning a numeric value to the quality. In contrast, a quantitative operation is different from numerical operation; "it has to do with the *comprehension* of the situation" (p. 187). The comprehension of quantitative situations gives rise to the construction of quantitative operations and relationships. In the context of this theoretical framework, he deals with the question of how can one "orient a student so that she would construct a scheme for speed that would be powerful enough that she would recognize (what we take as) more general rate situations as being largely the same as situations involving speed" (p. 183).

In Thompson's report of a teaching experiment with a fifth grade girl using a computer microworld of a rabbit and a turtle, an evolution of a quantitative concept of speed is presented. Two features stand out. First, the student works toward a scheme that seems to be composed of covarying accumulations of two segmented quantities, distance and time, and no quantity acts as the primitive, or the extensive quantity in Schwartz's terminology (1988), in that, as this student built her understanding, she actually viewed time as a speed per unit distance. "[H]er initial conception of speed was that it was a *distance*, and her initial conception of time was that it was a *ratio*" (p. 224). Second, Thompson points out several important issues emerging from this teaching experiment, two of which are (1) "the standard method for introducing speed in schools as 'distance divided by time'" before students have acquired a "mature conception of speed as quantified motion," would have little, if any, relevance to their initial understanding of speed. As a result, students would not be able to make a progress in their conception of speed as the student in his experiment did; (2) "what is called *dimensional analysis,* or the arithmetic of units (e.g., miles ÷ miles/hour = hour)" should be condemned, "at least when [it is] proposed as 'arithmetic of units,' and [we should] hope that it is banned from mathematics education" (p. 226).

In the second chapter in this part, Kaput and West report on their work with sixth grade students on missing-value proportional reasoning problems. They make a distinction similar to that made by Thompson between rate and ratio, which they term *rate-ratio* and *particular ratio,* respectively. Making this distinction allows them to describe what an understanding of rate must entail, including numerical and semantic equivalence and homogeneity of relation within the quantities being sampled.

The authors then propose a model of the development of a fuller or more mature understanding of proportional reasoning that goes from understanding particular ratios to rate-ratios through build-up strategies, abbreviated strategies, and finally the development of unit factor approaches. Going through a set of four problems, the authors articulate how a student must first identify the variables (the quantities that vary) and "The solver must then be able to form groups or segments that are the referents for the incrementing quantity steps and finally must be able to coordinate the two types of groups or segments to coordinate the dual incrementing acts" (p. 246–247).

In the two chapters, one sees similar approaches to quantity and computation, where the activity of conceptualizing the problem quantitatively precedes the computations. A second similar aspect of the two chapters is that in both approaches, one sees the beginnings of a function concept. The connections between early development of the function concept and proportional reasoning have also been articulated by Rizzuti (1991), who argued similarly to these authors, that covariation is a powerful way for students to approach contextual problems and organize a scheme to understand how quantities vary in relation to each other. This covariation approach to function resurfaces in the next chapters as the authors (Confrey and Smith) tackle the development of an understanding of exponential functions.

## Part IV. Multiplicative Worlds

Confrey's chapter, "Splitting, Similarity, and Rate of Change: A New Approach to Multiplication and Exponential Functions," offers a new primitive on which the concepts of both division and multiplication can be based. Based in actions such as sharing, folding, dividing symmetrically, growing, and magnifying, the *splitting* construct is contrasted to counting actions, which create addition (affixing, joining, annexing, removing). Confrey postulates that it has a complementary but equally significant role in the elementary curriculum, one neglected under current treatment.

Her approach to multiplication shares features with others. Schemes are important as in Steffe, Vergnaud, and Thompson, but the work reported on for the early grades is largely anecdotal. She offers a cycle of conceptual construction moving from "problematic to action to reflection" describing a scheme as a "cognitive habit of action," thus stressing the repetitious quality of schemes. Explicitly, she argues for attention to operations as evolving from actions and to numbers as evolving from operations. Therefore, she sees significant interactions between situational characteristics, the actions performed by students and their purposes and goals, the development of mental operations, and subsequently the construction of numbers.

In her argument for the recognition of splitting and counting as independent cognitive structures, she notes first that in counting the origin is zero, the successor action is adding one, the unit (as the invariance between predecessor and successor) is one, addition and subtraction are basic operations, intervals are made from differences via subtraction, and rate is difference per unit time. In splitting, she suggests, the origin is one, the successor action is splitting by $n$, multiplication and division are basic operations, ratio is used to describe the interval between two numbers (percentages, for example), and rate is the ratio per unit time.

Like the arguments in Greer, she offers an argument that the inclusion of additional representational forms is necessary to improve students' understanding. She includes among these the tree diagram, embedded similar figures, and a new similarity based plane. She shows how her analysis can lead to a more secure foundation for the development of the exponential function as its roots lie in repeated multi-plication.

The splitting conjecture is followed by the chapter by Smith and Confrey on "Multiplicative Structures and the Development of Logarithms: What Was Lost by the Invention of Function?" Modern notation and approaches can often act as lenses that limit as well as guide our observations and investigations. In an effort to discover other approaches to multiplication, the authors examined the history of logarithms. In doing so, they discovered that the assumptions about what would be the most primitive type of comparison between two quantities or magnitudes could vary depending on the observer and the task. A simple example of this is in asking someone to compare two distances, one of which is 10 feet and the other is 13 feet. One will most frequently describe the difference additively as 3 feet more. If asked to compare the distance to the moon (238,000 miles) to the distance to the sun (93 million miles), a multiplicative comparison will typically make more sense. (The sun is nearly 400 times as far away as the moon.)

In this chapter, the authors report on "the work of Thomas of Bradwardine and Nicole Oresme, who claim that, whereas the primitive action taken on magnitudes is addition, the primitive action taken on ratios is multiplication" (p. 334). They describe a world created by Oresme where the elements are ratios and the successor action is multiplication. In doing so, they recall an earlier theme, that of functions as covariations. By juxtaposing geometric and arithmetic sequences, a logarithmic-exponential function is created. The authors point out that this is a covariation approach to function, where the functional relation is not given as input-rule-output, but as a pair located in two covarying sequences, one varying additively and one varying multiplicatively. As in the Kaput and Thompson discussions of ratios and rates, one sees how tabular approaches can support a build-up, covarying rate of change approach to functions.

Going beyond the basic placement of these two sequences opposite each other, the authors argue how density in them is created through the insertion of additive and geometric means and how Napier eventually solved the problem of allowing spacing of the geometric sequence to be of any desired interval size. The chapter is an interesting addition to the volume for it locates the efforts of the authors in this book in a historical context.

## Part V. Intuitive Models

As has been indicated in the opening of this Introduction, research on the concepts of multiplication and division revived in the early 1980s with the work of Fischbein, Deri, Nello, and Marino (1985) on the constraints that primitive intuitive model impose on students' choice of the operation to solve multiplication and division word problems. With these models, multiplication and division have very restricted meanings; for example, multiplication is conceived as merely repeated addition. Greer devotes his chapter to extensions of these meanings of multiplication and division from their early conceptualization and the difficulties children have in making these extensions. Harel, Behr, Post, and Lesh's chapter in this section report on a study of teachers' limited conception of multiplication and division. More specifically, they address the impact of the number type on the solution of multiplicative problems by preservice and inservice teachers and reexamine the findings from other studies concerning this impact, with an instrument that controls for a wide range of confounding variables. They show that teachers' solutions of multiplicative problems strongly correlate with the intuitive rules derived from the models proposed by Fischbein et al. (1985) to explain children's solutions.

Harel et al. add several points to what is known in this domain. First they looked at the explanation suggested by Fischbein et al. (1985) and others to account for differences in subjects' performance on multiplicative problems with different non-whole-number operators, and show findings that are not consistent with this explanation. They theorize an alternative conceptual basis for these differences. Second, they suggest levels of robustness of the intuitive rules derived from Fischbein et al.'s models. Finally, they raise several open research questions with regard to the impact of the number type. For example, they point out that fractions and decimals may not have the same effect on the solution of multiplication and division problems, because the naming rule of fractions is different from the naming rule of decimals: "Under these naming rules, it is easier to identify the role of a problem quantity as an operator or operand if the quantity is a fraction than if it is a decimal; therefore it is easier to recognize its relation to other problem quantities." (p. 381).

## Summary

In a final discussion chapter, Kieran articulates ways to view the overriding ideas and issues in the volume. Drawing on an exhortation from Bereiter for larger objects of conceptual analysis, Kieran writes, "The growth of multiplicative structures . . . is critical for a person's conceptualizing or bringing forth the world in which he or she lives" (p. 387). His discussion highlights the distinctions in the book between the primitives of counting and splitting and elaborates on the strengths and assumptions of each. He identifies the new emphasis in the volume on units, iteration, building-up, distributing and norming, and he suggests that the development of these new ideas might help to alleviate the difficulties identified in the research in the volume on teachers' understanding of rational numbers.

In this book, we see another step in the community's understanding of the concepts of the multiplicative conceptual field. The authors unite in their attempts to create language that will allow them to describe what children and young adults do as they encounter problems from this domain. What we see is the emergence of a significantly new language that includes such terms as *schemes, units, norming, covariation, iteration, recursion, ratio, repeated splits, similarity, sequences, operations,* and *dimensions.* These words are not commonly used in the teaching of multiplication and division, and they promise to transform such instruction in rather significant ways.

The implications go beyond the field of multiplicative relations in that an agenda is set with quantity and scheme at the forefront. Using these terms, the authors seek to assist the student in "bringing forth the world in which he or she lives" in the words of Kieran. The book suggests that close attention must be paid to how students see the tasks, not only in individual teaching experiments, but in the rich, complicated, and noisy world of classroom instruction. The classroom work reported is limited, as would be expected at the start of any reconceptualization of the territory; but the work reported on teachers' weak understanding suggests that the need to apply this work to the classroom setting is pressing, for staff as well as students.

However, it should not be implied that the authors speak with a uniform voice. Those important differences that lead to further debate, articulation of views, and forms of evidence still are apparent in the texts. For instance, there are a variety of opinions on the relationships between quantity, operations, and numbers. At one end of the spectrum, Greer offers a description of "conservation of operation" and argues that a mathematically mature person will conserve the operation in relation to the quantities given across any number types. He suggests that this maturity is reached by overcoming the misconception that multiplication makes bigger and division makes smaller and might be facilitated by representations portraying smooth and continuous change in the magnitude of the numbers. Kaput and Thompson, in arguing for a separation of quantitative reasoning and numerical reasoning, seem to locate themselves implicitly toward the same end of the continuum, although in their empirical reports they give careful attention to cognitive jolts that occur when the segmentation of continuous quantity does not produce easy integral outcomes. These jolts, however, seem to be seen as computational disturbances and not as key in the development of the reasoning and relationships. Lamon, and Behr et al., in struggling to produce notation that can express the emergence of unit concepts, seem to be more interactional in their view of the matter. Lamon introduces new notation to help the reader avoid assuming the meaning of the division of quantity, and Behr et al. provide ample evidence of the multiple interpretations for any expression, such as 3/4. Steffe and Smith and Confrey are frankly explicit in their argument that numbers are constructed from actions and that therefore there is a necessary circularity in the number-operation relationship. The key seems to lie in one's approach to the unit. At the same time, quantities are segmented to produce units of quantity (Kaput) and units are descriptions of the invariance between a successor and a predecessor (Confrey), and both views of units are essential to the development of an understanding of multiplication and division.

The book is rich in ideas. Its implications for classroom practices are less clear. However, as a theoretical text with careful empirical support, mostly but not completely data of the interview-based or extended teaching experiment type, the book contributes to the ongoing dialogue about how to teach these most important topics.

## REFERENCES

Behr, M., R. Lesh, T. Post, and E. Silver. 1983. Rational number concepts. In *Acquisition of mathematics concepts and processes*, ed. R. Lesh and M. Landau, 91–126. New York: Academic Press.

Bell, Alan W., Efraim Fischbein, and Brian Greer. 1984. Choice of operation in verbal arithmetic problems: The effects of number size, problem structure and context. *Educational Studies in Mathematics* 15: 129–147.

Bell, Alan W., Malcolm Swan, and Gloria Taylor. 1981. Choice of operations in verbal problems with decimal numbers. *Educational Studies in Mathematics* 12: 29–41.

Confrey, Jere. 1988. Multiplication and splitting: Their role in understanding exponential functions. Proceedings of the tenth annual meeting of the North American chapter of the International Group for the Psychology of Mathematics Education. DeKalb, IL. November, 1988.

Dienes, Z. P., and E. W. Golding. 1966. *Sets, numbers and powers.* New York: Herder and Herder.

Fischbein, Efraim, Maria Deri, Maria Sainati Nello, and Maria Sciolis Marino. 1985. The role of implicit models in solving verbal problems in multiplication and division. *Journal for Research in Mathematics Education* 16 (1): 3–17.

Graeber, Anna O., and Dina Tirosh. 1988. Multiplication and division involving decimals: Preservice elementary teachers' performance and beliefs. *Journal of Mathematical Behavior* 7 (3): 263–280.

Greer, Brian. 1987. Nonconservation of multiplication and division involving decimals. *Journal for Research in Mathematics Education* 18: 37–45.

———. 1988. Nonconservation of multiplication and division: Analysis of a symptom. *The Journal of Mathematical Behavior* 7 (3): 281–298.

Hiebert, J., and M. Behr. 1988. Introduction: Capturing the major themes. In *Number concepts and operations in the middle grades*, ed. M. Behr and J. Hiebert. Reston, VA: National Council of Teachers of Mathematics.

Luke, Clifton. 1988. The repeated addition model of multiplication and children's performance on mathematical word problems. *Journal of Mathematical Behavior* 7 (3): 217–226.

Mullis, I., J. Dossey, E. Owens, and G. Phillips. 1990. *The state of mathematics achievement.* Washington, DC: National Center for Educational Statistics, Office of Educational Research and Improvement, U.S. Department of Education.

Nesher, Pearla. 1988. Multiplicative school word problems: Theoretical approaches and empirical findings. In *Number concepts and operations in the middle grades,* ed. M. Behr and J. Hiebert, 19–40. Reston, VA: National Council of Teachers of Mathematics.

Rizzuti, J. 1991. High school students' uses of multiple representations in the conceptualization of linear and exponential functions. Unpublished doctoral dissertation, Cornell University, Ithaca, NY.

Schwartz, Judah. 1988. Intensive quantity and referent transforming arithmetic operations. In *Number concepts and operations in the middle grades,* ed. M. Behr and J. Hiebert, 41–52. Reston, VA: National Council of Teachers of Mathematics.

Shulman. 1978. Invited address for the Research in Mathematics Education Special Interest Group (SIG/RME) of the American Educational Research Association annual meeting, Toronto, Ontario.

Simon, M., and G. Blume. 1992. Understanding multiplicative structures: A study of prospective elementary teachers. In *Proceedings for the sixteenth psychology of Mathematics education conference,* ed. William Geeslin and Karen Graham, vol. 3, 11–18. Durham, NH.

Sowder, J. 1988. Series forward. In *Number concepts and operations in the middle grades,* ed. M. Behr and J. Hiebert. Reston, VA: National Council of Teachers of Mathematics.

Steffe, L. 1988. Children's construction of number sequences and multiplying schemes. In *Number concepts and operations in the middle grades,* ed. M. Behr and J. Hiebert. Reston, VA: National Council of Teachers of Mathematics.

———, E. von Glasersfeld, J. Richards, and P. Cobb. 1983. *Children's Counting types: Philosophy, theory, and application.* New York: Praeger.

Vergnaud, Gerard. 1983. Multiplicative structures. In *Acquisition of mathematics concepts and processes,* ed. R. Lesh and M. Landau, 127–173. New York: Academic Press.

———. 1988. Multiplicative structures. In *Number concepts and operations in the middle grades,* ed. M. Behr and J. Hiebert. Reston, VA: National Council of Teachers of Mathematics.

# I
# THEORETICAL APPROACHES

# 1   Children's Multiplying Schemes

*Leslie P. Steffe*

The work on children's multiplying and dividing schemes in which I am engaged is based on two theoretical assumptions, both of which I have substantiated. The initial assumption is that children, when faced with their first arithmetical problems, use their current mathematical schemes to attempt to solve them. They seem to persist in doing this in spite of their teacher's explanation of accepted methods. Moreover, if the children's attempts are successful in producing an answer that is seen as "correct" by the teacher, the teacher often remains unaware of the fact that the children use their own methods. These methods, however, eventually become dysfunctional in current school mathematics programs because the children are discouraged from using them in favor of standard methods, imposed upon them by their teachers.

## CHILDREN'S METHODS

That children use their own methods when solving problems has been noted by various workers in the field (e.g., Booth, 1981; Erlwanger, 1973; Ginsburg, 1977; Hart, 1983; Resnick, 1982; Steffe, von Glasersfeld, Richards, and Cobb, 1983; Steffe and Cobb, 1988). In an investigation of the levels of understanding shown by British secondary school students, Booth (1981) found that "many children are not using the 'proper' mathematical methods taught them at school, but rather are relying upon naive, intuitive strategies . . . " (p. 36). Her coworker, Kathleen Hart (1983), said that child methods "are not teacher taught . . . and to a large extent involve counting . . . rather than the *four* operations" (p. 120). In a later paper, Booth (1984) stated that "The existence of these informal methods in mathematics must have profound consequences for teaching, curriculum development, and re-

3

search" (p. 159). Although Booth does not believe that the informal methods are incorrect, she does see a difficulty if they persist into the secondary school years—"The same methods often become extremely cumbersome and difficult to apply when the numbers involved are large, and often are inappropriate to the case of non-integers, being largely based on counting methods" (p. 159).

The views of Both and Hart stem from a large survey— 300 interviews linked to 10,000 written scripts (Hart, 1983, p. 120)—where no attempt was made to individually teach single children. Booth's current position is that

> Helping the child to learn these (formal mathematical) procedures means recognizing the methods that children actually use, and helping them both to understand the relationship between what they are doing and what the teacher is presenting, and to appreciate the *value* of making this connection, by helping them to recognize the limitations of their own approaches. (p. 160)

The expression "helping the children to recognize" is in obvious contrast with the notion that children can simply be *told*. They can, indeed, be told to do something, but they cannot be told to *understand*. This brings up the second theoretical assumption, which stems from the work of Piaget and especially the contemporary neo-Piagetians. It can be summarized simply by saying that any "knowledge" that involves carrying out actions or operations cannot be instilled ready-made into students or children but must, quite literally, be actively built up by them. "Human knowledge is essentially active. To know is to assimilate reality into systems of transformations. I find myself opposed to the view of knowledge as a copy, a passive copy of reality" (Piaget, 1970, p. 15). A current report on education expresses the same view: "All genuine learning is active, not passive. It involves the use of the mind, not just memory. It is a process of discovery, in which the student is the main agent, not the teacher . . . " (Adler, 1982, p. 50).

On the other hand, where mathematics teaching is concerned, I do agree with Menchinskaya (1969) when she says: "Neither scientific nor everyday concepts spring forth spontaneously; both are formed under the influence of adult teaching" (p. 79). But, from the perspective of the two theoretical premises mentioned, the process of education or teaching takes on the specific form of guiding the student's conceptual construction rather than imparting the "correct," adult way of doing things.

Lest this sound an extremist view, let me hasten to stress that I believe it to be a crucial attitude as children start to learn mathematics because little commonality can be assumed between the student's conceptual structures and those of the adult or the teacher. Piaget, among others, has shown the difficulty of concepts such as commutativity and reversibility, both of which are involved in understanding arithmetic. It is easy to forget that children, when they enter school, usually do not have operations of that kind and that it is not sufficient simply to tell them what they are. Hence, there is no reason whatever to assume that the child will interpret the formulation of a given problem in the way that seems "obvious" to the adult or the teacher; nor is there any reason to assume that the child could "see" that a particular way of proceeding must lead to a result that constitutes the solution.

From my perspective, then, the essential task at the beginning is not to provide "correct ways of doing" but, rather, to guide children to find ways of operating to reach their goals. Such guidance, necessarily, would have to start from points that are accessible to the child; and to establish these starting points it is indispensable to gain some insight into the child's conceptual structures and methods, no matter how wayward or ineffective they might seem. If there is any virtue at all in the notion that cognitive organisms *assimilate* their experiences to structures they already possess and *accommodate* these structures only when something does not lead to the expected result, it follows that the teacher will be far more successful in generating accommodation if he or she has gained some notion of what are the child's present structures and ways of operating.

The main thrust of my current research is based on these considerations and, therefore, aims at building up a model of children's initial, informal approaches to multiplying and dividing in order to derive from this "raw material" viable ways and means of guiding their conceptualizations and activities. The current notion of school mathematics is based almost exclusively on formal mathematical procedures and concepts that, of their nature, are very remote from the conceptual world of the children who are to learn them. Yet, only through their own conceptual powers can they acquire understanding and make mathematical progress. Recent work has shown that the generative power of children as they approach numerical problems is extremely exciting (Booth, 1984; Confrey, 1991; Streefland, 1984; Steffe and Cobb, 1988). The problem, as I see it, is to find a way of harnessing that power. I agree with Booth that this cannot be done unless we have a detailed model of how children construct mathematical concepts in situations and contexts *as they see them.*

To this end, my effort is aimed at mapping schemes of action and operation involving composite units as they are elaborated and reorganized by children. Because I am ultimately interested in fostering and guiding these changes and elaborations, this part of the research comprises two tasks. First, I am specifying the schemes of action and operation involving composite units that can be inferred from the actual mathematical behavior of the children who participated in my teaching experiment. Second, I am investigating the construction of these schemes by the children.

## CHILD-GENERATED ALGORITHMS

To distinguish notions and methods children appear to have developed on their own in out-of-school contexts (Saxe, 1990) from those that seem to be developed in the context of indirect teaching, I use the term *child methods* for the former and *child-generated algorithms* (Hatfield, 1976) for the latter. Child-generated algorithms are constructed by children when trying to solve tasks they meet in school and are, in fact, elaborations of child methods acquired elsewhere. As such, I consider them nothing other than what Piaget (1980) calls *scheme:* "All action that is repeatable or generalized through application to new objects engenders . . . a 'scheme'" (p. 24).

There is a distinct difference in how I view child-generated algorithms and the view that they are the child's method of performing standard algorithmic tasks like long division. The difference is manifest in the conception of indirect teaching that follows from the second theoretical assumption of the neo-Piagetians. In a teaching experiment, I do not model or demonstrate an algorithm for multiplying or dividing and then study how the child might have interpreted my actions. Rather, through interactive communication with the child, I present situations without modeling or demonstrating solution methods the child might imitate. My goal is to study the child's assimilating operations, goals, and intentions, the activity used to achieve these aims, and the results of acting.

If I observe regularity in the child's solution attempts across several of my situations, this is an occasion for abstracting a scheme that might explain my observations. On the other hand, if I observe independent changes in how the child operates or changes as a result of our interactive communication, this is an occasion for abstracting an accommodation in the child's functioning schemes that might explain my observations. In these ways, I do not start with predetermined methods that I aim to teach the child and expect the child to learn but, rather, I seek to learn the child's schemes and how the child might modify those schemes in the context of solving my situations.

Schemes consist of three parts (von Glasersfeld, 1980). First, there is an experiential situation (i.e., a "trigger" situation, as perceived or conceived by the child, with which an activity and its result have been associated); second, there is the child's specific activity or procedure; and third, there is a result (again, as perceived or conceived by the child). A child's current numerical knowledge can be viewed as the coordinated schemes of action and operation the child has currently constructed. Investigating children's schemes, therefore, is tantamount to investigating the building blocks of mathematical thinking and opens new vistas for school mathematics (Steffe, 1983a).

The counting scheme, for example, is fundamental in the construction of the whole numbers (cf. Brownell, 1928; Ginsburg, 1977; Saxe, 1982; Steffe et al., 1983; Carpenter and Moser, 1984; Steffe and Cobb, 1988). It is a scheme that undergoes continual development and refinement as the child progresses, and it is not limited to counting by ones. Its elaboration to include composite units (i.e., units whose numerosity is greater than one) is crucial in learning multiplication and division. Only when teachers take into account and build on schemes a student already has constructed can they bring new material within reach of the student in an informed way;

and conversely, being able to construct new ways and means of operating fosters a sense of power and control in the student that more than anything else enhances motivation.

How powerful assimilation and accommodation can be is exemplified by Tyrone, a 7-year-old participant in one of my past teaching experiments, which briefly dealt with multiplication problems (Steffe and von Glasersfeld, 1980).

> Tyrone was presented with a problem he himself had asked to do: 20 × 20. He sequentially put up fingers twenty times in synchrony with saying "20, 40, 60, 80, . . . , 340, 380, 400". He put up each of his ten fingers ("20, 40, . . . , 200") and then put up his ten fingers again ("220, 240, . . . , 400"). He stopped counting without suggestion, indicating that he fully intended to repeat 20 twenty times. The next task given was 30 × 20. He began counting from "four hundred"— "420, 440, 460, 480, 500, 520, 540, 560, 580, 600!" He again put up fingers sequentially and, on his own, stopped counting at "600" because he considered it the result. (Steffe, 1982, p. 4)

Tyrone's sense of competence and control was demonstrated by the fact that he kept asking for more problems. His counting scheme was vastly different from the standard algorithms that he might meet in the school curriculum. His method for multiplying was an elaboration of his more primitive counting scheme—counting by ones. He had a strong sense of control because he assimilated the tasks using *his* counting scheme and solved the tasks using *his* strategies and methods. There was no discontinuity between his own mathematical reality and what he was asked to do. It was not difficult to encourage Tyrone because he independently sought situations that he could solve.

It is a drastic mistake to ignore child-generated algorithms in favor of the "standard" paper and pencil algorithms currently being taught in the elementary schools. Other than the work already cited, there is solid evidence that imposing the standard algorithms on children yields discontinuities between children's methods and their algorithms (Easley, 1975; Brownell, 1935; McKnight and Davis, 1980). Even when they are to some extent based on operative arithmetical concepts, the standard algorithms become essentially instrumental for the children (Skemp, 1978) and pose a serious threat to the retention of insight (Freudenthal, 1979; Erl-wanger, 1973).

The numerical knowledge represented by the elabora-tions and refinements of children's schemes is operative rather than instrumental and should be used to foster in-sight into creative problem solving. In fact, the whole view of school mathematics changes when the mathematical activity of the child is stressed. Easley (1975) has pointed out that the "notion that mathematics at the elementary school level could be a constructive or even a creative discipline is diffi-cult . . . for most American teachers and educators to ac-cept" (p. 31). Yet, the young child, in a state of mathematical naivete but innate curiosity, in many respects resembles the mathematician working on the frontiers of mathematics. Viewing mathematics as a human activity qualifies children's numerical schemes as part of the creative discipline that we all know mathematics comprises.

## MEANING THEORY IN ARITHMETIC

Acknowledging the gap between current school mathematics programs and children's methods is but an expression of the constructivist principle that each child must construct his or her own representations of reality (Piaget, 1937). Mathemat-ics instruction, however, has not yet benefited from this con-structivist principle in spite of the current debates on mathe-matics education (Steffe and Wood, 1990; *Everybody counts,* 1989; *Curriculum and Evaluation Standards for School Mathematics,* 1989). Researchers in the Second Internation-al Study of Mathematics found that current mathematics teaching can be characterized almost universally as the for-mal, symbolic presentation of mathematical rules and proce-dures in lecture formats (McKnight, 1987). Scant attention is given to mathematics as a human activity and the mathemati-cal activity of the students is essentially ignored. The ves-tiges of meaning theory that remain in current approaches to mathematics teaching hark back to two historically impor-tant schools of thought. One principal contributor to *the structural school* believed that "meaning is to be sought in the structure, the organization, the relationships of the sub-ject itself" (Brownell, 1945, p. 481). The origin of a second school, *the operational school,* can be traced to Percy Bridgman's operational analysis of the fundamental concepts of modern physics (Bridgman, 1927). Van Engen (1949), the principal contributor to this school, believed that the meaning of a symbol, such as 4, is "an intention to act and . . . the act need not, in itself, take place. However, if the individual is challenged to demonstrate the meaning of the symbol, then the action takes place" (p. 324). Van Engen viewed semantics as interpretations of operational definitions. These operational definitions were taken to be universal and, therefore, identical for all children. Especially in mathematics, the main concepts seemed transparent and were expected to become "self-evident," provided they were properly explained.

Since Piaget's (1936; 1937; 1945) revolutionary publications, contemporary epistemology has made us aware of the highly complex processes of abstraction that underlie "understanding" in general and mathematical understanding in particular. Although a great many everyday concepts can be abstracted from sensory experiences, this is not the case with the mathematical ones. If we believe, as did Van Engen, that the meaning of a mathematical symbol is essentially an intention to act, then it follows that mathematical concepts will be formed from operations whose material includes the results of actions. Indeed, Piaget (1970) characterized "reflective abstraction" as "the mode of abstraction that derives its knowledge . . . from actions and from the subject's operations. . . . Thus defined, reflective abstraction is necessarily constructive and enlarges and enriches the structures from which it starts . . . " (p. 221). The realization that schemes can function on different levels of abstraction makes it plausible to think of mathematical concepts in the context of schemes. The word *concept* refers to "any structure that has been abstracted from the process of experiential construction as recurrently usable. . . . To be called 'concept' these constructs must be stable enough to be re-presented in the absence of perceptual input" (von Glasersfeld, 1982, p. 194).

Schemes that have been constructed through the process of reflective abstraction fit this view of concepts. They can be used in assimilation in experiential contexts, in establishing goals and intentions, and in attempts to reach the goals and intentions they make possible. In Piaget's (1971) words, "the response is there first, or . . . at the beginning there is structure" (p. 15). This becomes apparent as soon as one considers how children use schemes purposefully to achieve their goals. Tyrone, for example, intentionally counted by twenties to solve 20 × 20. He anticipated that he could solve his task by counting by twenties and knew what he was going to do before he did it. Schemes serve as anticipatory structures that guide activity within a goal-directed framework. In general, "anticipation is nothing other than . . . application of the schema (or scheme) . . . to a new situation before it actually happens" (Piaget, 1971, p. 195). In other words, upon recognition (assimilation) of a situation, a scheme is anticipatory if the child can use the scheme in representation of the activity or result prior to executing the response of the scheme.

Without Piaget's (1973) insight that children see ideas differently, the early meaning theorists' emphasis on operations and structures led them to create classic works that emphasized standard algorithmic procedures dictated by adult conventions (Brownell, 1945; Van Engen and Gibb, 1956). In contrast, I look at the child's mathematical activity to infer their meaning of arithmetical words, numerals, and procedures. Attempts to understand the child's perspective in teaching experiments have already led to novel understanding of children's mathematics (Confrey, 1991; Kieren and Pirie, 1991; Kieren, 1990; Steffe et al., 1983; Steffe and Cobb, 1988).

## MATHEMATICS LEARNING

Mathematics learning consists of the accommodations children make in their functioning schemes as a result of their experiences. Although possible, there is no necessity that such accommodations occur on the spot in a flash of insight (Wertheimer, 1942). They may occur on the spot or during periods of rest or reflection with no major reorganization of the involved scheme (Steffe and Cobb, 1988; Cobb and Steffe, 1983). Social interaction is a primary context for accommodation and provides the basis for my study of mathematics learning in constructivist teaching experiments (Steffe, 1983b; Cobb and Steffe, 1983; Steffe, 1991a). For convenience, I call social interactions in mathematical contexts *mathematical interactions*, and these mathematical interactions that transpire between myself and a child serve as a basis for the accommodations we both make. As Schubert and Lopez Schubert (1981) point out, teaching in this sense is rooted in action:

> It is in the subtly powerful interaction of some teachers . . . with their students. It is in the daily striving of teachers who try to understand their students' sources of meaning . . . their personal "theories" or sense-making constructs. It exists in attempts made by teachers to determine how their experience and knowledge can bolster their students' quest for meaning. (p. 243)

In the course of a mathematical interaction, I act with intended meaning and the children interpret my actions using their mathematical schemes, creating the actual meanings they perceive. I infer their schemes based on their language and actions and make decisions about what new knowledge the children might construct, how they might construct it, and what aspects of old knowledge they need to refine or consolidate. This dynamic and flexible quality of the teaching experiment is the very thing that recommends it for creative investigation of mathematics learning. A statement by David Hawkins (1973) at the Second International Congress of Mathematical Education provides a sharp characterization of my approach.

> The really interesting problems of education are hard to study. They are too long-term and too complex for the laboratory, and too diverse and non-linear for the comparative method. They require longitudinal study of individuals, with intervention a dependent variable, dependent upon close diagnostic observation. The investigator who can do that and will do it is, after all, rather like what I called a teacher. So, the teacher himself is . . . the best researcher . . . (p. 135)

The teaching experiment is crucial in my work because mathematical schemes are never acquired by a child in one piece, but must be built up over a rather extended period of time; and learning how children learn mathematics must be accomplished as it actually takes place. The teaching experiment essentially replaces the Piagetian clinical interview (Ginsburg, 1981; Opper, 1977) as the fundamental research methodology in mathematics education. Skemp (1979), in a review of the teaching experiment methodology, commented that "I believe that by these means, better understanding of the learning and teaching of mathematics is likely to be reached than by any other way that has been devised so far" (pp. 28–29).

Children's mathematical learning must be observed in contexts where we as teachers can intervene with the intention of influencing it, examining its limits, and examining the child's generative power and flexibility. This provides a new perspective on reflective abstraction. No longer should reflective abstraction be restricted to the most general reorganizations that accompany discontinuities in mathematical learning and are considered inaccessible to observation in experiential contexts (see Thom, 1973, for an expression of this view). Rather, I cast it into functional forms and use it to explain how children modify their mathematical experiences. In the context of scheme theory, I operationalize reflective abstraction as those accommodations children make in their schemes in experiential contexts to neutralize perturbations that are introduced through their actions (von Glasersfeld, 1991; Steffe, 1991b).

## PROBLEMS UNDER INVESTIGATION

In a teaching experiment on children's multiplication and division (Steffe and von Glasersfeld, 1985a), I took the children's number sequences as my starting point because it was my assumption that whatever multiplication and division might become for these children, the operations would be constructed as modifications of their number sequences. Univocal expressions of multiplication that stress the adult's voice, like the following, are not based on this essential assumption. "Multiplication is most commonly linked to the process of combining equal groups. Children also need to see that it relates to array, 'times as many' and 'combination'" (*Curriculum and Evaluation Standards for School Mathematics*, 1989, p. 42). In my view, these meanings of multiplication are based on the mathematical concepts of the authors' of the report cited rather than on their attempts to modify those concepts to construct a multiplication of children.

### Number Sequence Types

In past teaching experiments, I isolated three distinctly different number sequences—the initial, the tacitly nested, and the explicitly nested sequences (Steffe and Cobb, 1988). In general, a child's number sequence can be thought of as a mental record of counting (Steffe, 1991b). Understanding how these records are formed is essential in understanding my definition of a child's number sequence as a sequence of unit items that contain records of counting acts (Steffe and von Glasersfeld, 1985b).

An act of counting consists of the co-occurrence of uttering a number word and producing a countable item of one kind or another. If a child takes a sequence of such acts of counting as an occasion for review, using its unitizing operation, the child may create a sequence of unit items where each unit item contains records of the experiential counting act that was used in its formulation. In a review of counting, if there are no perceptual records of counting available, the child may regenerate the experience of the acts of counting without overtly executing the counting acts. If these imagined acts of counting are each used as material of the unitizing operation, the child creates a sequence of abstracted unit items—"holes" or "slots"—that contain what are now *interiorized* records of the counting acts. These records may be of sensory material produced in the auditory channel, the visual channel, the kinesthetic channel, or combinations thereof.

To use a number sequence in re-presentation means to regenerate the experience of counting that the records point to without generating actual sensory material. If auditory records are involved, the child creates a *verbal number sequence.* A lexical item of a verbal number sequence serves as a symbol for an initial segment of the number sequence—for the sequence of unit items symbolized by *one* up to and including the lexical item. When a child uses a number sequence in assimilation, the results are easy to observe. Given, for example, the following task: There are seven marbles in this cup (rattling marbles in the cup). Here are four more marbles (placing four marbles in another cup). How many marbles are there in all? If the child says there are seven in the cup, and proceeds to count on "8, 9, 10, 11—eleven!" it suggests that in uttering "seven" the child knows that the number word stands for a collection of perceptual unit items that, if counted, would be coordinated with the number words from *one* to *seven*. The child knows this and therefore does not have to run through the counting activity that is implied because the initial segment of the child's number sequence is symbolized by *seven*.

To designate the result of the operation of taking the sequence of unit items one through seven as a single unit, I use the term *composite unit*, a unit that itself is composed of units. A child who has constructed only the initial number sequence is yet to establish this operation. For such a child, *seven* can be a symbol for a number sequence but not for a unit containing that number sequence. In the preceding example, upon saying "eleven," if the child had just constructed the initial number sequence, the number word would refer to the sequence of involved counting acts but not to a unit containing that sequence. This restriction of the initial number sequence can be understood by reflecting on the process that produces it. In the review of counting that I explained, which yielded interiorized records of counting acts, the child operated on individual counting acts rather than on a *segment* of counting acts.

Children who have constructed only the initial number sequence can learn to unite sequences of counting acts into composite units. But this operation has to be actually performed and is not symbolized by the lexical items of the initial number sequence. That is, the child cannot take composite units as givens, but must *make* them in experiential contexts. One principal advancement in the tacitly nested number sequence is that a number word does symbolize the uniting operation used to take its initial segment as a unit. *Seven*, for example, refers to the operations necessary to take the lexical items of its initial segment *as material for making a composite unit*. The ability to create a unit containing an initial segment of a number sequence is a crucial step in the construction of whole numbers.

Another principal advancement is that the tacitly nested number sequence is a reversible counting scheme. That is, the child can take the number sequence as both a result of counting and as a situation of counting in the same counting episode. Although it is not currently interpreted in this way, double counting is simply the manifestation of reversibility of the number sequence and an indication that the child has constructed the tacitly nested number sequence.

I understand the construction of reversibility as a product of reinteriorization of the initial number sequence that occurs when a child takes an associated verbal number sequence as material in a review of counting. After constructing the number sequence at two levels of interiorization, the child can take the contents of the first level as the material of the operations at the second level, which is the constitutive operation of reversibility. However, there is still a restriction in the operations that are symbolized by the lexical items of the tacitly nested number sequence. Given an initial segment corresponding to, say, nine, a composite unit, say, seven, can be distinguished in nine, disembedded from nine, and treated as a number separate from nine *while leaving it in nine.* These operations are not symbolized for the child by the lexical items of the tacitly nested number sequence, even though the child can learn to carry them out.

When a child's part-to-whole numerical operations are symbolized, I call the child's number sequence *explicitly nested.* The inclusion relation for numbers is now symbolized and the child's number sequence is graduated in the sense that a lexical item of the sequence, say, seven, refers to a composite unit containing one through seven as well as a unit of one that can be iterated seven times. Iterating by one also sums by one, and the child understands that one is included in two, two in three, three in four, and so forth (Sinclair, 1971). I have found that these differences in children's number sequences have profound consequences for what they can establish as multiplication and division (Steffe, 1992).

## Composite Units as Iterable

One of the problems on which I am currently working is to investigate the construction of the first few whole numbers as iterable units. I am in the process of investigating how children make these constructions, so my comments about possible constructive itineraries should be taken as provisional.

In my experience, only the most enterprising children construct iterable composite units without having been presented with specific problem situations designed to encourage those constructions. However, the children who have constructed only the initial number sequence are far from constructing composite units as iterable. Surely, we would not expect such children to construct composite units as iterable anytime in the near future because they can at best make such units of units in experiential contexts. I know, for example, that two and three do not emerge as iterable units in one fell swoop. Early on, these number words refer to an assemblage of units that I call a *numerical pattern* for children who have constructed the initial number sequence. A numerical pattern is a sequence of abstract unit items that contain records of the elements of a counted figurative pattern. Although I did not mention them, figurative patterns serve a crucial role in the construction of the initial number sequence. They serve as a basis for constructing numerical patterns, which constitute the meaning of the first few number words of the initial number sequence. In fact, the initial number sequence itself is nothing but an elaborated numerical pattern. The conceptual leap that must be made to construct composite units as iterable is bridged by intermediate unit types (Steffe and von Glasersfeld, 1985b). Before becoming iterable, they emerge as *experiential composite units*, as *abstract composite units*, and as a unit type that I call *iterating composite units*.

An experiential composite unit is formed by taking an implementation of a numerical pattern as one thing, using a uniting operation. A numerical pattern is considered a dynamic and not a static entity. To implement or use a numerical pattern means to instantiate the records that compose it. The records of a pattern do not make a picture of the pattern, but they constitute a program or recipe whose enactment constitutes a sensory pattern.

An abstract composite unit is established by taking the elements of a *re-presented* numerical pattern as one thing. The differences in an experiential and an abstract composite unit are consequential because the experiential composite unit is made at the same level of interiorization as the involved numerical pattern. It appears to the child *in experience,* and the only modification involved is that the records of the uniting operation become recorded in the numerical pattern. An abstract composite unit is formed at a level of interiorization *above* the involved numerical pattern. Having no actual sensory material involved is an essential step in "leaving it behind." A re-presentation is already an abstracting operation and removes the child one step from its sensory-motor world. Separating a ground from background in re-presentation using the uniting operation is another act of abstraction, which means that the processes that produce experiential and abstract composite units are one level of abstraction apart—the level introduced by re-presentation.

Outlining various unit types provides a model of how the units of two through nine might become iterable. Children for whom a number word symbolizes an abstract composite unit can use this concept when finding how many elements are in so many of the composite units. Metaphorically, counting the elements of so many composite units is like jumping from one ladder of so many rungs to the next ladder, when the ladders are already in place with the intention of counting the rungs. *Iterating* a unit of ten, on the other hand, is like repeatedly laying down a ladder of so many rungs end-to-end with the intention of finding how many elements are generated. In this case, the ladder represents a resource one has in re-presentation that one can anticipate using for counting.

In their analysis of the iterable unit, Richards and Carter (1982) stated that for children to use *ten* as an *iterable* unit the child "must be able to strip it of its composite quality" (p. 16). I believe this is precisely what is started when a child makes an iterating unit. When a child can count "40 is 1; 50 is 2; 60 is 3; . . . ; 90 is 6" when finding how many tens there are from thirty to ninety, this is an indication that ten has become iterable; especially if thirty pennies are hidden, some more are put with them, and the child is told that there are now ninety pennies hidden and is asked to find how many dimes could be fairly exchanged for the added pennies. There are other uses of counting by tens that are indicators. For example, a child might independently coordinate a count by tens with a count by ones—"52, 62, 72, 73, 74, 75, 76"—to solve an arithmetical equation like "42 + —— = 76" (Steffe, 1984, pp. 107–109).

Having given this brief account of number sequence types

and composite unit types, I turn to illustrating how I am using these constructs to understand children's multiplying schemes (see Steffe, 1992, for a more complete account).

## A Premultiplying Scheme

For a situation to be established as multiplicative, it is necessary to at least coordinate two composite units in such a way that one of the composite units is distributed over the elements of the other composite unit. An 8-year-old child who began the teaching experiment having constructed only the initial number sequence was incapable of doing this. The modifications he made in his number sequence did not produce a multiplicative scheme, although he was able to solve some situations I would call *multiplicative*. One of his most advanced solutions occurred in the following protocol where one row of three blocks was visible and five rows were hidden. *T* stands for "teacher," and *Z* for "Zachary." *T:* If you had six rows of blocks, I wonder how many blocks that would be? *Z:* (using his index finger, he traces a segment on the table and then touches the table three times, where his points of contact form a row; he makes six such rows) Eighteen.

The sweeping motion over the table coupled with three pointing acts is solid indication that Zachary used his linear spatial pattern (a numerical pattern) in enacting his concept of three, creating a composite unit of three. Its unitary nature is solidly indicated by his keeping track of the six times that he used it (he made no visual records that were apparent to me), so I infer that Zachary's concept of three at this point contained records of uniting acts that he had performed in the past. My assumption is that Zachary re-presented a row of three, and in an attempt to count its elements, he enacted his re-presented concept by sweeping a finger to make something tangible to count. The enactment created a figurative composite unit of three "out there" that was as real for him as the row of three in his visual field. It was an act of externalization of his mental processes and the resulting figurative composite unit served the same role as a finger pattern, in that it provided Zachary a "tangible" composite unit whose elements could be counted.

We can see that Zachary enactively distributed his concept of three across the elements of his concept of six. Because there was no indication whatsoever that he made this units coordination prior to the activity of his scheme, I refer to the whole enactive units-coordinating scheme as an *enactive concept of multiplication*. But I would not call it a *concept* of multiplication nor would I consider calling the scheme a *multiplying scheme* because the units coordination was only enacted.

## A Multiplying Scheme

I had no indication that Zachary coordinated two units of different ranks prior to executing the activity of his units-coordinating scheme at any time during the teaching experiment. For that reason, his concept of multiplication did not progress beyond being enactive. Maya, a child who has constructed the tacitly nested number sequence, coordinated units prior to counting the first time that I worked with her. The situation in which I observed her coordinating units was as follows. I placed a red piece of construction paper in front of her and several congruent rectangular blue pieces cut so that six blue pieces fit exactly on the red piece. Then I asked her to find how many blue pieces fit on the red piece. After she said that six of them would fit on the red piece, I removed the 3 blue pieces she had placed on the red piece and placed two orange squares on one blue piece (they fit exactly). I then asked her how many orange squares would go on the red piece without allowing her to actually make the placements— "figure it out using the blue ones." After looking straight ahead, she subvocally uttered number words and said, "Twelve." In explanation, she tapped the table twice with each of six fingers synchronously with uttering the number words, "1, 2; 3, 4; 5, 6; 7, 8; 9, 10; 11, 12."

Tapping the table twice with each of six fingers indicates that Maya interpreted the request to "figure it out using the blue ones" numerically. Operating as she did was possible because she could use her number sequence in two ways—to create an extensive meaning for *six*, which I take as a composite unit generated by her using her numerical concept of six in re-presentation, and to count the elements of that composite unit, which in this case were symbols for pairs of orange pieces. The content of her figurative composite unit of six is, of course, unknowable, but whatever it was I take it to be a sequence of symbols for abstract units that she could "fill" with material of any kind; in this case symbols for a unit of two. To solve the problem as she did, Maya would have to distribute her units of two across each of the six units of one which were in re-presentation prior to counting.

It is critical in understanding the nature of Maya's two-for-one units coordination to remember that it was produced as a functional accommodation of the operations involved in double counting and her assimilating operations, and it was situational. As a product of reversibility of her number sequence, her two-for-one units coordination could be said to be implicit in double counting and this is the reason I refer to her complete two-for-one units-coordinating scheme as an *implicit concept* of multiplication.

Even though Maya counted in pairs "1, 2; 3, 4; . . . ; 11, 12" in explanation for how she found "twelve," when using her two-for-one units-coordinating scheme, I do not consider her implicit concept of multiplication to be repeated addition in the sense of taking twos as six addends. Coordination as I have explained it in Maya's case is a root structure that can yield what one might want to call a *repeated addition,* but it should not be confused with repeated addition. In fact, it is in nature like Confrey's (this volume) "one to many" action that she identified in the case of her splitting structure. "*Splitting* can be defined as an action of creating simultaneously multiple versions of an original, an action often represented by a tree diagram" (pp. 292). Of course, Maya made only the trunk of a tree and six branches, but that does not deter me from interpreting her coordination as being multiplicative in its constitution.

Confrey (this volume) seems to believe that the number sequence leads *necessarily* into interpreting multiplication as repeated addition—"we have argued that the majority of the current approaches to multiplication . . . rely too exclusively on repeated addition and its underlying basis in counting" (p. 291). I agree with Confrey that we should abandon encouraging multiplication only as repeated addition if repeated addition means "the identification of a unit then counting consecutively instances of that unit" (p. 292). Maya's units-coordinating scheme taught me how to make a distinction between multiplication and addition in the case of children's number sequences. In fact, Maya counted in precisely the same manner that Confrey (p. 301) explained people do when naming the results of three two splits—"1,2 . . . 3,4 . . . 5,6 . . . 7,8". So, like Confrey, I make an appeal that people not interpret multiplication of whole numbers only as repeated addition based on what they can observe children do, which may be to implement the activity of their units-coordinating scheme. It is what goes on before counting that constitutes a true units coordination.

## An Iterating Concept of Multiplication

Tenryn, a child who could perform the operations involved in reinteriorization, established a repeated adding scheme about one month after I began working with him during the fall term of his third grade in school. He had just made nine rows of blocks with three per row, arranging them into a rectangular array with nine rows and three columns.

> T: How many blocks do you have (covering the blocks)?
> Tn: (looks at the teacher for approximately 8 seconds and then puts up his left index, middle, and ring fingers. He then subvocally utters for approximately 75 seconds. As he subvocally utters, he alternately looks into space and at his finger pattern) Twenty-seven. T: How in the world did you figure that out?
> Tn: My fingers. T: OK, tell me what you did. Tn: (puts up the same finger pattern and moves it synchronous with uttering number words) 3, 6, 9, 12, 15, . . . T: (interrupting) how did you keep track? (Tn: smiles uncertainly as if he does not know.) Did you use that hand (pointing to Tenryn's right hand)? (Tn: shakes his head "no".) T: (points to his head) How did you keep track in your mind? How did you know how to keep count? Tn: I counted by threes (holding up his finger pattern). T: You counted by threes? How many times did you count? Tn: Nine. T: How did you keep track of nine? How did you know when to stop? Tn: At twenty-seven. T: Show me how you kept track. Tn: (puts up his finger pattern once again and moves it synchronous with uttering) 3, 6, 9, 12, . . . T: (interrupting) how many times is that? Tn: Three! (Pauses, then puts up his finger pattern to continue) Four (meaning that it was four instead of three. He then continues on) five would be 12; six, 18 (he had his finger pattern active and moves it with each count); seven, 21; eight, 24; nine, 27!

During the 8 seconds after I asked Tenryn how many blocks he had, he was actively engaged in assimilating the situation. What went on during these 8 seconds is indicated by the activity that followed. Because he used only *one finger pattern* as he subvocally uttered for approximately 75 seconds, I take his numerical finger pattern of three as a unit that he could use a specific number of times and call it *an iterating unit*. Although it is not indicated in the protocol, it was crucial that he could form the goal of using it a specific number of times for it to be available for use.

My argument is that Tenryn had a scheme he could use to establish *nine units of three*. There was no need to distribute a unit of three across the nine units of one prior to initiating the activity, because he had a unit of three—his finger pattern—that he could use in iterating. Tenryn did make an enactive units coordination when keeping track of each unit of three until reaching nine such units. In this sense, he established a unit of three as a countable unit and used his number sequence for one to keep track of making these composite units. It qualifies Tenryn's scheme as an enactive units-coordinating scheme. I call his entire scheme an *iterating concept* of multiplication as repeated addition and take it as a root of repeated addition.

Tenryn's iterating concept of multiplication only *seemed* to be more advanced than Maya's implicit concept of multiplication. Maya was aware of six twos as a result of her units coordination but she did not constitute a unit of two as an iterating unit until later on in the teaching experiment. Rather, she counted the elements of six twos using her number sequence. The unit of two was not a countable unit item, but the elements of the six twos were countable.

What went on prior to Tenryn's executing the activity of his iterating concept of multiplication was quite different. His concept of three could be used in re-presentation and coordinated with his number sequence, not distributed across the elements of his number sequence, but rather implemented using those elements as material. Tenryn's numerical finger pattern for three was like an activated tuning fork in that he fully intended to repeat it nine times, continuing to count the fingers he used after each implementation. The experience of "nine threes" did not appear to him prior to his implementations of his numerical finger pattern even though he intended to make nine threes. He could anticipate what the results would be because he could count to nine. But the experience of nine threes depended on his actual implementations. In other words, nine threes was one of the results of the activity of his iterating concept of multiplication as repeated addition, whereas "six twos" was the result of the assimilating operations of Maya's implicit concept of multiplication.

In the discussion of Maya's two-for-one units-coordinating scheme, I made an appeal that people not interpret multiplication of whole numbers only as repeated addition based on what they can observe children do, which may be to implement the activity of a units-coordinating scheme. Based on Tenryn's case, however, we see that such an interpretation might be appropriate. Because it was Tenryn who contributed his iterating concept of multiplication to the situations that I presented, we should not dogmatically discourage multiplication as repeated addition if repeated addition means "identifying a unit and then counting consecutively instances of that unit" (Confrey, this volume, p. 292), because that is essentially what Tenryn did. We should not make a priori decisions to teach multiplication either as repeated addition or as distribution of a composite unit across the elements of another composite unit, because either decision excludes the other. Rather, we should develop alternative interpretive constructs concerning *children's multiplication* and abandoning encouraging any single interpretation for all children. Therefore, I find, as does Confrey, that repeated addition should not be the *only* model for multiplication.

## A Part-to-Whole Units-Coordinating Scheme

Johanna, a child in the teaching experiment who had constructed the explicitly nested number sequence, made accommodations in her number sequence in constructing multiplying schemes that went beyond anything either Maya or Tenryn was able to do. Two primary ways of operating were available to Johanna that were not available to either Maya or Tenryn. Johanna could unite a bounded figurative plurality of composite units into another composite unit, forming a unit of units of units; and she could engage in numerical part-to-whole reasoning. For example, when investigating her operations, I asked her to take twelve blocks, told her that together we had nineteen, and asked her how many I had. After sitting silently for about 20 seconds, she said, "Seven," and explained, "Well, ten plus nine is nineteen; and I take away the two—I mean, ten plus two is twelve, and nine take away two is seven!"

Johanna first decomposed nineteen into two parts, ten and nine. She then transformed the parts into twelve and seven by adding two to ten and compensated by subtracting two from nine. To operate in this way, it would be necessary for her to not only decompose nineteen into the two units, ten and nine, but to also disembed both of these subunits from nineteen and use them as the material of operating independent of nineteen. The compensation involved is solid indication that, after she disembedded the two units from nineteen, she still viewed them as constitutive parts of nineteen. This example indicates to me that Johanna could disembed numerical parts from a numerical whole, use these parts as material for further operating, and reconstitute *that* result as the original numerical whole.

Like Maya, when there was a definite numerosity of units of three hidden, Johanna could use her concept of three in a units coordination. In contrast to Maya, she could then take those results as material for further operating in the restricted situation that I call a *missing rows* situation. We see its consequences in the following two situations. In the first situation, I asked Johanna to make four rows of three, leaving them in her visual field. I then covered some other rows and then told her that there were now seven rows of three and asked her to find how many were hidden. She said "Three," almost immediately. Johanna could treat the rows as if they were singleton units but still maintain their numerosity. The latter is indicated in another situation, where she found that four rows of three were hidden in the same explicit way (seven were visible). I then asked her how many blocks were hidden and, after about 10 seconds, she said, "Sixteen," using units of four rather than three. I do not view her taking each row as containing four rather than three blocks as a *necessary* error. She seemed to be explicitly aware of the coordination between her number sequence and units of three.

The power of Johanna's operations were revealed in a situation where I hid five rows of blocks with four per row, which Johanna had made, and asked her how many blocks were in the first three rows. She sat silently for about 25 seconds with her hands resting in her lap in deep concentration, and then said, "Twelve." So, I then asked her how many more rows she had and she said, "Two." She added that there were eight blocks in the two rows. Quite encouraged, I then asked her how many blocks were in all five rows. She again sat silently for about 15 seconds and replied, "Twenty!" In explanation, she said, "Because I added up. Twelve plus four is sixteen, and sixteen plus four is twenty!"

What was significant in the situation is that Johanna used her concept of four in reprocessing the results of re-presenting five rows of four, counting out the first three units of four and uniting them into another composite unit. There is also solid reason to believe that she then went on, counting the remaining elements of the five units of four that she had made in re-presentation and united these two composite units together, forming another unit of units of four. These inferences are based on the 24 seconds she sat in deep concentration before she said *twelve,* on the fact that she immediately knew that two more rows of four were left, and the fact that she progressively united these two composite units with twelve to find the number of blocks. This modification in operating (making units of units of units) was permanent and can be called a *functional accommodation* in her units-coordinating scheme.

The progressive uniting operations ("twelve plus four is sixteen, and sixteen plus four is twenty") cannot be taken as a prima facie indication that four was permanently established as an *iterable* unit. Based on the protocol alone, the most that can be said is that her concept of four seemed to be permanently established as an iterating unit. So, I took eight blocks and asked her, "If I make rows out of my blocks, and add them to your rows, how many rows would you have?" After 10 seconds, she said, "Twenty-eight." So, I said, "Twenty-eight blocks, right? How many rows would you have?" Johanna then said, "Oh! Hm—seven rows." In explanation she said, "Because, you have four blocks in each row and you have five of them, and five plus two is seven!" Immediately after, I put seven rows together, took twelve more blocks, and asked the same question. Without actually making the additional rows, Johanna said, "Ten." She explained, "I counted the number of rows of four in twelve and three rows, and seven plus three is ten!"

Johanna had just found that there were twenty blocks in five rows of four using her units-coordinating scheme. So she could count the eight blocks onto the twenty blocks without experiencing a conflict between two ranks of units as being countable. Combining the blocks together into twenty-eight blocks, she formed a composite unit containing the twenty-eight blocks, the first twenty of which she had already organized into a unit of units of four. She could view the last eight blocks from the vantage point of this unit of units of four and continue to use her concept of four in re-presentation, forming a *continuation* of the five units of four. This is the first time that I would say that Johanna used an iterable unit more than once in re-presentation to establish a bounded figurative plurality of composite units of indefinite numerosity when solving my composite units situations. The consequences of establishing this structure are immediately obvious, because she mentally made units of four from the eight blocks and added those two units onto the preceding five.

When finding ten rows in the very next task, Johanna simply found how many more rows of four could be made using the twelve blocks and mentally added these three rows onto the preceding seven. She could drop the step of counting twelve onto twenty-eight because she could take the results of using her part-to-whole operations as a given. That is, she could take the twenty blocks and the twelve blocks together into a unit of indefinite numerosity and still work with the parts without making the indefinite numerosity definite. She eliminated it as a redundancy.

## ON ESTABLISHING A REVERSIBLE UNITS-COORDINATING SCHEME

The modification in operating that Johanna made was permanent, because later in the teaching experiment I asked her to place twelve blocks into a liter container and asked her how many groups of three she could make using the twelve blocks. After she found four by counting by threes, I placed some more blocks into another container and asked Johanna how many groups of three were in the other container if there were twenty-seven blocks when the blocks were poured together. Because Johanna stood in deep concentration for about 30 seconds subvocally uttering number words before answering, "Five.", my inference is that she counted from twelve up to twenty-seven by three, counting the units of three that she made. I also infer that the number word *twelve* symbolized a unit containing four units of three if for no other reason than she found the sum of five and four when I asked her how many groups there were altogether.

Finding the number of groups of three she could make from twelve blocks was a critical part of her solution. She took the results, four composite units of three elements each, as material for making a unit. I believe that she united these four composite units, making a unit of units of units. After performing this operation, she had established a situation analogous to what I have already explained in the case of her functional accommodation of her units-coordinating scheme. Three was now an iterable unit that she could use in re-presenting a continuation of the unit containing four units of three and in counting beyond a unit containing four units of three using her units-coordinating scheme.

Johanna was in the process of constructing a system of operations that would feed the results of using her units-coordinating scheme back into its "situation." This system of operations can be used to establish the units-coordinating scheme as reversible. The system also makes it possible for a child to use his or her concept of three in re-presentation more than once, creating a bounded plurality of composite units of indefinite numerosity and the involved composite unit as iterable. However, this structure of a unit of units of units was not available to Johanna prior to using her schemes.

## Iterative Multiplying Schemes

Johanna had not constructed what I would call an *iterative multiplying scheme*, because her composite units were not yet iterable nor was her scheme reversible. That is, the operations involved in using a composite unit in re-presentation more than once, creating a bounded figurative plurality of composite units of indefinite numerosity, were not symbolized. She could perform these operations in a context, but they were not available to her prior to operating. I explained how she could re-present five rows with four blocks each, take that bounded figurative plurality as the material of her uniting operation, establishing a unit of five units of four, and then separate the five units of four into three units and two units, using the same concept of four that she used in the re-presentation. Reprocessing the elements of a unit of units of four with one of its elements strips the units of four of their sensory quality and reinteriorizes them at a new level. Records of the motion involved in reprocessing also become contained in the concept of four that is used in reprocessing. If they are permanent records, the concept of four takes on a characteristic of being iterable. As an iterable unit, it is at the same level of interiorization as the composite units to which it was applied.

If Johanna had completed the construction of her units-coordinating scheme as a reversible scheme, then she could take its situation and results as being interchangeable. In this case, she would have completed construction of what I call a generalized number sequence—"one is three, two is six, three is nine, . . . " The generalized number sequence symbolizes the operations involved in the units coordination that Maya taught me and that I now regard as the roots of multiplication. Its construction serves as the basis for composite units to be iterable. I call it an *iterative multiplying scheme* and with it expressions of multiplication like "four three times" begin to appear.

## Abstract Concepts and Properties

My interest in iterative schemes should not be interpreted as implying that I view the child as "nothing but a counter" or that I intended to mold the children so that their primary response to arithmetical problems would be always to count. Rather, my main hypothesis is that iterative multiplying and dividing schemes are essential material in the construction of more abstract concepts of multiplication and division. The progress children make in constructing the explicitly nested number sequence must be recapitulated after they construct the generalized number sequence, in that it must be used as material for part-to-whole operations. This process transforms the generalized number sequence into a measurement scheme. I cannot elaborate on this constructive process, but it has its consequences for mathematics education in middle and late childhood. It is here that I put the concepts of quotitive and partitive division, commutativity, distributivity, and forms of strategic reasoning.

*Commutativity.* Experiential regularities of the results of multiplying can lead to an empirical generalization that the multiplier and multiplicand can be interchanged without changing the results. For example, after Tyrone had solved a problem by iterating four nine times, he started counting "nine, eighteen, . . . " He then said, "There's two ways to do it," and explained, "nine times four and four times nine." In Skemp's (1978) terms, Tyrone's use of commutativity was "instrumental" rather than "relational." Baroody, Ginsburg, and Waxman (1983) made a similar analysis when discussing first-graders' use of an additive commutative principle.

> The development of a secure knowledge of this principle so early is probably the result of informal experience. Young children's addition gives the appearance of a primitive notion of commutativity ("proto commutativity"): the order in which the addends are dealt with does not affect the correctness of the sum. . . . With additional experience, the proto concept may be elaborated to form a genuine concept of commutativity. (p. 186)

Attaining the internal *logical necessity* of commutativity requires at least partitioning and recombining the results of multiplying in creative ways. Iterable units must be extracted from their context and become arithmetical objects to which further conceptual operations can be applied. If the child runs through the activity of repeating the elements of the unit of four nine times in thought ($9 \times 4$), each unit of one within the unit of four would be instantiated nine times. Taking the instantiations of each unit of one as a new composite unit yields the product $4 \times 9$. In this case, the commutative principle would be an object of awareness and would involve a creative reorganization of the abstract multiplicative concept. I am investigating commutativity and its use by children because of its obvious importance.

*Curtailment.* The shortening of a process in solving problems is called *curtailment* (Krutetskii, 1976, p. 263).

Curtailment is shown in only a "very elementary form in the primary grades" (p. 337). Nevertheless, curtailments of iterative schemes are important in the progress of the child. One of the most obvious reasons for seeking possible curtailments of iterative schemes springs from the necessity to acquire knowledge of the basic multiplication facts.

A curtailment of an iterative scheme refers to shortening the second part of the scheme. The results of, say, repeating six four times can provide the child with a "functional" connection between "4 × 6" and twenty-four. Curtailment of the response via reflection and abstraction leads to a solid numerical connection. That these curtailments can happen is unquestionable (Gates, 1942); *how* they happen is yet to be explored. I believe that there are moments when it seems natural to the child not to count, and I am accounting for this seemingly natural change in the child's knowledge. I am particularly interested in curtailments of reversible schemes; that is, in investigating acquisition of the division facts based on knowledge of the multiplication facts and upon reversible schemes.

## Modeling Mathematical Knowledge

I contend that the progress in children's schemes of actions and operations that are functioning reliably and effectively can be modeled. In El'konin's (1967) assessment of Vygotsky's research, the essential function of the teaching experiment was the production of models: "Unfortunately, it is still rare to meet with the interpretation of Vygotsky's research as *modeling*, rather than empirically studying, developmental processes" (p. 36).

A viable way to investigate the nature and development of the child's schemes is to attempt to reconstruct the steps that led the child to whatever schemes it has acquired. Such a reconstruction is necessarily hypothetical, because

> another person's conceptions are, by definition, not observable. In this connection it is crucial to remember that conceptual structures (knowledge) are, in our view, not transferable. . . . Even if a child was aware of its own conceptions and could reflect upon and verbalize them, even if it told the observer (teacher, experimenter) what it believes to be, say, its concept of numbers, that observer could only interpret

that verbal message in terms of his or her own experience. (Steffe et al., 1983, p. xvi)

The essential virtue of the teaching experiment is that it allows the study of constructive processes—those critical moments when restructuring take place and is indicated by alterations in the child's behavior. To El'konin (1967), the method was an essential part of Vygotsky's experimental-genetic method. He described it as allowing "a dissection in abstract form of the very essence of the genetic process of concept formation" (p. 36). This agrees with my own understanding of the modeling process. However, a child develops in mathematics under the influence of teaching, and if one wants to observe mathematical conceptual development, one has to observe the child in the teaching context. The successful conceptual models of the construction of the counting scheme (Steffe et al., 1983; Steffe and Cobb, 1988) were produced as a result of prior teaching experiments.

My emphasis on model building is consistent with the views of Dewey (McLellan and Dewey, 1908) and Skemp (1981). A quote from Skemp (1981) is particularly significant:

> The mathematical schemas which we want children to construct over a period of ten years or so are formed from many concepts, each of which has itself much interiority. Micromodels of this kind are needed not only to enable us to teach these concepts successfully, but also repeatedly to bring home to us, by a variety of examples, the great concentration of knowledge which mathematics contains. (p. 2)

My effort to derive "micromodels" of children's operations will not only classify how children may solve the specific problems, but also serve to formulate ways of proceeding for teachers in elementary schools.

## EDUCATIONAL SIGNIFICANCE OF THE WORK

It cannot be stressed too much that teaching in a teaching experiment occurs under laboratory-type conditions. The teachers were experienced mathematics educators, the number of students was small, and the interaction of the teachers and students was intensive as well as extensive. Laboratory-type conditions are essential if the child's construction of mathematical concepts and operations is ever to be explained in any significant depth. Observing manifestations of children's constructive processes firsthand provides a crucial experiential basis for formulating models of their construction.

Research in mathematics education has turned to providing a useful understanding of how the child's mathematical concepts may be built up piece by piece. As a result, there are indications of how children's spontaneous methods can be educationally exploited to the best advantage. Ultimately, children must construct their own ways and means of operating; as a mathematics educator, I believe that we can play a leading role in that development, but only on condition that we learn to bring our teaching strategies into harmony with what children can do at any given point. If we want them to learn more than collections of "mathematical facts," if we want them to understand the arithmetical operations, we must provide a sequence of steps that explicitly lead from their first intuitive attempts to the practice of more sophisticated methods.

I believe that adults can help children learn mathematics; but to do this successfully, teachers must take careful stock of the conceptual "raw material" that is there to start with. As I have had ample opportunity to observe, the generative power of children is impressive, but as yet essentially uncharted. This project is an attempt, on the basis of conceptual analysis and modeling, to find out more about their conceptual raw material and the possibilities of its transformation. The insights gained about children's methods and their progressive change under the influence of experimental teaching can provide powerful new guidelines to the educator whose aim is to foster the development of genuine mathematical understanding.

The generalizability of the results hinges on the theoretical constructs used to explain children's behavior. Skemp (1979) discussed this issue with regard to the teaching experiment. His two reasons for expecting results to be "generalizable" are as follows:

> First, the mathematical structures to be learnt are the same, or nearly so, for all the children whose learning of them it is hoped eventually to help. Second, though chil-

dren themselves vary both in their learning abilities and in the schemas which they have available . . . there are regularities in the learning process itself . . . (pp. 28, 29)

I concur with Skemp. Long ago, Piaget (1964) identified experience as one crucial factor of intellectual development. This applies with equal force to the researcher who aims at acquiring knowledge concerning children's conceptual structures and their evolution. No amount of sporadic testing and statistical elaboration of test results can possibly provide insights into such evolution that would be comparable to the insights gained in the longitudinal study of individual children's progress. Only a protracted effort of observation, interaction, and further observation can supply the intensity and the continuity of experience that are indispensable if substantive results are to be obtained.

## ACKNOWLEDGMENTS

This chapter is based on NSF Grant No. MDR-8550463. All opinions and findings are those of the author. I would like to thank Ernst von Glasersfeld for his insightful contributions to NSF Grant No. MDR-8550463 on which this chapter is based. I would also like to thank Jere Confrey for her helpful comments and Guershon Harel for his encouragement and support.

## REFERENCES

Adler, M. J. 1982. *The Paideia proposal: An educational manifesto:* New York: Macmillian Publishing Company.

Baroody, A. J., H. P. Ginsburg, and B. Waxman. 1983. Children's use of mathematical structure. *Journal for Research in Mathematics Education,* 14(13): 156–168.

Booth, L. R. 1981. Child methods in secondary school mathematics. *Educational Studies in Mathematics* 12: 29–41.

———. 1984. Children's conceptions and procedures in arithmetic: some implications. In *Proceedings of the eighth international conference for the psychology of mathematics education,* ed. B. Southwell et al. Matcham NSW: D. Prescott.

Brownell, W. A. 1928. *The development of children's number ideas in the primary grades.* Chicago: University of Chicago Press.

———. 1935. Psychological considerations in the learning and the teaching of arithmetic, in *Teaching of arithmetic*, ed. W. D. Reeve. *Tenth yearbook of the National Council of Teachers of Mathematics*. New York: Bureau of Publications, Teachers College, Columbia University.

———. 1945. When is arithmetic meaningful? *Journal of Educational Research* (March).

Bridgman, P. W. 1927. *The logic of modern physics*. New York: Macmillan Co.

Carpenter, T. P., and J. M. Moser. 1984. Acquisition of addition and subtraction concepts. *Journal for Research in Mathematics Education* 15: 179–202.

Cobb, P., and L. P. Steffe. 1983. The constructivist researcher as teacher and model builder. *Journal for Research in Mathematics Education* 14 (12): 83–94.

Confrey, J. 1991. The concept of exponential function: A students' perspective. In *Epistemological foundations of mathematical experience*, ed. L. P. Steffe. New York: Springer Verlag.

Easley, J. A. 1975. Thoughts on individualized instruction in mathematics. In *Schriftenreihe des IDM*, ed. H. Bauersfeld, M. Otte, & H. G. Steiner, vol. 5, 21–48. Universität Bielefeld, Institut für Didaktik der Mathematik.

El'konin, D. B. 1967. The problem of instruction and development in the works of L. S. Vygotsky. *Soviet Psychology*. 5 (3): 34–41.

Erlwanger, S. H. 1973. Benny's concept of rules and answers in IPI mathematics. *Journal of Children's Mathematical Behavior* 1: 7–26.

Freudenthal, H. 1979. Learning processes. In *Some theoretical issues in mathematics education: Papers from a research presession*, ed. R. Lesh and W. Secada. Columbus: Ohio State University ERIC/SMEAC.

Gates, A. I. 1942. Connectionism: Present concepts and interpretations. In *The Psychology of learning, forty-first yearbook of the NSSE*, ed. N. B. Henry. Chicago: University of Chicago Press.

Ginsburg, H. 1977. *Children's arithmetic: The learning process*. New York: D. Van Nostrand Company.

———. 1981. The clinical interview in psychological research on mathematical thinking: Aims, rationales, techniques. *For the Learning of Mathematics*, 1(3): 4–11.

Hart, K. M. 1983. I know what I believe, do I believe what I know? *Journal for Research in Mathematics Education* 2: 119–125.

Hatfield, L. 1976. Child-generated computational algorithms: An exploratory investigation of cognitive development and mathematics instruction with grade II children. Unpublished manuscript.

Hawkins, D. 1973. Nature, man, and mathematics. In *Developments in mathematical education: Proceedings of the second International Congress on mathematical education*, ed. A. G. Howson. Cambridge: Cambridge University Press.

Kieren, T. 1990. Children's mathematics/mathematics for children. In *Transforming children's mathematics education: International perspectives*, ed. L. P. Steffe and T. Wood. Hillsdale, NJ: Lawrence Erlbaum Associates.

———— and S. E. B. Pirie. 1991. Recursion and the mathematical experience. In *Epistemological foundations of mathematical experience*, ed. L. P. Steffe. New York: Springer Verlag.

Krutetskii, V. A. 1976. *The psychology of mathematical abilities in school children*, ed. J. Kilpatrick and I. Wirzup; trans. J. Teller. Chicago: University of Chicago Press.

McKnight, C. 1987. *The underachieving curriculum*. Champaign, IL: Stipes Publishing Company.

———— and R. B. Davis. 1980. The influence of semantic content on algorithmic behavior. *The Journal for Mathematical Behavior* 3: 39–79.

McLellan, J. A., and Dewey, J. 1908. *The psychology of number*. New York: Appleton (originally published, 1895).

Menchinskaya, N. A. 1969. The psychology of mastering concepts: Fundamental problems and methods of research. In *Soviet studies in the psychology of learning and teaching mathematics*, vol. 1, ed. J. Kilpatrick and I. Wirzup. Stanford, CA: School Mathematics Study Group.

National Council of Teachers of Mathematics. 1989. *Curriculum and Evaluation Standards for School Mathematics*. Reston, VA.

National Research Council. *Everybody counts: A report to the nation the future of mathematics education.* Washington, DC: National Academy Press, 1989.

Opper, S. 1977. Piaget's clinical method. *Journal of Children's Mathematical Behavior* 1(4): 90–107.

Piaget, J. 1936. *La naissance de l'intelligence chez l'enfant*. Neuchatel, Switzerland: Delachaux et Niestle.

————. 1937. *La construction du reel chez l'enfant*. Neuchatel, Switzerland: Delachaux et Niestle.

————. 1945. *La formation du symbole chez l'enfant*. Neuchatel, Switzerland: Delachaux et Niestle.

————. 1964. Development and learning. In *Piaget rediscovered: A report of the conference on cognitive studies and curriculum development*, ed. R. E. Rippe and V. N. Rockcastle. Ithaca, NY: Cornell University Press.

————. 1970. *Genetic epistemology.* New York: Columbia university Press.

————. 1971. *Biology and Knowledge.* Chicago: University of Chicago Press.

————. 1973. Comments on mathematical education. In *Developments in mathematical education: Proceedings of the second international congress on mathematical education,* ed. A. G. Howson. Cambridge: Cambridge University Press.

————. 1980. The psychogenesis of knowledge and its epistemological significance, In *Language and learning: The debate between Jean Piaget and Noam Chomsky,* ed. Massimo Piattelli-Palmarini. Cambridge, MA: Harvard University Press.

Resnick, L. B. 1982. Syntax and semantics in learning to subtract. In *Addition and subtraction: A cognitive perspective,* ed. T. Carpenter, J. Moser, and T. Romberg. Hillsdale, NJ: Lawrence Erlbaum Associates.

Richards, J., and R. Carter. 1982. The numeration system. In *Proceedings of the fourth Annual Meeting of the Psychology of Mathematics Education,* ed. S. Wagner. Athens: University of Georgia Press.

Saxe, J. 1982. Developing forms of arithmetical thought among the Oksapmin of Papua, New Guinea. *Developmental Psychology* 18 (4): 583–594.

————. 1990. *Culture and cognitive development: Studies in mathematical understanding.* Hillsdale, NJ: Lawrence Erlbaum Associates, Inc.

Schubert, W. H., and A. L. Lopez Schubert. 1981. Towards curricula that are of, by, and therefore for students. *Journal of Curriculum Theorizing* 3(1): 239–251.

Sinclair, H. 1971. Number and measurement. In *Piagetian cognitive-development research and mathematical education.* ed. M. F. Rosskcopf, L. P. Steffe, and S. Taback. Reston, VA: National Council of Teachers of Mathematics.

Skemp, R. 1978. Relational understanding and instrumental understanding. *Arithmetic Teacher* 26: 9–15.

————. 1979. Theories and methodologies. Paper presented at the wingspread conference on the initial learning of addition and subtraction skills.

————. 1981. New directions in the psychology of mathematical education: A personal view. Third annual meeting of the North American chapter of PME.

Steffe, L. P. 1982. Conceptual structures for arithmetical operations. *Problem solving* 4(8).

————. 1983a. Children's algorithms as schemes. *Educational Studies in Mathematics* 14: 109–125.

————. 1983b. The teaching experiment in a constructivist research program. In *Proceedings of the fourth International Congress on Mathematical Education*, ed. M. Zweng et al. Boston: Birkhauser.

————. 1984. Communicating mathematically with children. In *Theory, research, and practice in mathematical education: Working group reports and collected papers of the fifth international congress on mathematical education.* ed. A. Bell, B. Low, and J. Kilpatrick. Shell Center for Mathematical Education, University of Nottingham.

————. 1991a. The constructivist teaching experiment: Illustrations and implications. In *Radical constructivism in mathematics education*, ed. E. von Glasersfeld. Boston: Kluwer Academic.

————. 1991b. The learning paradox: A plausible counterexample. In *Epistemological foundations of mathematical experience*, ed. L. P. Steffe. New York: Springer Verlag.

————. 1992. Schemes of action and operation involving composite units. *Learning and Individual Differences: A Multidisciplinary Journal in Education*, 4(3): 259–309.

Steffe, L. P. and P. Cobb. 1988. *Construction of arithmetical meanings and strategies.* New York: Springer-Verlag.

Steffe, L. P. and E. von Glassersfeld. 1980. Analysis of the child's construction of whole numbers. NSF Grant No. SED-80 16562.

———— and E. von Glasersfield. 1985a. Child generated multiplying and dividing algorithms. NSF Grant No. MDR-8550463.

———— and E. von Glassersfeld. 1985b. Helping children conceive of number. *Recherches en Didactique des Mathematiques* 6 (2–3): 269–303.

Steffe, L. P., E. von Glasersfeld, J. Richards, and P. Cobb. 1983. *Children's counting types: Philosophy, theory, and application.* New York: Praeger Scientific.

Steffe, L. P. and T. Wood, ed., 1990. *Transforming childrens' mathematics education: International perspectives.* Hillsdale, NJ: Lawrence Erlbaum Associates.

Streefland, L. 1984. Unmasking N-distractors as a source of failures in learning fractions. *Proceedings of the eighth international conference for the psychology of mathematics education.* Matcham NSW: D. Prescott.

Thom, R. 1973. Modern mathematics: does it exist? In *Developments in mathematical education: Proceedings of the second international congress on mathematical education*, ed. A. G. Howson. Cambridge: Cambridge University Press.

Van Engen, H. 1949. Analysis of meaning of arithmetic. *Elementary School Journal* 49: 321–329, 395–400.

—— and E. G. Gibb. 1956. *General mental functions associated with division.* Cedar Falls: Iowa State Teachers College.

von Glasersfeld, E. 1980. The concept of equilibration in a constructivist theory of knowledge. In *Autopoiesis, communication and society,* ed. F. Benseler, P. M. Hejl, and W. K. Kock. New York: Campus.

——. 1982. Subitizing—The role of figural patterns in the development of numerical concepts. *Archives de Psychologie* 50: 191–218.

——. 1991. Abstraction, re-presentation, and reflection: An interpretation of experience and Piaget's Approach. In *Epistemological foundations of mathematical experience,* ed. L. P. Steffe. New York: Springer Verlag.

Wertheimer, M. 1942. *Productive thinking.* New York: Harper and Row.

# 2 Multiplicative Conceptual Field: What and Why?

## Gerard Vergnaud

### INTRODUCTION

It is a trivial idea to consider that children (and students) develop more and more complex competencies and conceptions by using their former knowledge to make sense of new situations and try to grasp them. This process can be viewed as a general process of adaptive behavior: assimilation and accommodation, as Piaget first stressed it.

But most psychologists have tried to theorize about the progressive complexity to children's competencies within content-free frameworks: logic, information processing, linguistics, or factor analysis. Piaget himself has sometimes paid attention to the conceptual components of children's knowledge (space, time and speed, probability . . . ) and sometimes tried to reduce the conceptual complexity progressively mastered by children to some kind of general logical complexity, represented in his theory by such structures as those of grouping, the INRC group, and combinatorics.

The conceptual field theory asserts that a more fruitful approach to children's cognitive development is provided by using a framework referring to the contents of knowledge themselves and to the conceptual analysis of the domain. This approach has already provided enlightening results for the acquisition of elementary arithmetic (additive structures, multiplicative structures); for elementary physics, biology, or economics; for elementary algebra and geometry; and for different technological domains. As far as MCF (multiplicative conceptuals field) is concerned, it is now clear that one cannot reduce proportional reasoning, or the concepts of fraction and ratio, or the algorithms of multiplication and division, to any logical, information processing, or linguistic reasoning. Logic, computer science and linguistics do not provide us with concepts sufficient to conceptualize the world and help us meet the situations and problems that we experience. This is true for the acquisition of rational knowledge as well as for science itself. This epistemological point of view is in line with, but goes far beyond the consequences one can draw from, Godel's theorem.

The conceptual field theory also stresses that the acquisition of knowledge is shaped by the situations and problems first mastered and that knowledge has therefore many local features. All concepts have a restricted domain of validity, which varies with experience and cognitive development. The conceptual field theory is a pragmatist theory, though that does not mean it is empiricist. A problem is not a problem for an individual unless the individual has concepts enabling him or her to consider it as a problem for him or herself; the process of simulating a problem goes far beyond the abstraction of regularities from the observable world. Problems are practical and theoretical, and not merely empirical, even for young children. When a class of problems is solved by an individual (this means that he or she has developed an efficient scheme to deal with all or nearly all the problems of the class), the *problematic* character of that specific class passes away. This new power enables one to tackle new situations and objects and try to understand new properties and relationships, and therefore to pose and recognize or consider new problems for oneself. Thus, this process is a continuing cycle.

From my point of view, the most general features of the multiplicative conceptual field are the following:

- Its framework is mathematical, in a wide sense of mathematics. But the mathematics used to analyze MCF takes account of the contrasts and interconnections among the conceptual operations needed to progressively master this field. I will illustrate this point later.
- The situations and problems that offer a sound experiential reference for MCF are not purely mathematical, especially at the elementary and early secondary levels. The child's early experience of buying goods and sharing sweets, and his or her first understanding of speed, concentration, density, similarity, or proba-

bility are essential. The didactical consequences of this are important.

- The identification of the bulk of concepts needed to analyze the bulk of MCF situations is an essential theoretical problem for researchers. I use the word *bulk* instead of *set*, because the frontiers of MCF are not strictly defined, neither in terms of the concepts nor the situations and problems. There is also a dialectical tie between situations and concepts in the sense that each bulk depends heavily on the other. The concept of a conceptual field represents some kind of equilibrium between the need to classify the objects we have to study in psychology and mathematics education and the need to understand something about learning, development, and teaching. Obviously, students move in the whole repertoire of their mathematical competencies, yet I consider MCF as a reasonable-sized object for research and theory.

- Finally, one must not minimize the role of language and symbols in the development and the functioning of thinking. This is, of course, true for MCF as well as for additive structures, algebra, or mechanics. It is therefore essential to classify and analyze the variety of symbolic and linguistic signifiers that we may use when communicating and thinking about MCF, even though signifiers are not concepts or conceptual operations but only stand for them. An essential theoretical and empirical task for researchers is to understand why a particular symbolic representation can be helpful, under which conditions and when and why it can be profitably replaced by a more abstract and general one. The necessity to educate all students to a reasonably proficient level in algebra makes these considerations essential.

The conceptual field theory is therefore a complex theory. This complexity is inevitable because we need to embrace, in one single theoretical glance, the whole development of the situations progressively mastered, of the concepts and theorems required to operate efficiently in those situations, and of the words and symbols that can effectively represent these concepts and operations to students, depending on their cognitive levels.

Teaching is essential. Situations, verbal and symbolic mediations, and scaffoldings of all kinds are the usual ways by which teachers help students learn. Therefore one can hardly hold a radical constructionist approach. I prefer to speak of the appropriation process by which students make social knowledge their own, personal knowledge, with the help of teachers, parents, and peers (Vygotsky, 1962). This point of view is a constructionist point of view, but in a restricted meaning of the word: nobody, in place of the student, can grasp the meaning of a problem (and eventually its solution), make sense of a mathematical sentence, or develop a new mathematical scheme to be part of the student's repertoire. The role of teachers is nevertheless essential, but I will not develop this point in this chapter.

## Intuitive Knowledge and Formal Knowledge

Because it contains an explicit reference to the idea of *concept,* some researchers consider that the conceptual field theory concerns the learning and teaching of explicit and formalized concepts. This is not true. Its first aim is rather to account for the knowledge contained in most ordinary actions, those performed at home, at work, at school, or at play by children and adults. It also refers to the knowledge involved in problem solving. Specifying the complete meaning of the theory requires several kinds of clarification.

The conceptual field theory asserts that one needs mathematics to characterize with minimum ambiguity the knowledge contained in ordinary mathematical competences. The fact that this knowledge is intuitive and widely implicit must not hide the fact that we need mathematical concepts and theorems to analyze it. I have introduced the ideas of *concepts in action* and *theorems in action* for that very purpose. The expression *intuitive knowledge* clarifies nothing, except that the subject uses his or her knowledge spontaneously, without reflecting much on its contents and groundings. A cognitive approach requires a more precise analysis, which has to be mathematical for mathematical competencies.

However, and this is a second point to be clarified, one cannot actually achieve this analysis using an a priori framework. The thesis that the analysis must be mathematical does not mean that it can be found ready-made in mathematics. The classification of relationships and situations, and the distinctions concerning the formation of the concepts of fraction and rational number that the reader will find in this chapter, have not been a primary focus of mathematicians. However, these areas of study interest psychologists and researchers in mathematics education. Great attention must be devoted to the comparative difficulty of different classes of problems and procedures and the different verbal expressions and writings produced by students. Psychology has led me, for example, to stress the fact that multiplication is not usually conceived by children as a binary operation, that the isomorphic properties of the linear function are more easily grasped than the constant coefficient properties, and that many ways of reasoning concern relationships between magnitudes or quantities, rather than pure numbers.

Another point deserves clarification. Today, the literature is full of papers concerning real-life mathematical competencies and real-life learning, as opposed to school mathematics and school learning. This opposition is misleading in the sense that no mathematical procedure observable in real-life situations cannot potentially be found in the classroom, provided students are offered a wide variety of situations to deal with, rather than stereotyped algorithms. It is a real problem that school (especially in some countries) does not offer students a variety of meaningful situations and problems. But this problem must not be confused with my claim that we need formal mathematics to characterize real-life competencies.

The last point that may require clarification concerns the need to establish more clearly the kinds of relationships that connect the formation of intuitive knowledge with consciousness, and make it more explicit. Vygotsky made a useful distinction between consciousness before and consciousness after, showing examples in which widely automated and unconscious competencies could be developed first, followed by some reflection and analysis, whereas in other examples, consciousness and explication, were conditions for the emergence of new competencies. The explication, symbolization, and even the formalization of mathematics may be more crucial for certain competencies than for others, even at the elementary and early secondary levels. It is a crucial point for a theory of teaching.

## What Is MCF?

MCF is simultaneously a bulk of situations and a bulk of concepts. A concept is made meaningful through a variety of situations, and different aspects of the same concepts and operations are involved in different situations. At the same time, a situation cannot be analyzed with the help of just one concept; at least several concepts are necessary. This is the main reason that researchers should study conceptual fields and not isolated situations or isolated concepts.

Another reason comes from the fact that students master certain classes of situations before they master others; it may take up to ten years for a student to go from the simplest to the most complex ones. During that process, she or he will have to deal with a variety of things: situations, words, algorithms and schemes, symbols, diagrams and graphs . . . and will learn sometimes by discovering, sometimes by repeating, sometimes by representing and symbolizing, sometimes by differentiating, sometimes by reducing different things to one another. Because the landscape of knowledge acquisition is so complex, the theoretical framework of researchers must also be complex.

From a *conceptual* point of view, MCF has the following essential ingredients:

- multiplication and division;
- linear and bilinear (and $n$-linear) functions;
- ratio, rate, fraction, and rational numbers;
- dimensional analysis;
- linear mapping and linear combinations of magnitudes.

From a *situational* point of view, MCF comprises a rather large number of situations that need to be classified and analyzed carefully, so that one may describe a hierarchy of possible competencies developed by students, inside and outside school. It is the problem of analyzing the cognitive tasks underlying these situations, and the procedures used by students to deal with them, including erroneous procedures, that has pushed me to use such sophisticated concepts as those just listed.

By considering the situations used in the classroom to introduce multiplication and division, one is first compelled to consider that multiplication and division are only the most visible part of an enormous conceptual iceberg. School overestimates explicit knowledge and underestimates or even devalues implicit knowledge; and one cannot readily analyze simple multiplication or division problems within the panoply of MCF. And yet the simple multiplication involved in the calculation of the price of five miniature cars at the cost of $4 each, raises crucial questions.

1. The result is given in dollars, not miniature cars. Why?

2. One can understand multiplication of $4 \times 5$ as the interaction of paying $4, 5 times; but it would be impossible to explain to 7- or 8-year-olds that multiplication of $5 \times 4$ is 4 iterations of 5: one cannot add miniature cars and find dollars, and there is no reason to iterate 5, as only 5 miniature cars have been bought.

3. "5 times more" is meaningful, as it is a scalar relationship and has no dimension. "4 times more" is meaningless. Of course the multiplication of $5 \times 4$ is meaningful, but it represents a functional relationship between different possible quantities of cars and their costs.

4. These two multiplications rely upon different theorems:

(a)     scalar $f(5) = 5f(1)$

It is usually introduced through iterated addition and therefore relies upon the additive isomorphism property

$$f(1 + 1 + 1 + 1 + 1 + 1) = f(1) + f(1) + f(1) + f(1) + f(1)$$

from which the multiplicative isomorphism property $f(n \cdot 1) = nf(1)$ is conceptually derived.

(b)     functional $f(5) = 4 \cdot 5$

It uses the constant coefficient property $f(x) = ax$, instead of the previous isomorphism property.

5. The constant coefficient represents neither cars nor dollars but dollars per car. Dimensional analysis is implicitly present.

I have already analyzed this example elsewhere (Vergnaud, 1988), and I have also explained about the classification of problems in earlier papers (Vergnaud 1983; 1988). Therefore, here I will just review the main categories of multiplicative structures and stress the epistemological points that appear to me to be the most essential (for more details, see Vergnaud, 1983).

- Simple proportion:

$$
\begin{array}{cc}
M_1 & M_2 \\
a & b \\
c & d
\end{array}
$$

Calculate one of these four magnitudes knowing the other three.

- Concatenation of simple proportions:

$$
\begin{array}{ccc}
M_1 & M_2 & M_3 \\
a & b & \\
 & c & d \\
f & & g
\end{array}
$$

Calculate one of these six magnitudes knowing the other five.

- Double proportion: calculate one of the six magnitudes in the following table, knowing the other five.

| $M_2$ | $M_1$ | $a$ | $b$ |
|---|---|---|---|
| | $M_3$ | | |
| $c$ | | $f$ | |
| $d$ | | | $g$ |

- Comparison of rates and ratios:

$$
\begin{array}{cccc}
M_1 & M_2 & \quad & M_1 & M_2 \\
a & b & & c & d
\end{array}
$$

Which rate ($b/a$, $d/c$) is bigger? Or which ratio ($c/a$, $d/b$)?

It is clearly impossible to analyze the procedures used by students in these situations without the framework of linear and bilinear functions and the clear identification of the magnitudes involved: elementary, quotient, and product magnitudes.

The following example of a procedure used by some 10- to 13-year-olds exemplifies the need for a sophisticated mathematical framework to theorize about the intuitive knowledge of students.

*Given:* The consumption of flour is on average 3.5 kg per week for ten persons. *Question:* What quantity of flour is needed for fifty persons over twenty-eight days? *Answer:* 5 times more persons, 4 times more days, 20 times more flour; therefore 3.5 × 20 = 70 (kg)

It is impossible to give account of that reasoning without making the hypothesis of the following implicit theorem in the subject's head:

$$f(n_1 x_1, n_2 x_2) = n_1 n_2 f(x_1, x_2)$$

Consumption (5 × 10, 4 × 7) = 5 × 4 Consumption (10, 7)

Of course, this theorem is available because the ratio of 50 persons to 10 persons, and the ratio of 28 days to 7 days are simple and visible. It would not be so easily applied to other numerical values. Therefore, its scope of availability is limited. Yet it is a mathematical theorem, and can be expressed in different ways:

1. In words: The consumption is proportional to the number of persons when the number of days is held constant; it is proportional to the number of days when the number of persons is held constant.

2. By a double-proportion table, as shown in Figure 2.1.

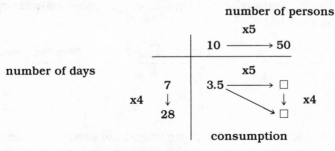

Fig. 2.1
*Double-proportion table.*

3. By a formula:

$$C = KP \cdot D$$

where
C = consumption,
P = number of persons,
D = numbers of days,
K = C(1, 1), consumption per person and per day.

C is proportional to P when D is constant and to D when P is constant. Therefore, it is proportional to the product.

It is clear that these different modes of expressing the same reasoning are not cognitively equivalent: The last one is more difficult. These modes rather are complementary and illustrate different ways of making explicit the same hidden mathematical structure at different levels of abstraction.

### Situations, Schemes, Concepts, and Symbols

The conceptual field theory is a psychological theory of cognitive complexity. There are several ways to gain cognitive complexity and several ways to fail at an attempt to gain cognitive complexity.

The first, and very essential, way to make progress is to learn to manage a new class of situations. The hierarchical classification of multiplication and division problems, which takes into account the conceptual structure, the domain of experience used, and the numerical values, therefore, is important to the study of the growth of cognitive complexity. For instance, the distinction between multiplication, division 1 (partition), and division 2 (quotation) is commonly accepted as the first basis of MCF.

| Multiplication | | Division type 1 | | Division type 2 | |
|---|---|---|---|---|---|
| $M_1$ | $M_2$ | $M_1$ | $M_2$ | $M_1$ | $M_2$ |
| 1 | a | 1 | ? | 1 | a |
| b | ? | b | c | ? | c |

But this is true only when the domain of experience referred to is conceptually easy (sharing discrete objects, buying goods) and when the numerical values are whole numbers (small whole numbers for *b*). These situations provide the first meaning for multiplication and division:

*b* times more, *b* times less

Division 2 already demands two more cognitive steps: Find how many times *a* goes into *c*, or apply to *c* the inverse functional coefficient /*a*. These two operations are equivalent mathematically, but not conceptually: The first one consists of finding a scalar ratio, and the second one an inverse quotient of dimensions.

Along similar developmental considerations, the primitive conception of fractions comes from the partition structure and is usually available for very simple values: 1/2 first, 1/4 one or two years later, and 1/*n* (for *n* < 10) by the end of elementary school. Archimedian fractions are therefore viewed as both operators and quantities: 1/*n* is first viewed as dividing by *n* some discrete or continuous quantity—it is therefore an operator—but the result is a fractionary quantity 1/*n*.

At the same time, young students use scalar ratios and functional rates, as shown in fig. 2.2 They can therefore combine Archimedean fractions and scalar ratios into non-Archimedean fractions, *p/q*, provided *p* < *q*. Some examples would include the following:

- Sharing a pastry cut into eight parts. What fraction is eaten by 5 children who each eat one part?
- Sharing a bag of twenty-four sweets, divided into eight parts. What fraction is eaten by five children? How many sweets does each child eat? How many sweets are eaten by all five children?

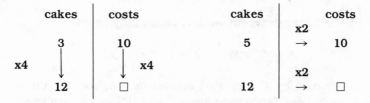

Fig. 2.2
*Scalar ratios and functional rates.*

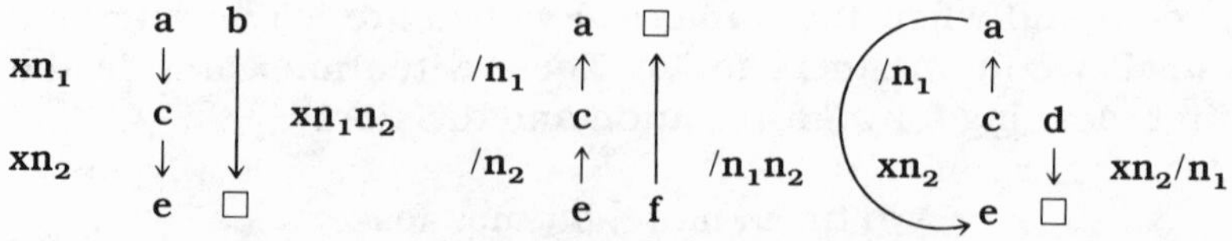

$$
\begin{array}{cc}
& a \quad b \\
xn_1 & \downarrow \\
& c \\
xn_2 & \downarrow \quad \downarrow \\
& e \quad \square
\end{array}
\quad xn_1n_2
\qquad
\begin{array}{c}
a \quad \square \\
/n_1 \quad \uparrow \quad \uparrow \\
c \\
/n_2 \quad \uparrow \\
e \quad f
\end{array}
\quad /n_1n_2
\qquad
\begin{array}{c}
a \\
/n_1 \quad \uparrow \\
c \quad d \\
xn_2 \quad \downarrow \quad xn_2/n_1 \\
e \quad \square
\end{array}
$$

*Note: $n_1$ and $n_2$ are any whole numbers.*

Fig. 2.3
*Combining scalar ratios.*

They also combine scalar ratios in several different ways, as shown in figure 2.3. The main conceptual difficulties met by students in these combinations concern the commutativity of division and multiplication

$$/n_1 \text{ and } xn_2 = xn_2 \text{ and } /n_1$$

and the need to multiply when one combines two divisions

$$/n_1 \text{ and } /n_2 = /(n_1 \times n_2)$$

It is worth noticing that during the elementary school years, students are also introduced to double proportion in two different domains:
as combinatorics

> How many possible different colored houses can be painted with three colors for the roof and four colors for the walls?
> How many possible couples of dancers can be formed with five boys and 7 girls?

and as area and volume

> What is the area of a room 5 meters long and 4 meters wide?
> In the latter case, formulas are usually taught

$$A = L \times W \qquad V = L \times W \times H$$

Therefore, by the end of elementary school, students have already been faced with some essential aspects of multiplicative structures. They have had to deal with different problems of proportion, with different kinds of operations involving ratios and rates, and with different types of symbolisms. However, they can master only a small part of the conceptual field; they still have a long way to go to understand it fully. For instance, they have yet to build such high-level concepts as those of rational number, function, and variable, dependence and independence.

They also have to achieve more modest steps, such as extending the scope of validity of their intuitive knowledge to complex ratios and rates and to nonwhole numbers. It is now well known that there are strong epistemological obstacles to such extension: The beliefs that one cannot divide a number by a larger one, multiplication makes bigger and division smaller, and so on. Students also have to extend their knowledge of multiplicative structures to such difficult domains as geometry (similarity and homothety), physics (density, mechanics), probability, and so forth. This extension raises a sharp theoretical issue. From a cognitive point of view, applying multiplicative structures to new domains of experience is both necessary to conceptualize them properly, and made possible only if some specific conceptualization of the domain has taken place. It seems to be a vicious circle.

The vicious circle can be disrupted only if one develops a reasonably complex theory of cognitive development and learning, especially of the relationship between schemes, concepts, and symbols.

What is a scheme? A scheme is defined as the invariant organization of action for a certain class of situations. This dynamic totality, introduced by Piaget (after Kant), to account for both "sensory-motor skills" and "intellectual skills" requires a strict and deep analysis if one wishes to understand the relationship between competences and conceptions.

A scheme is finalized; goals imply expectations. A scheme generates actions; it must contain rules. A scheme is not a stereotype, as the sequence of actions depends on the parameters of the situation: Its application involves hic et nunc computations. A scheme also involves operational invariants: categories to pick up relevant information (concepts in action) and propositions from which inferences are made (theorems in action). All these aspects of a scheme are illustrated in Figure 2.4.

This analysis makes it clear that no action is possible without operational invariants that enable the subject to pick up information and compute what to do and expect.

The theory of conceptual fields offers a way to under-

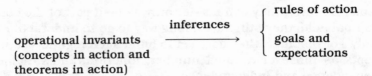

Fig. 2.4
*What does a scheme consist of?*

stand how the intuitive knowledge contained in behavior works. Intuitive knowledge is made essentially of operational invariants, that is, concepts in action and theorems in action. They are the "conceptual" part of schemes, however implicit or explicit, conscious or unconscious they may be. If a scheme addresses a class of situations, it must contain invariants that will be relevant over the whole class. This is especially visible when a scheme is extended to a larger class of situations. Transfer presupposes invariants as well as differentiation and restriction.

Among the most important theorems in action developed by students, one finds the isomorphic properties of the linear function

$$f(x + x') = f(x) + f(x')$$
$$f(x - x') = f(x) - f(x')$$
$$f(c_1x_1 + c_2x_2) = c_1f(x_1) + c_2f(x_2)$$

and the constant coefficient properties of the linear function

$$f(x) = ax$$
$$x = 1/af(x)$$

and some specific properties of bilinear functions

$$f(c_1x_1, c_2x_2) = c_1c_2f(x_1, x_2)$$

Among the most important concepts in action developed by students, one find those of quantity and magnitude, unit value, ratio and fraction, function and variable, constant rate, dependence and independence, quotient and product of dimensions.

Concepts in action are necessary ingredients of theorems in action, in the same way that propositional functions and arguments are necessary ingredients of propositions. But concepts are not theorems. They allow no derivation (or inference, or computation); a derivation requires propositions. Propositions can be true or false; concepts can be only relevant or irrelevant. Yet, there are no propositions without concepts.

Reciprocally, there are no concepts without propositions, as it is the need to derive action from the representation of the world and have true (or at least truer) conceptions of the world that makes concepts necessary. A computable model of intuitive knowledge must comprise concepts in action and theorems in action as essential ingredients of schemes.

Schemes play the most essential part as they generate actions. (Intellectual operations also are actions.) They can generate actions because they contain operational invariants, which constitute the core of representation.

But a concept is not fully a concept unless it is explicit. Moreover, the process of making concepts and theorems explicit helps one identify the relevant or irrelevant invariants. Therefore, linguistic expressions, symbols, and symbolic representations that may accompany, at the signifier level, the formation of concepts and theorems, must also be studied. Explication and symbolization are an important path through which cognitive complexity is gained.

Not only is it important that students be faced with a variety of occasions to extend or restrict the scope of validity and availability of their schemes and to develop new schemes, but also they be helped by external means, like linguistic and extralinguistic signifiers, in recognizing the invariant structure of different problems and therefore the possibility of using the same schemes or similar ones. Not only is it important that situations be clearly and exhaustively classified from the point of view of their conceptual structure, but also that the invariants (concepts and theorems) be worded, symbolized, diagramed, or graphed so that they become elements of explicit rational conceptions and do not remain elements of only implicit schemes. This is probably a necessary condition for the transference of concepts and theorems to any numerical values and to any domain of experience.

As a matter of fact, transfer and generalization necessarily require the recognition of the "same" structure in different situations. Does it help to associate specific words and
sentences or specific symbols to similar problems and relationships? In other words, is it helpful for students that hidden invariants and structures be made explicit? There is
probably no universal answer to this question, but it is likely
that making some relationships explicit can be helpful. This
is the reason why the conceptual field theory considers language and symbols important. For instance, it is important to
express and eventually symbolize the structure of data and
questions, and to use words and symbols that can be used by
students. In the previous example of the consumption of
flour by fifty students over twenty-eight days, the double-
proportion table can be handled easily by a 10- to 15-year-old,
whereas the algebraic notation cannot.

The "table-and-arrow diagram" offers many advantages.
It uses the properties of the two-dimensional space to represent some relevant properties of simple proportion and double proportion. A list of its strengths for each type of proportion follows.

- Simple proportion:
  Parallelism is used to represent different kinds of
    quantities or magnitudes.
  Vertical arrows indicate ratios.
  Horizontal arrows indicate functions and quotients of
    magnitudes.
  Combinations of arrows represent products of opera-
    tors.
- Double proportion:
  Orthogonality is used to represent independence.
  Parallelism line to line or column to column repre-
    sents proportion.
  Margins are used to represent the values of the ele-
    mentary magnitudes, and the inside of the table is used
    to represent the values of the product magnitude.
  Arrows are used to represent ratios, functions, and
    combinations of ratios.

Words are important, but mathematical sentences are
usually complex when expressed in natural language. Alge-

braic symbols are economical and powerful, but they make theorems very abstract for children. The table-and-arrow diagram is a prealgebraic representation that is less abstract than algebra while still representing the essential relationships. This is why I find it important to communicate about MCF with the help of such a symbolic system. Students find it fairly easy to use at the primary and early secondary level, and many adults find it easier to use than algebra. This does not mean that algebra is not necessary. Most students seem to find both algebra and proportion tables useful for a long while.

The status of an explicit and symbolized theorem is different from the status of an implicit local operational invariant. But the former has no meaning if it is not grounded in the latter, and the latter is available only in a limited range of situations. Moreover, scientific concepts and theorems are debatable and public. Implicit concepts and theorems by their nature cannot be explicitly debated. Therefore an enormous amount of the discussion that is expected from students in the learning of mathematics could not take place if mathematics consisted only of schemes. There is a need for symbolizing and formalizing, which makes mathematics different from a bulk of schemes addressing a bulk of situations. The concept of linear function cannot emerge from dealing with proportion problems only. It has to be worded and analyzed as a general and comprehensive concept.

And yet situations and schemes are essential from a psychological point of view. They are also essential for didactics, as the capacity to invent complex and meaningful didactic situations is probably the most genuine activity of professional teachers. As it is presented in this chapter, MCF is reduced to its main elementary components. It is another enterprise to combine these components to provide students with more challenging and motivating situations. I have not addressed that problem here.

## CONCLUSION

The multiplicative conceptual field can be viewed as

- a set (bulk) of situations that require multiplication, division, or combination of such operations;

- a set (bulk) of schemes that are needed to deal with these situations. Schemes are invariant organizations of behavior for well-defined classes of problems; but they can also be evoked to solve new problems;
- a set (bulk) of concepts and theorems that make it possible to analyze the operations of thinking needed: linear and nonlinear functions, fraction, ratio, rate and rational number, dimensional analysis, vector space theory. (These three concepts maybe explicit, but they are very often implicit only in schemes.)
- a set (bulk) of formulations and symbolizations.

All four sets are necessary to understand how students master more and more complex situations, more and more profoundly and reliably, and to understand how teachers can help them by presenting appropriate situations to them and giving them appropriate explanations.

Especially important is the choice of situations that can make new concepts or new aspects of a concept more meaningful. We need to develop a powerful theory of teaching situations tied to both the epistemology of mathematics and the psychology of learning mathematics.

Operational knowledge is an answer to genuine practical and theoretical problems. This is apparent when one considers the history of science and the history of techniques and technology. How much of this idea can we transpose into the classroom?

Even if it is each individual student's cognitive decision to recognize or discover a new property of a concept, there is a large set of possible ways for teachers to help students: organize interesting and mathematically fruitful situations and activities, focus attention, explain and symbolize the relevant relationships and operations of thinking, or reduce the gap between the problem and its solution.

The practical competence of teachers must be analyzed in terms of a strong cognitive theory of learning and teaching. This is what the conceptual fields theory tries to provide. This theory asserts that the core of cognitive development is conceptualization. Therefore, we must devote all our attention to the conceptual aspects of schemes and to the conceptual analysis of the situations for which students develop their schemes, in school or in real life. Words and symbols are nevertheless essential. Therefore we must also devote our attention to the adequacy of linguistic and extralinguistic means by which we help students identify invariants and recognize them as mathematical objects.

## REFERENCES

Behr, M., and J. Hiebert (eds.). 1988. *Number concepts and operations in the middle grades.* Reston, VA: National Council of Teachers of Mathematics. Hillsdale, NJ: Lawrence Erlbaum.

Lesh, R., and M. Landau (eds.). 1983. *Acquisition of mathematics concepts and processes.* New York: Academic Press.

Piaget, J. 1967. *Biologie et connaissance.* Paris: Gallimard.

Vergnaud, G. 1983. Multiplicative structures. In *Acquisition of mathematics concepts and processes,* ed. R. Lesh and M. Landau, 127–174. New York: Academic Press.

———. 1988. Multiplicative structures. In *Research agenda in mathematics education: Number concepts and operations in the middle grades,* ed. J. Hiebert and M. Behr, 141–161. Hillsdale, NJ: Lawrence Erlbaum.

——— et al. 1990. Epistemology and psychology of mathematics education. In *Mathematics and cognition,* ed. J. Kilpatrick and P. Nesher, 2, 17. Cambridge: Cambridge University Press.

Vygotsky, L. S. 1962. *Thought and language.* Cambridge, MA: MIT Press.

# 3 Extending the Meaning of Multiplication and Division

## *Brian Greer*

In discussing how multiplication, defined within the integer domain in terms of repeated addition, can be extended to fractions, De Morgan (1910, originally 1831) stated:

> If we could at once take the most general view of numbers, and give the beginner the extended notions which he may afterwards attain, the mathematics would present comparatively few impediments. But the constitution of our minds will not permit this. . . .
>
> In the limited view which we first take of the operations which we are performing, the names which we give are necessarily confined and partial; but when, after additional study and reflection, we recur to our former notions, we soon discover processes so resembling one another, and different rules so linked together, that we feel it would destroy the symmetry of our language if we were to call them by different names. *We are then induced to extend the meaning of our terms* . . . (p. 33, emphasis added)

Similar remarks were made by Cajori (1917):

> That, in the historical development, multiplication and division should have been considered primarily in connection with integers, is very natural. The same course must be adopted in teaching the young. First come the easy but restricted meanings of multiplication and division, applicable to whole numbers. *In due time the successful teacher causes students to see the necessity of modifying and broadening the meanings assigned to the terms.* (p. 183, emphasis added).

This chapter is concerned with the extension of the meaning of multiplication and division from their early conceptualizations. The "vertical" dimension of this extension is the progress through successively more general number systems;[1] the "horizontal" dimension involves the introduction of a wide range of situations modeled by the operations. In the first part of the chapter, current taxonomies of multiplicative situations are criticized for failing to differentiate sufficiently, particularly along the vertical dimension, and a more differentiated scheme is proposed. This is followed by a discussion of the conceptual difficulties pupils have in extending multiplication and division beyond the integer domain. Finally, some suggestions are made as to how the extension of meaning for multiplicative concepts may be promoted.

## CLASSIFICATIONS OF SITUATIONS MODELED BY MULTIPLICATION AND DIVISION

### Limitations of Existing Taxonomies

Somewhat overlapping classifications have been proposed by, among others, Vergnaud (1988), Schwartz (1988), and Nesher (1988). These classifications differentiate, in each case, between three very broad classes. A slightly more differentiated alternative has been put forward by Greeno (1987, p. 64), who suggested that "the problems used in the elementary grades all can be understood" using six classes for multiplicative relations (and three for additive relations). Marshall has proposed that *all* story problems (in the sense of all those commonly encountered in schools) can be classified within five "semantic profiles," labeled *change, group, compare, restate*, and *vary* (Marshall, Barthuli, Brewer, & Rose, 1989).

Is this degree of parsimony desirable? To reduce the variety of situations to such a small number of classes implies a focus on commonality of structure at an abstract level. It is not clear, however, that it is useful to ignore distinctions with important psychological and pedagogical implications. In particular, from the perspective of this chapter, a crucial limitation of the classifications just considered is that they do not differentiate between applications involving: (1) only integers derivable by counting discrete objects or entities, (2) integers and fractions derived by divisions of integers by integers, (3) decimals as measures of quantities. (For convenience, it will be implicit throughout that only positive numbers are being referred to, and the terms *fractions* and *decimals* will refer to positive rational numbers expressed in the corresponding notations).

In all of the classification schemes listed, the following three problems would be put in the same category:

1. If three children each have four oranges, how many oranges do they have altogether?
2. A boat moves at a steady speed of 4.2 meters per second. How far does it move in 3.3 seconds?
3. An inch is 2.54 cm. How long in cms is 3.3 inches?

Representations of these problems using the schematic diagrams proposed by Vergnaud (1988) and by Marshall et al. (1989) are shown in Figure 3.1.

A more differentiated classification is shown in Table 3.1 (see Greer, 1992, for further details). Within this scheme, Problem 1 is classified as an *equal groups* problem, whereas 2 is classified as a *rate* problem. Problem 3 is placed in the separate *measure conversion* class. This distinction reflects

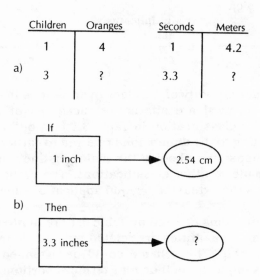

Fig. 3.1
*Schematic representations of multiplicative problems used by (a) Vergnaud (1983) and (b) Marshall et al. (1989).*

TABLE 3.1.  PROPOSED CLASSIFICATION OF SITUATIONS
MODELED BY MULTIPLICATION AND DIVISION

|  | Rate | Multiplicative comparison | Rectangular array/area | Product of measures |
|---|---|---|---|---|
| Integers | Equal groups | Multiplicative comparison | Rectangular array | Cartesian product |
|  |  |  | Rectangular area |  |
| Integer multiplier | Equal measures |  |  |  |
| Fractions | Rational rate | Multiplicative comparison | Rectangular area |  |
|  |  | Part-whole |  |  |
| Decimals | Rate | Multiplicative comparison | Rectangular area | Product of measures |
|  | Measure conversion | Part-whole |  |  |
|  |  | Multiplicative change |  |  |

the particular property of the class of situations in which the measures involved are alternative measures of the same quantity. The classification in Table 3.1 is two-dimensional, corresponding to the dimensions referred to earlier. The horizontal dimension embodies the major distinctions similar to those made in other classifications. The vertical dimension reflects the extension beyond applications involving integers.

The situations covered by Table 3.1 are limited to those in which three quantities are related by a single multiplicative relationship. The scheme could be extended in many ways; for example, to include four-term proportionality problems, and the class of situations that Vergnaud (1988) calls *multiple proportions*, in which one quantity is simultaneously proportional to two or more other quantities. A further extension would cover multiple proportionality to powers of quantities (integral or fractional, positive or negative), and then to the important application domain of dimensional analysis in physics and biology (McMahon and Bonner, 1983). Another extension would be to exponentiation (reducible, in its simplest form, to repeated multiplication) and related topics (Confrey, this volume).

## Instructional Representations

In conjunction with analysis and classification activity, various instructional representations have been devised (Greeno, 1987). For example, Nesher (1988) referred to findings reported by Mechmandarov (1987) that good results were achieved when pupils were trained to use a tabular layout similar to that of Vergnaud (see Figure 3.1) to schematically represent problems. (In fact, this type of layout has a long history, having been used at least as long ago as the sixteenth century (Cajori, 1917)).

Marshall et al. (1989) have developed a computer-based teaching system based on her semantic analysis. Students learn to match situations to the appropriate schematic representations and to match elements of the problem statement to components of those representations. (The representation for "vary" problems is illustrated in Figure 3.1.)

Another approach is the type of system devised by Shalin and Bee, as described by Greeno (1987). The representation of a simple multiplication problem within this system is illustrated in Figure 3.2. Thompson (1989) has developed this style of representation further in his Word Problem Assistant, which is based on a theory of quantity-based reasoning and generalizes naturally from arithmetic to algebraic problems.

A number of misgivings have been raised about the general approach illustrated by these examples. With reference to similar schemes devised for addition and subtraction problems, Verschaffel and De Corte (1993) cautioned against overvaluing the positive effects reported for training experiments on the grounds that they may be attributable to the standardized and stereotyped nature of the problems used in the posttests. In a similar vein, Confrey (1987, p. 84) expressed doubt about "the optimistic conjecture that we are doing something significant if we take problems that we give in tests and teach people how to solve them."

There is plenty of evidence that current teaching allows

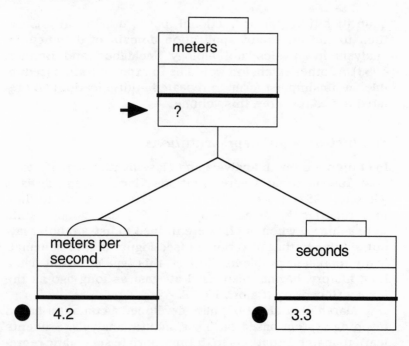

Fig. 3.2
*Schematic representation of a simple problem (after Shalin and Bee, 1985).*

and reinforces the use of superficial strategies to solve word problems without constructing a representation of the situation described in the problem (Nesher, 1980; Reusser, 1988). There is a risk that the use of instructional representations could likewise lead to superficial, procedural strategies that bypass such an in-depth analysis. It would be of interest to test children trained on tabular or schematic representations of rate problems by giving them a problem such as, If three men can dig a hole in four days. How long will it take six men? A different set of issues (Greer, 1992) is raised by problems such as, A man can run a mile in 4 minutes. How long will it take him to run 3 miles?

The apparent assumption that all the problems that matter can be handled by a small number of representations is a source of concern. For example, Treffers (1989) argues that forcing children to solve word problems by selecting a specified diagram from a limited set of alternatives destroys their "natural" problem-solving skills; instead, Treffers advocates that children should construct their own diagrams for representing and solving word problems.

There is experimental evidence that the diversity of problems encountered by students has a big effect on their ability to solve novel types of problems (Marshall, 1985). The view taken in this chapter is that for children to understand the relationship between the "algorithmic simplicity" and the "phenomenal variety" (Freudenthal, 1983, p. 117) of multiplication and division, attention must be given to extension of meaning for both dimensions of Table 3.1. The next section concentrates on the conceptual obstacles to extension along the vertical dimension.

## EXTENSION OF MULTIPLICATION AND DIVISION BEYOND THE INTEGER DOMAIN

### The Extension of Mathematical Concepts

For most children, multiplication of $m$ by $n$ is at first encountered in a context of $n$ equal groups, each containing $m$ discrete objects. Mathematically, this can be represented as repeated addition (a definition found in Euclid, for example) or, more precisely, as the operation of a "repetition number" on a cardinal number (Fowler, 1987, pp. 14–16). In response to both practical problems and formal analysis, the conceptual complex of numbers and operations on them is progressively extended to include rationals, negative numbers, real numbers, and beyond. The extension of operations, functions, relations, and so forth, defined in a "natural" way for the integer domain, is characteristic of mathematics—other familiar examples include the gamma function as a generalization of *n factorial* and the extension of exponents from natural numbers to expressions such as $3^{-2}$ and $e^{i\pi}$. There is a parallel between the extension of multiplication beyond its initial definition as repeated addition and exponentiation beyond its initial definition as repeated multiplication. Descartes liberated exponentiation from its roots in three-dimensional Euclidean geometry by writing $2x^3$ in place of 2*A cubus*. De Morgan (1910, p. 185) quotes Laplace as follows:

> Newton extended to fractional and negative powers the analytical expression which he had found for whole and positive ones. You see in this extension one of the great advan-

tages of algebraic language which expresses truths much more general than those which were at first contemplated, so that by making the extension of which it admits, there arises a multitude of new truths out of formulae which were founded upon very limited suppositions.

As Sinclair (1990) has pointed out, reflection on the historical perspective also reminds us that in teaching mathematics, we are expecting children to cover in a few years developments that, on the cultural scale, took centuries. Insight into the conceptual difficulties experienced by children may be obtained by reflecting on historical parallels (Fischbein, 1987; Piaget and Garcia, 1989). Of particular interest historically (though outside the scope of this chapter) is the extension of multiplication and division to negative numbers (Fischbein, 1987, Chapter 8; Semadeni, 1984).

For any mathematical concept, it is inevitable that the early conceptualizations of that concept (whether in historical or ontogenetic terms) will be based in concrete activity where the relationship between the aspects of reality being modeled and the mathematical objects used to model them is relatively direct (as in the equal groups situation). Fischbein (1987) has argued that when the concept is extended, the intuitive models established earlier continue tacitly to affect thinking and that helping children to become aware of and control these covert influences on their thinking is a major pedagogical challenge.

## Children's Difficulties: Summary of Evidence

1. Misconceptions, such as that multiplication always makes bigger and division smaller (MMBDMS) and that division is always division of the larger number by the smaller, have been familiar to teachers and researchers for a long time. Such misconceptions are manifested both on purely numerical tasks and on word problems, and they are often explicitly stated by subjects (see, for example, Af Ekenstam and Greger, 1983; Bell, Fischbein, and Greer, 1984; Graeber and Tirosh, 1990; Greer, 1987, 1988). Nesher (1987) commented that word problems in textbooks rarely expose students to counterexamples.

2. In a number of studies, children have been asked to

construct word problems corresponding to given multiplication and division calculations (e.g., Af Ekenstam and Greger, 1983; Bell et al., 1984; Fischbein, Deri, Nello and Marino, 1985; Greer and McCann, 1991). In general, children have been found to perform poorly on such tasks unless the numbers are "easy."

3. Many recent studies have used the choice-of-operation methodology, in which a word problem is presented and the task is to decide which operation would be appropriate to find the answer, *without having to carry out the calculation.* (This methodology is not new; Thorndike (1922, p. 43) cites a "reasoning" test devised by Courtis that used it.) It has consistently been found that the numbers used drastically affect the ease with which the appropriate operation can be identified, even though, logically speaking, they have no bearing on it (Bell, Greer, Grimison, and Mangan, 1989; De Corte, Verschaffel, and Coillie, 1988; Fischbein et al., 1985; Graeber and Tirosh, 1990; Greer and Mangan, 1986; Luke, 1988).

4. An even more striking set of findings shows that, when problems that differ only in terms of the numbers involved are juxtaposed, children often do not realize that the appropriate operation must be the same in both cases. This phenomenon has been labeled *nonconservation of operations,* by analogy with Piagetian nonconservations (Greer, 1987, 1988).[2]

It may be concluded that, for many children who have been taught multiplication and division of decimals, understanding of the applications of the operations in modeling situations is weak. More specifically, it has been demonstrated that generalization from the integer domain cannot be taken for granted. The investigations of choice-of-operation and conservation tasks show that although, from a logical point of view, the numbers in a problem carry no information regarding the appropriate operation, the types of numbers still exert a very strong influence on the ease or difficulty with which the appropriate operation can be identified. Clearly, word problems do not generally evoke general schemas into which the numbers can simply be "plugged," and instructional representations based on this assumption must be suspect. To the extent that children may be said to possess schemas for such problems, they are "weak schemas" (Anderson, 1984) or "theorems-in-action" (Vergnaud, 1988). For these to be developed into "strong schemas" in Anderson's sense, or "theorems, in action" (Kieren, 1987) requires an appreciation of the invariance of the operation over the numbers involved, which is the keystone of the extension of meaning for multiplication and division.

## Focus: When Multiplication Makes Smaller and Division Bigger

When an extension of meaning of a mathematical concept is under consideration, the particular choice from those possible is guided, in general, by two considerations—maintenance of properties (Semadeni, 1984) and usefulness (matrix multiplication is a good example of the latter). By definition, extension to a more general domain implies that not all properties will be maintained. In the case of multiplication and division, the lapsing of the MMBDMS rule entails a major conceptual reconstruction, the difficulty of which for children has been illustrated.

This conceptual obstacle has historical counterparts. Cajori (1917, pp. 182–183) records how Pacioli, an Italian mathematician of the fifteenth century, was "greatly embarrassed by the use of the term 'multiplication' in case of fractions, where the product is less than the multiplicand." Thorndike (1922, p. 72) remarked that "some of my readers will probably confess that even now they feel a slight irritation or doubt in saying or writing that $16/1 \div 1/8 = 128$."

The links between multiplication and increase, division and decrease, forged during experience in the natural number domain, are liable to be strengthened by early work with fractions. It is natural enough that the first fractions to be introduced are those of the form $1/n$ where $n$ is a positive integer and the emphasis is on division by $n$. Multiplication by $3/4$ is taking "three-quarters of," that is, dividing by 4 and multiplying by 3. Consequently, the basis for mindful experience of counterexamples to the MMBDMS rule is limited.

The association between $1/n$ and division is reflected in the common mistake whereby, for example, children write $6 \div 1/3 = 2$. Here there are two cues for division by 3 (compare the common error $(6 - (-3) = 3)$. Likewise, the difficulty of construing $1/n$ as a single-entity multiplier is reflected in the error made by preservice teachers reported by Graeber and Tirosh (1988, p. 274):

$$\tfrac{1}{2} \text{ of } 70 \text{ is } 2\sqrt{70}$$

$$\text{so} \quad .65 \text{ of } 30 \text{ is } .65\sqrt{30}$$

Even with decimals, the impact of counterexamples can be limited if emphasis is put on paper-and-pencil computation; to put it crudely, calculations with decimals can be done in the same way as calculations with integers, with the addition of some rules for the placing of the decimal point. In their studies of preservice teachers, Tirosh and Graeber (1989) demonstrated how misapplied procedural knowledge could be detrimental to conceptual knowledge. For example, the procedural rule for division "change the divisor to a whole number" is transformed into beliefs that "you cannot divide by a decimal" and "the divisor must be a whole number" (Tirosh and Graeber, 1989, p. 92).

Use of a calculator offers the chance to detach the procedural from the conceptual meaning of multiplying or dividing by a decimal, thus providing a way for students to experience counterexamples having a greater impact. You enter 16, press the multiplication key, enter 0.85, press the equals key—and the answer is less than 16! Various games using calculators have been devised at the Shell Centre in Nottingham to promote the insight that any number can be obtained from any other number by multiplying or dividing by the appropriate operand (Bell, 1986).

Sowder (1988) quotes a student as follows: "sometimes when you multiply, like let's say you're multiplying by point seven five. The number comes out smaller than the original number. *So it's kind of like dividing, when you multiply by decimals*" (p. 232, emphasis added). Even though this student correctly stated the rule for multiplication, the link between making smaller and division was still compelling.

Natural language both reflects and reinforces the bifurcation between multipliers greater than and less than 1. As Confrey (this volume) points out, we say "three times a number" or "3.2 times a number" but ".85 of a number". The differences between natural language and "the mathematical register" in this context is also referred to by Pimm (1987): "A common mathematical practice is that of expanding the range of applicability of a particular word. . . . an enlargement by a scale factor of a half looks remarkably like a contraction, yet the mathematical term applied in certain texts is still *enlargement*" (p. 91). Students need to be made aware of these differences between natural and mathematical language and shown the power of the mathematical style—a calculator or a computer program does not distinguish between "3.2 *times* 6" and "0.85 *of* 6" and there is a single setting on a photocopier for enlarging or reducing by a required factor. This uniformity is *useful* and parallels the structural unity of the group of positive rational (or real) numbers under multiplication, within which you can get from any number to any other number by either multiplication or division.

In summary, a straightforward suggestion for the tackling of the conceptual obstacles discussed here is to improve the experiential base on which inductions are made by students. In the final section, which follows, further suggestions for promoting extension of meaning of multiplicative concepts more generally are offered.

## PROMOTING EXTENSION OF MEANING

### Taking a Long Perspective

As Vergnaud has pointed out, the development of the conceptual field of multiplicative structures takes many years. Applications of multiplication and division span many levels, including

1. Informal procedures of children and adults involving context-embedded calculations with whole numbers.

2. Applications involving nonintegers derived from integers by division (notably intensive quantities).

3. A range of applications involving positive decimals as measures—measure conversion, multiplicative comparison, and so on.

4. Extension to algebra, necessitating (for closure under division and subtraction, respectively) extension to the rational numbers and to the negative numbers (which can be, and were for centuries, managed without in practical arithmetic).

5. Extension to complex numbers and their applications in modeling, notably in physics.

Attention needs to be paid not only to this vertical extension of meaning, but also to exposing children to an appropriately wide variety of situations embodying multiplicative relationships. To some extent this is indicated in Table 3.1; a fuller list would include the following conceptual complexes:

1. Rectangular arrays, rectangular area, Cartesian product, combinatorial problems, product of measures.
2. Proportionality, geometrical similarity, multiple proportionality, dimensional analysis.
3. Multiplicative comparisons, part-whole relationships, multiplicative change, growth, exponentiation.
4. Change of unit, measure conversion.
5. Intensive quantity, rate, linear function.

Obviously, the map of interrelated concepts could be drawn differently, and whatever groupings of related concepts are made, there will be multiple links between them. Individual researchers have emphasized particular aspects and, to a greater or lesser extent, have implied that a *single* particular conceptualization offers the key to understanding multiplicative concepts. In the long run, however, it will be necessary to look for some sort of balanced synthesis.

In terms of the relative scales of the educational problems at different stages, it could be argued that a disproportionate amount of research has been concentrated on the earliest stages of the development of multiplicative concepts, within the integer domain. There seems to be an assumption that some essential "core" notion—"*the* concept of multiplication"—holds the key, and all further development is mere elaboration.

By contrast, the view taken here is that

1. Epistemological and psychological questions of comparable interest and importance are raised at every stage of reconceptualization throughout the long development of the conceptual field. For example, Tirosh and Almog (1989) investigated the difficulty for students of the extension from real to complex numbers; they found that students were re-

luctant to accept complex numbers as numbers and incorrectly attributed to them the ordering relationship that holds for real numbers.

2. The extension of concepts, relations, functions, and so forth from one domain to a more general domain is a characteristic mode of development within mathematics and an appropriate subject for study, both in terms of specific examples (as in the main focus of this chapter) and in terms of the general processes involved (with ontological-historical parallels).

As the domain of multiplicative concepts is progressively extended, the relationship between the mathematical structure and real situations becomes less directly perceivable. Fischbein (1987) has argued that the limitations of intuition necessitate at some point acceptance of mathematics as a formal, deductive system. Likewise, Freudenthal (1983, p. 81) stated that, "Mathematics is characterised by a tendency which I have called *anontologisation:* cutting the bonds with reality." However, to the extent that mathematical structures model aspects of the world, bonds with reality remain. The notion of modeling as the link between mathematics as providing descriptions of aspects of the real world and mathematics as formal structures offers a unifying perspective for multiplicative concepts from their origins to the most advanced applications.

The operations may be seen as modeling a variety of situations, which is the perspective taken in Greer (1992). However, the term *modeling* is often used in the opposite direction, when concrete representations are said to provide models for the operations (e.g., Vest, 1976)—as when a rectangular array is used to illustrate commutativity. This latter view is often associated with the pedagogical aim of legitimizing calculation procedures and promoting understanding of the formal properties of number and operations. Concentration on this aim can leave students with impoverished conceptualizations, ill-equipped to handle applications outside a narrow range. Vest (1976, p. 398) gave the example of a textbook that, at a stage when rectangular area was the only model for multiplication of the fractions that had been presented, posed the problem: "John lives 1/3 mile from school. Harry lives only half that distance from school. How far from

the school does Harry live?" Instances of children being expected to force problems into inappropriate representations
are also referred to by Verschaffel and De Corte (1993).

In the case of decimals, the situation may be even worse
than for fractions, because computational proficiency is perceived as attainable through purely syntactical training, with
little or no reference to situations modeled.

By contrast, the view taken here that the aim of teaching
arithmetic is to enable children to mathematize situations
and understand the nature of the modeling process that relates aspects of the real world and mathematical structures.
The contrast between the two aims is suggested schematically in Figure 3.3.

## Invariance of Operations: A Powerful Idea

Semadeni (1984) judged that formal considerations for the
extension of mathematical systems (such as "the principle of
permanence of laws") are too sophisticated to be used directly. Instead, he proposed the "concretization permanence
principle":

> Rather than forcing problematic concretizations for the ex-
> tended operations or proceeding formally, the teacher is ad-
> vised to start with some sound concretization which is fa-
> miliar to the children within the original number range and
> which is capable of extension. Keeping the concretization
> fixed and varying numbers, students get to examples involv-
> ing numbers from the broader domain. Since the context is
> the same (or similar) and only numbers are changed, the
> same name—by analogy—is used for the extended opera-
> tion, though its intuitive meaning may appear changed.
> (p. 381)

For example, for the division of fractions, he suggests as a
starting point the situation in which 6 kg of sugar is packed
into bags holding 2 kg each (p. 386). A sequence of related
problems is then introduced by varying the capacity of the
bags: 1 kg, 1/2 kg, 1/3 kg, 2/3 kg. "When the divider becomes
1/2, the crucial point is to accept the validity of the previous
schema, to agree . . . that the operation should be called *division* (in spite of the fact that now the quotient is greater than
the dividend)." This situation could be exploited further, by

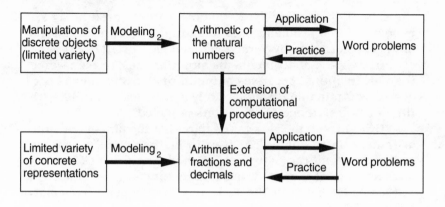

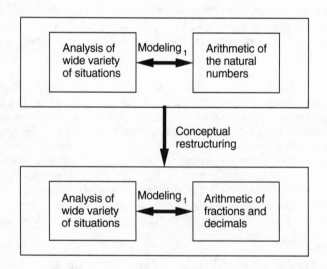

Fig. 3.3

*Contrasting views of the aims of teaching arithmetic.* Modeling₁ *refers to the operations as modeling situations;* Modeling₂ *refers to situations as models for the operations (see text).*

varying the size of bag through 3 kg, 6 kg, 12 kg, and so forth, to introduce cases where the divisor is greater than the dividend and the quotient, consequently, is less than 1.

Greer (1988) suggested a situation involving a bucket

being drawn up a well by use of a winch, where the rope or cable does not wrap around itself (Figure 3.4). Using a physical model or a computer simulation (ideally both), attention can initially be focused on how far the bucket moves as the result of 1, 2, 3, . . . complete turns of the winch; the extension to the distance raised by $3^{1}/_{2}$, $2^{1}/_{4}$, 3.34 turns is then natural. By using a dynamic simulation with the number of turns and the distance the bucket has been raised being indicated, the idea of systematic variation exemplified by the sugar bags example can be extended to continuous variation.

The invariance of multiplication and division over the numbers involved is a powerful idea that potentially can be harnessed to overcome the limitations of intuition. Given a problem in which the numbers are "hard," that is, do not allow the appropriate operation to be intuitively grasped, the strategy can be employed of replacing the numbers with "easy" ones. The operation having been identified, it is applied to the original numbers. However, Bell, Swan, and Taylor (1981), in attempting to make this strategy available to pupils, found that the difficulty was that most pupils assumed that changing the numbers could change the operation. If this strategy is to be available to students—or if, as suggested by Sinclair (1990), problems should be analyzed structurally before numbers are introduced; or if general schematic frames such as those illustrated in Figures 3.1 and 3.2 are to be totally effective—work has to be done on establishing the principle of invariance.

Semadeni's approach just discussed is one way of tackling this problem. Another possible way of doing this would be take a problem such as the "explorer problem," devised by Greer (1987), in which the time required for each of four sections of a journey is to be found:

| 25 | miles at | 20 | miles per hour |
|----|----------|-----|----------------|
| 3  | miles at | 6   | miles per hour |
| 12 | miles at | $^{3}/_{4}$ | miles per hour |
| 15 | miles at | 3   | miles per hour |

Children typically use a variety of methods to work out the answers, masking the availability of a uniform procedure. Now suppose a student is put behind a screen with a calculator, with the instruction to take the numbers that are shown through two slots in the screen, marked *distance* and *speed*,

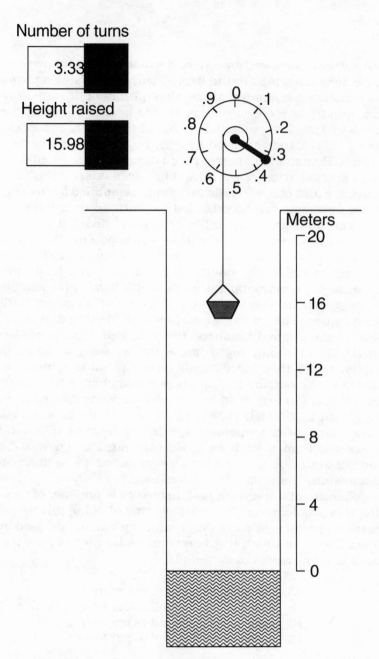

**Number of turns**

3.33

**Height raised**

15.98

.9  0  .1
.8      .2
.7      .3
.6  .5  .4

Meters
20

16

12

8

4

0

Fig. 3.4

*Raising the bucket: A context for the extension of multipliers from integers to nonintegers. Tabular and graphical representations could be added. (The shaded rectangles represent covers that could be dragged back by a mouse to reveal more decimal places as required).*

respectively; divide the first by the second; write the answer on a piece of paper; and put it through a third slot marked *time*. This activity could form the basis for discussion, including predictions—will this "computer simulation" produce the right answer for different values of distance and speed?

As Semadeni pointed out, invariance is no accident; rather it is a consequence of the principle that, in extending mathematical concepts, maintaining invariance of properties is an overriding consideration. A radical pedagogical strategy would be to make pupils aware of this and sell the powerfulness and utility of the principle.

Computer-supported schematic representations were referred to earlier, but there is considerable potential for using the image-manipulating capabilities of the computer to represent multiplicative relationships more directly, through the use of mediating representations such as number lines and through dynamic simulations appealing to intuitions about continuity and to general conceptions of growth.

Kaput has developed considerable software aimed at promoting understanding of intensive quantity; more recently, he has turned his attention to possible extensions to the continuous world (Kaput, in press). The most powerful mediating representation is undoubtedly the number line. Two such lines can represent, in a sense, an extension of the tabular representation illustrated in Figure 3.1 to the continuous case. Figure 3.5 shows how software under development would be used to represent the solution of the problem: "How many liters of soda will be needed for ten children if every two children drink three liters?" The first step is to put a line through some pair of values, one on each number line (number of children, number of liters), conforming to the intensive quantity three liters per two children (Fig. 3.5(a)). The software would then allow the line to be dragged into a perpendicular position, changing the scale (compressed or stretched) so that corresponding numbers of children and liters would be aligned (Fig. 3.5(b)).

A different use of a transformation of scale is based on the idea of an "elastic measure," which could either be constructed as a physical model or simulated. The use of this to represent multiplication and partitive and quotitive division is shown in Figure 3.6.

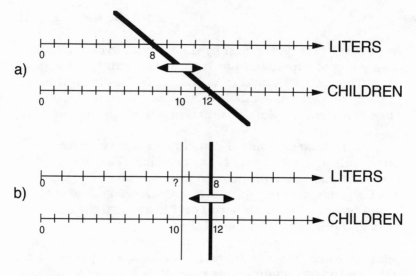

Fig. 3.5
*Representing isomorphism between measures in a continuous environment (Kaput, in press); see text for details.*

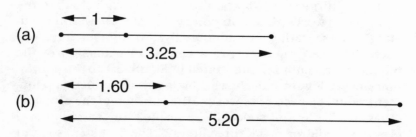

Fig. 3.6
*Multiplication and division using an elastic measure. Multiplication of 3.25 by 1.60 is represented by starting with a measure of length 3.25, with 1 marked accordingly, (a). The end of the measure is extended until the point originally corresponding to 1 is at 1.60, (b); the result (5.20) is then read off. Quotitive division of 5.20 by 3.25 is represented by starting in the same way, (a), then extending the end of the measure until the length is 5.20, (b); the position of the marked point (1.60) is then read off. Partitive division of 5.20 by 1.60 is represented by starting with (b), then letting the measure contract until the point originally measuring 1.60 is measuring 1, at which point the total length (3.25) is read off, (a).*

With the simulated version, the stretching of the measure could be done by grabbing and dragging the end using a mouse. Moreover, as the stretching is carried out, the numbers indicating the lengths could change simultaneously, thus emphasizing the complete, continuous transformation, rather than simply the end points. In general, there is considerable potential for representing continuous change, and relating it to very general concepts of growth.

Continuous change could also be embodied in a simulation of the well situation shown in Figure 3.4. A further feature of this example is its potential link with the kinaesthetic sense (either through manipulation of a physical model or if the turning of the winch in a simulation was controlled by corresponding rotary movements of a mouse or other device).

Future developments, such as Mathcars, under development by Kaput (1988), are likely to draw on a powerful combination of linked representations of multiple types drawing on multiple sensory and information-processing modes within a real-time simulation.

## SUMMARY

It has been argued that the meaning of multiplication and division, in the sense of the gamut of ways in which they can be applied in modeling aspects of reality, should be extended so that the operations are regarded not merely as a part of computational mathematics, but as related to knowledge structures in physics, biology, and so on and to social contexts (cf. Vergnaud, 1988, and this volume). Vertical extension of the meanings of the operations is beset with cognitive obstacles; elucidating these and finding means to help children overcome them more effectively is a major challenge for research in this field.

## NOTES

1. See Confrey, this volume, for a discussion of the conceptual development and integration of different number systems.

2. Confrey, this volume, pp. 303–304, gives an alternative interpretation of such findings.

## ACKNOWLEDGMENT

This research was funded under grant CRG.890977 from NATO.

## REFERENCES

Af Ekenstam, A., and K. Greger. 1983. Some aspects of children's ability to solve mathematical problems. *Educational Studies in Mathematics* 14: 369–384.

Anderson, R. C. 1984. Some reflections on the acquisition of knowledge. *Educational Researcher* 13(9): 5–10.

Bell, A. W. 1986. Diagnostic teaching, 2: Developing conflict-discussion lessons. *Mathematics Teaching* 116: 26–29.

Bell, A. W., E. Fischbein, and B. Greer. 1984. Choice of operation in verbal arithmetic problems: The effects of number size, problem structure and context. *Educational Studies in Mathematics* 15: 129–147.

Bell, A. W., B. Greer, L. Grimison, and C. Mangan. 1989. Children's performance on multiplicative word problems: Elements of a descriptive theory. *Journal for Research in Mathematics Education* 20: 434–449.

Bell, A. W., M. Swan, and G. Taylor. 1981. Choice of operations in verbal problems with decimal numbers. *Educational Studies in Mathematics* 12: 399–420.

Cajori, F. 1917. *A history of elementary mathematics*. New York: Macmillan.

Confrey, J. 1987. Contribution to discussion following Greeno (1987).

De Corte, E., L. Verschaffel, and V. Van Coillie. 1988. Influence of number size, problem structure and response mode on children's solutions of multiplication word problems. *Journal of Mathematical Behavior* 7: 197–216.

De Morgan, A. 1910. *Study and difficulties of mathematics*. Chicago: University of Chicago Press (originally published, 1831).

Fischbein, E. 1987. *Intuition in science and mathematics*. Dordrecht: Reidel.

Fischbein, E., M. Deri, M. S. Nello, and M. S. Marino. 1985. The role of implicit models in solving verbal problems in multiplication and division. *Journal for Research in Mathematics Education* 16: 3–17.

Fowler, D. H. 1987. *The mathematics of Plato's academy: A new reconstruction*. Oxford: Oxford University Press.

Freudenthal, H. 1983. *Didactical phenomenology of mathematical structures*. Dordrecht: Reidel.

Graeber, A. O., and D. Tirosh. 1988. Multiplication and division involving decimals: Preservice elementary teachers' performance and beliefs. *Journal of Mathematical Behavior* 7: 263–280.

———. 1990. Insights fourth and fifth graders bring to multiplication and division with decimals. *Educational Studies in Mathematics* 21: 565–588.

Greeno, J. G. 1987. Instructional representations based on research about understanding. In *Cognitive science and mathematics education*, ed. A. H. Schoenfeld 61–88. Hillsdale, NJ: Lawrence Erlbaum.

Greer, B. 1987. Nonconservation of multiplication and division involving decimals. *Journal for Research in Mathematics Education* 18: 37–45.

———. 1988. Nonconservation of multiplication and division: Analysis of a symptom. *Journal of Mathematical Behavior* 7: 281–298.

———. 1992. Multiplication and division as models of situations. In *Handbook of research on mathematics teaching learning* ed. D. Grouws 276–295. Reston, VA: National Council of Teachers of Mathematics; New York: Macmillan Publishing Co.

Greer, B., and Mangan, C. 1986. Understanding multiplication and division: From 10-year-olds to student teachers. In *Proceedings of the tenth international conference for the psychology of mathematics education*, ed. L. Burton and C. Hoyles, 25–30. London: London Institute of Education.

——— and M. McCann. 1991. Do children know what calculations are for? Paper presented at the fifteenth annual meeting of the international group for the psychology of mathematics education, Assisi, Italy.

Kaput. J. 1988. Looking back from the future: A history of computers in mathematics education from 1978 to 1998. Paper given at the invitational conference on mathematics education and technology, Northern Illinois University, DeKalb.

———. In press. The role of concrete representations of multiplicative structures: Creating cybernetic and pedagogical ramps from the concrete to the abstract. In *Making sense of the future*, ed. D. Perkins, J. Schwartz, and M. Stone Wiske, Oxford University Press.

Kieren, T. 1987. Remarks made at eleventh annual meeting of the international group for the psychology of mathematics education, Montreal.

Luke, C. 1988. The repeated addition model of multiplication and children's performance on mathematical word problems. *Journal of Mathematical Behavior* 7, 217–226.

Marshall, S. P. 1985. An analysis of problem solving in arithmetic textbooks. Paper presented at the meeting of the American Psychological Association, Los Angeles.

Marshall, S. P., K. E. Barthuli, M. A. Brewer, and F. E. Rose. 1989. *Story problem solver: A schema-based system of instruction* (Tech. Rep. No. 89-01). San Diego: San Diego State University, Center for Research in Mathematics and Science Education.

McMahon, T. A., and J. T. Bonner. 1983. *On size and life.* San Francisco: W. H. Freeman.

Mechmandarov, I. 1987. The role of dimensional analysis in teaching multiplicative word problems. Unpublished manuscript. Tel-Aviv: Center for Educational Technology.

Nesher, P. 1980. The stereotyped nature of school word problems. *For the Learning of Mathematics* 1(1): 41–48.

———. 1987. Towards an instructional theory: The role of student's misconceptions. *For the Learning of Mathematics* 7(3): 33–40.

———. 1988. Multiplicative school word problems: Theoretical approaches and empirical findings. In *Number concepts and operations in the middle grades,* ed. J. Hiebert and M. Behr, 19–40. Hillsdale, NJ: Lawrence Erlbaum Associates; Reston, VA: National Council of Teachers of Mathematics.

Piaget, J., and R. Garcia. 1989. *Psychogenesis and the history of science.* New York: Columbia University Press.

Pimm, D. 1987. *Speaking mathematically.* London: Routledge & Kegan Paul.

Reusser, K. 1988. Problem solving beyond the logic of things: Contextual effects on understanding and solving word problems. *Instructional Science* 17(4): 309–338.

Schwartz, J. 1988. Intensive quantity and referent transforming arithmetic operations. In *Number concepts and operations in the middle grades,* ed. J. Hiebert and M. Behr, 41–52. Hillsdale, NJ: Lawrence Erlbaum Associates; Reston, VA: National Council of Teachers of Mathematics.

Semadeni, Z. 1984. A principle of concretization permanence for the formation of arithmetical concepts. *Educational Studies in Mathematics* 15: 379–395.

Shalin, V., and N. V. Bee. 1985. *Analysis of the semantic structure of a domain of word problems.* Pittsburgh: University of Pittsburgh, Learning Research and Development Center.

Sinclair, H. 1990. Learning: The interactive re-creation of knowledge. In *Transforming children's mathematics education,* ed. L. P. Steffe and T. Wood, 19–29. Hillsdale, NJ: Lawrence Erlbaum.

Sowder, L. 1988. Children's solutions of story problems. *Journal of Mathematical Behavior* 7: 227–238.

Thompson, P. 1989. A cognitive model of quantity-based algebraic reasoning. Paper presented at the annual meeting of the American Educational Research Association, San Francisco.

Thorndike, E. L. 1922. *The psychology of arithmetic.* New York: Macmillan.

Tirosh, D., and N. Almog. 1989. Conceptual adjustments in progressing from real to complex numbers. In *Proceedings of the thirteenth annual conference of the international group for the psychology of mathematics education*, vol. 3, ed. G. Vergnaud, J. Rogalski, and M. Artigue, 221–227. Paris: G. R. Didactique.

Tirosh, D., and A. O. Graeber. 1989. Preservice elementary teachers' explicit beliefs about multiplication and division. *Educational Studies in Mathematics* 20: 79–96.

Treffers, A. 1989. Personal communication, cited by Verschaffel & De Corte (1993).

Vergnaud, G. 1988. Multiplicative structures. In *Number concepts and operations in the middle grades*, ed. J. Hiebert and M. Behr, 141–161. Hillsdale, NJ: Lawrence Erlbaum; Reston, VA: National Council of Teachers of Mathematics.

Verschaffel, L., & De Corte, E. (1993). A decade of research on word-problem solving in Leuven: Theoretical, methodological, and practical outcomes. *Educational Psychology Review*, 5(3), 1–18.

Vest, F. R. 1976. Teaching problem solving as viewed through a theory of models. *Educational Studies in Mathematics* 6: 395–408.

# II
## THE ROLE OF THE UNIT

# 4 Ratio and Proportion: Cognitive Foundations in Unitizing and Norming

*Susan J. Lamon*

A major contribution of cognitive science to mathematics education has been knowledge of the cognitive processes students use in mathematical problem solving. In several well-defined mathematical domains (Carpenter and Moser, 1983; Hiebert and Wearne, 1985; Post, Wachsmuth, Lesh, and Behr, 1985), the rich descriptions of children's thinking processes in relation to specific content bear considerable potential for enhancing the classroom teacher's pedagogical content knowledge and ability to assess and teach for meaningful learning. Knowledge of the interplay between content and thinking frameworks may, in fact, influence the teacher's entire instructional decision-making process (Carpenter & Fennema, 1988), thus providing an interface between research on children's thinking and research on teaching (Fennema, Carpenter, and Lamon, 1988).

The current psychological zeitgeist provides further impetus to increase the number of domains in which detailed analyses of children's thinking are available. Constructivism, the belief that people construct knowledge for themselves, has cast an aura of illegitimacy over the prescription of content strictly according to mathematical structure and has advanced the viewpoint that individual students' existing interpretation and organization of mathematical experiences determines the appropriateness of further instruction.

## RESEARCH GOALS IN COMPLEX DOMAINS

There is a question as to whether such detailed characterizations of children's thinking are possible within complex mathematical domains. The domain of ratio and proportion, for example, is one where such a body of knowledge is desirable. The domain represents a critical juncture at which many types of mathematical knowledge are called into play and a point beyond which a student's understanding in the mathematical sciences will be greatly hampered if the conceptual coordination of all the contributing domains is not attained. Proportional reasoning plays such a critical role in a student's mathematical development that it has been described as a watershed concept, a cornerstone of higher mathematics, and the capstone of elementary concepts (Lesh, Post, and Behr, 1988).

In most complex mathematical domains, however, it is impossible to designate a single, linearly ordered path through the many content domains essential to its understanding. Development looks more like a tree with an intricate branching system. By the time a child engages in proportional reasoning, prior knowledge and experience are so extensive and interactive that to ask how that student arrived at any given point in his or her learning is not to ask the important question. The compelling questions related to complex domains are these: What do children know before instruction? What can instruction do to facilitate learning? And how do we recognize learning when it occurs? Thus, rather than trying to map out optimum routes through the content of elementary school mathematics à la Gagne, a more rewarding enterprise, if successful, will be the identification of mathematical processes that facilitate growth in mathematical thinking both in horizontal and vertical directions.

An important consideration is connectivity. The logical status of many topics within rational numbers, for example, is of such complexity that it necessitates not only an understanding of many prerequisite domains, but also an ability to relate them to each other (Streefland, 1983). Therefore, one of the goals of current research is to identify important mathematical processes, themes, or connections by which thinking becomes progressively more sophisticated from early childhood through early adulthood. We might call these *mechanisms* by which more advanced reasoning evolves. This pursuit involves the investigation of complex domains with the goal of discovering continuities or discontinuities with earlier learned mathematical structures and processes. Under this approach, for example, one might ask the following: "How can we view ratio, and eventually proportional reasoning, as an extension of some basic mathematical idea(s)?" "What arithmetic knowledge may be useful in developing the more complex domains encountered in middle school?" "What intuitive, informal, or existing knowledge aids the learning of rational number concepts?" That is, researchers are seeking basic conceptual schema that constitute part of the backbone of mathematics, processes that are recurrent, recursive, or of increasing complexity across mathematical domains.

In a search for some viable way to construe the multiplicative conceptual field, two valuable lines of research have been seeking such mechanisms for mathematical growth. Research within complex mathematical domains is being conducted using both top-down and bottom-up approaches. Theoretical mathematical and semantic analyses of rational number topics (Behr, Harel, Post, and Lesh, 1990; Harel, Behr, Post, and Lesh, 1990) are based on the assumption that the current curriculum provides a limited perspective of the multiplicative conceptual field and that understanding will require a broader range of experiences. Determining what experiences might be most important to foster understanding requires a thorough analysis of the quantities (both units of measure and magnitudes) germane to multiplicative situations. A second methodology examines children's knowledge in clinical interviews either before formal instruction (Lamon, 1989) or after limited instruction (Mack, 1990) or in teaching experiments (Steffe, this volume) to determine which informal (intuitive, child-generated, previous, or existing) knowledge forms a useful foundation upon which instruction might be built.

This chapter examines sixth grade children's thinking in the process of solving ratio and proportion problems before these children had received any instruction in the domain. Instead of merely documenting the types of solution strategies children use, it analyzes them in terms of a framework that holds some promise for providing explanatory power for increasingly sophisticated reasoning. Specifically, it explores unitizing and norming, one of the processes to which both content analyses (Behr et al., 1990; Harel et al., 1990) and research on children's thinking (Steffe, this volume; Lamon, 1989) are pointing as a mechanism for the growth of mathematical thinking. After a discussion of unitizing and norming, children's use of these mechanisms in relation to ratio and proportion problems will be examined. The results will then be analyzed as a counterpoint to other research on the development of units in both the additive and multiplicative conceptual fields.

## UNITIZING

The ability to construct a reference unit or a unit whole, and then to reinterpret a situation in terms of that unit, appears critical to the development of increasingly sophisticated mathematical ideas. This process, which begins in early childhood (Steffe and Cobb, 1988), involves the progressive composition of units to form increasingly complex quantity structures.

The notion that higher order grouping allows a student to view aggregates and individual members of a set simultaneously, and thus to gain mathematical power, has been demonstrated in many areas, including the acquisition of early counting strategies (Steffe and Cobb, 1988), children's partitioning strategies (Pothier, 1981), and the acquisition of early addition and subtraction strategies (Carpenter and Moser, 1983).

The process of forming composite units probably begins visually in early childhood quantifying activities such as subitizing and is then extended into counting activities. When a child is counting on his or her fingers and begins to substitute the counting word *five* for the fingers on one hand, the child has adopted a more powerful grouping technique; likewise, when the child counts on from the first number in an addition problem or counts by tens, we can interpret his or her strategy an indication of higher level conceptual organization.

Research in the natural development of language hierarchies (Callanan and Markman, 1982; Markman, 1979) has substantiated this perspective, suggesting that more sophisticated thinking results when one reframes a situation in terms of a more collective unit because it invokes a part-whole schema that allows the student to think about both the aggregate and the individual items that compose it.

Advancing from addition and subtraction into multi-

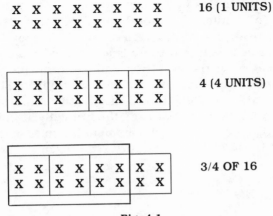

Fig. 4.1
*A simple multiplicative structure.*

plicative structures requires the conceptual coordination of multiple compositions (see Behr, Harel, Post, and Lesh, this volume). Some of the simplest multiplicative structures require a three-tiered composition of units. One such structure, illustrated in figure 4.1, involves the following example: Find 3/4 of 16 objects.

1. Consider the 16 objects as 16 units. You then have 16 one units.

2. Create units of units, that is, 4 composite units, each consisting of 4 one units. You then have 4 four units.

3. Create units of units of units, that is, create one three unit consisting of three of the four four units.

In addition to the change in the magnitudes of quantities, most multiplicative situations are referent-transforming compositions (Schwartz, 1988). That is, multiplicative structures combine two magnitudes with different labels to produce a quantity whose label is not the same as either the multiplicand or the multiplier. For example, five bags of candy with six candies per bag yields thirty candies (not bags of candy and not candies per bag). Sometimes the result is an intensive quantity, a new unit of measurement, a special relationship between two extensive quantities. This new quantity needs to be conceptualized as an entity in itself, different from its constituent measures. For example, if a car travels a distance of 207 miles in 3 hours, it averages about 69 miles per hour (not miles and not hours). Thus, multiplicative structures involve several layers of cognitive complexity. Although it is not clear that there is a continuous sequence of development from the counting and iteration processes of addition and subtraction into the more complicated intensive measurements units used in multiplicative situations (e.g., miles per hour, candies per bag), theoretically, it is impossible to avoid some discussion of composite units in the development of these complex quantity structures.

## NORMING

Composite units are usually formed in the pursuit of some larger goal. Often, a composite unit is formed and then a given situation is reinterpreted in terms of that unit. Freudenthal (1983) used the term *norming* to describe this process of reconceptualizing a system in relation to some fixed unit or standard. This norming process, or the adoption of some framework of units in which to conceptualize a situation, is prevalent in mathematical thinking. For example, imagine that the earth is the size of a pin's head (about 1 mm diameter) and then reconceptualize the solar system in terms of that definition. Then the sun would appear as a sphere with a diameter of 10 cm at a distance of 10 m from the earth (Freudenthal, 1983). Other roles of unitizing and norming in the world of rational numbers are examined next.

## UNITIZING AND NORMING IN RATIO AND PROPORTION

Norming, or the reconceptualization of a system in relation to some fixed unit or standard, is a convenient construct by which to analyze many rational number processes. Two of those processes are the within and between strategies used to solve proportions.

### Norming in the Measure Space

One of the simplest examples of norming occurs in the determination of a scale factor within a measure space. Within a

$M_1$
(measures of length)

$4\phantom{)}$
$7$         $7 = 1(4) + 3/4(4)$

Fig. 4.2
*A scalar operator transposes a measure within a single measure space.*

measure space, one element is a scalar multiple of another; that is, a scalar operator, or a nonvariable factor, operates upon one element of the space to produce another and vice versa. This process entails a reinterpretation of one measure in terms of the other, using the process of scalar decomposition. This idea is illustrated in figure 4.2, using Vergnaud's (1983) schema for representing multiplicative relationships.

    In this case, the four was chosen as the unit whole and seven was written in terms of whole four units and a fractional part of that four unit. Similarly, if seven had been chosen as the unit, four would be $(4/7)$ $(7)$.

## The Within Strategy for Solving Proportions

One of the critical numerical relationships in a proportion is that the scalar operator that links the measures in the first measure space is the same operator that links the measures in the second measure space. One of the methods of solving for a missing value in a proportion involves this important relationship. A student is said to be using a within strategy or a scalar method when he or she equates two internal or within-measure space ratios and uses the sameness of scalar operators to determine the missing term. This process is shown in figure 4.3, which presents a schema for the follow-

$M_1$                  $M_2$
(numbers of shirts)      (number of yards)

x3    $5$             $7$
         $15$            $x$    x3

Fig. 4.3
*The scalar method for finding a missing value in a proportion: an instance of norming.*

ing problem: If I can make five team shirts with 7 yards of material, how many yards of material will I need to make a team shirt for each of fifteen children on the soccer team?

If we think of five as the unit whole, we can think of fifteen as three of those units. Therefore, the scalar operator that transforms five into fifteen is three.

The numbers of yards of material in the second measure space are related by the same scalar operator. Therefore, the missing value can be found by operating with the scalar three upon the unit whole of seven to get twenty-one.

## The Between Strategy for Solving Proportions

In a proportion, another critical relationship is the functional relationship between measure spaces. We say that a student is using a *between strategy* or a *functional method* when he or she equates two external or between-measure space ratios and relies on the functional relationship between the measure spaces to find the missing term. These are ratios whose first terms form $M_1$ and second terms form $M_2$ or vice versa. The functional relationship enables the solution for a missing value by scalar decomposition of a function, another norming process. Figure 4.4 illustrates this process as it applies to the following problem: The pharmacist gave you 7 ounces of medicine for $8.75. What would you expect to pay for a bottle containing 4 ounces?

In this example, the function operator, 1 1/4, represents the coefficient of the linear function from $M_1$ to $M_2$. If you were solving this problem mentally, for example, you might think of the $8.75 in terms of whole dollar amounts and partial dollar amounts. $8.75 is the cost of 7 ounces at $1.00 per ounce plus $.25 per ounce. Therefore, 4 ounces would cost 4 whole dollars plus 4 quarter dollars.

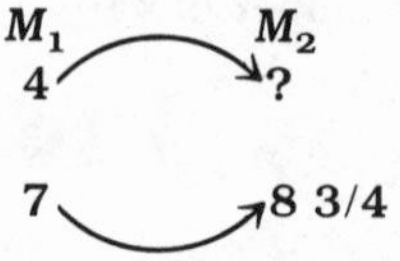

$M_1$      $M_2$
4     ?

$8\ 3/4 = 1(7) + 3/4(7)$
$f(x) = 1(x) + 3/4(x)$

7     8 3/4

$1(4) + 3/4(4) = 5$

Fig. 4.4
*Finding the function operator through a norming process.*

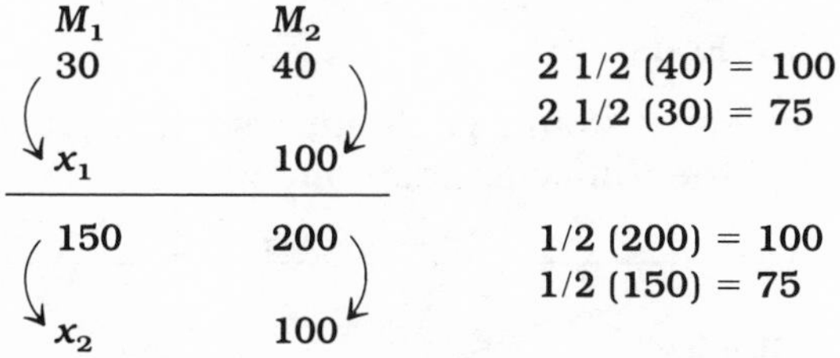

$$M_1 \qquad M_2$$

$$\begin{array}{cc} 30 & 40 \\ \left( \begin{array}{c} \\ \downarrow x_1 \end{array} \right. & \left. \begin{array}{c} \\ 100 \end{array} \right) \end{array} \qquad \begin{array}{l} 2\ 1/2\ (40) = 100 \\ 2\ 1/2\ (30) = 75 \end{array}$$

$$\begin{array}{cc} 150 & 200 \\ \left( \begin{array}{c} \\ \downarrow x_2 \end{array} \right. & \left. \begin{array}{c} \\ 100 \end{array} \right) \end{array} \qquad \begin{array}{l} 1/2\ (200) = 100 \\ 1/2\ (150) = 75 \end{array}$$

Fig. 4.5
*Percentages: norming to a standard.*

## Percentages as Norming

The process of norming can achieve yet another level of sophistication when, for reasons of uniformity or convenience, an independent unit is chosen as the standard for norming. In calculating percentages, for example, 100 is always chosen as the second number in a ratio. To standardize the relationship between 30 of 40, and 150 of 200, for example, we use the sameness of scalar operators. This process is shown in figure 4.5.

## Norming and Other Rational Number Topics

The processes just examined are well-documented solution strategies to missing-value proportion problems. We have merely examined them under a different interpretive framework. Changing one's perspective often facilitates some observations and connections that may not have been forthcoming otherwise. For example, the proposed unitizing and norming framework provides a theoretical backdrop for other rational number topics.

## Norming and the Meaning of Fractions

One of the most critical understandings in the fraction world is that the numerical symbol (1/3, for example) can represent different amounts, depending on what the unit or the unit whole happens to be. How many or how much is meant by 1/3 is ambiguous unless we know (among other things) to which whole it refers. The notions of unitizing and norming play a major role in the development of the fraction concept. For

a. Cuisenaire Rods

Let *brown* be equal to the whole number 1. What color represents each of the following fractions?

$$1/8 = \underline{\hspace{2cm}} \qquad\qquad 1\ 1/4 = \underline{\hspace{2cm}}$$

b. Pattern Blocks

If  = 1, draw 1/8.

If ⬡ = 1, what fraction does △ represent?

c. If ●●<br>●● = 1/3, draw 1/2.

d. If ○<br>○○ = 1, what fraction is shaded below?

●● ●● ○<br>●● ●● ●○

Fig. 4.6
*Norming in the development of the fraction concept.*

example, the student activities shown in figure 4.6 require that the student understand composite units when the unit whole is either continuous or discrete and further reinterpret another quantity in terms of that composite whole.

## Norming in the Division of Fractions

The process of dividing fractions may also be interpreted as a norming process. It requires the selection of the divisor as a norming unit and the reinterpretation of the dividend in terms of the divisor. This process is shown in figure 4.7.

In short, it appears that the analysis of ratio and proportion through a framework of unitizing and norming may provide some useful insights into the critical relationships we would like children to understand. The ability to reinterpret information in terms of different unit wholes—sometimes several times within the same situation—appears to be essential to understanding ratios. The question remains as to how closely this rational analysis conforms to what children actually do as they encounter the domain. To examine this question, the remaining sections of the chapter will report on a study of the intuitive understanding of ratio and proportion

$$3/4 \div 1/2 = ?$$

How many 1/2s are in 3/4 ?

1/2 is the unit whole and 3/4 is reinterpreted in terms of that unit.

$$3/4 = 1(1/2) + 1/2(1/2)$$

Answer: 3/4 = 1 1/2 of the unit whole

Fig. 4.7

*Fraction division as a norming process.*

of twenty-four sixth-grade children. Before presenting this study, I will set the stage with a brief review of pertinent literature.

## RESEARCH IN RATIO AND PROPORTION

Though research with preadolescent students is not extensive, it appears that student representations of situations involving ratio and proportion occur on an informal, qualitative basis long before those students are capable of treating the topic quantitatively (Tourniaire, 1986; Treffers and Goffree, 1985; Van den Brink and Streefland, 1979; Steffe, 1969; Steffe and Parr, 1968) and that many student strategies and processes develop independent of instruction (Post, Behr, Lesh, and Wachsmuth, 1985). That is, children naturally employ a form of mathematical intuition, an informal knowledge system (Kieren, 1983; Treffers & Goffree, 1985; Streefland, 1984, 1985). This informal knowledge includes a visual understanding of ratio and proportion (Steffe and Parr, 1968), especially of congruence and similarity (Van den Brink and Streefland, 1979). In addition, children use self-invented solution strategies, which are primitive, context bound, relatively symbol free, and based upon counting, adding, and halving (Hart, 1984). Some fairly sophisticated student-invented methods have been documented (e.g., see Rupley, 1981), but analysis of those strategies has been principally in relation to within (internal or scalar) and between (external or functional) frameworks (e.g., Lybeck, 1978; Noelting, 1980a, 1980b; Hart, 1981; Karplus, Pulos, and Stage, 1983;

Vergnaud, 1983; cf. Tourniaire and Pulos, 1985). The chief thrust of empirical research efforts has been to document the difficulties students have with proportional reasoning and identify the subject, task, and context variables that affect problem difficulty. The consensus is that preference for between or within strategies is highly dependent on task variables. The influence and interaction of these variables is so complex that we are left with few implications for instruction.

Although previous studies indicate that children's thinking in relation to ratio and proportion problems can be reasonable, productive, and even insightful, they have provided little information as to how, when, and in response to which situations a child's predominantly additive view of the world changes to include a multiplicative perspective. Basic questions remain unanswered: At what point can we say a child reasons proportionally? What kinds of relationships does someone need to have grasped before we can claim that he or she understands ratio and proportion? Could an interpretation of children's thinking in terms of a unitizing-norming framework provide any new insights?

### Children's Preinstructional Thinking

Twenty-four sixth-grade children, who had not yet received formal instruction in ratio and proportion, participated in clinical interviews in which five problems were designed to elicit information regarding the children's power of unitizing and norming. The five problems are shown in Figure 4.8.

The balloon problem and the subscription problem investigated students' preference for simple units (one units) or composite units. The apartment problem examined their ability to handle multiple compositions. The pizza and the alien problems were intended to check preferences for between or within strategies in the presence of unit ratios and divisibility relationships. Key solution strategies will be indicated for each of these problems, analyzed in terms of a unitizing and norming framework, and then related to other research on unit formation.

### Results

The study indicated that children come to instruction in ratio and proportion with some informal understanding. Al-

*The balloon problem.* Ellen, Jim, and Steve bought 3 helium filled balloons and paid $2.00 for all three. They decided to go back to the store and get enough balloons for everyone in their class. How much did they have to pay for 24 balloons?

*The subscription problem.* The following subscription card was taken from a magazine.

**WHY IT HAPPENED. WHAT IT MEANS.**

\* Now you can receive Newsweek at **46¢\*\* an issue**—that's **77% off** the cover price. And you can pay in 3 easy installments!

| Term | 6 mos. | 9 mos. | 12 mos. |
|------|--------|--------|---------|
| Cover Price | $52.00 | $78.00 | $104.00 |
| Your Cost | $11.97 | $17.97 | $ 23.97 |
| 3 Payments Each Only | $ 3.99 | $ 5.99 | $ 7.99 |
| | 4.00 | 6.00 | 8.00 |

Please check:
- 6 months (26 Issues)
- 9 months (39 Issues)
- 12 months (52 Issues)
- Full payment enclosed
- Bill me in 3 installments
- Bill me in full

Offer good in U.S. only and subject to change.
Basic Rate 79¢
\*\*Rates Rounded
\*Per mo. installment

761527 13-10

Reprinted with permission from *Newsweek*, Inc.

Do you get a better deal if you buy the magazine for a longer period of time?

*The apartment problem.* In a certain town, the demand for rental units was analyzed and it was determined that, to meet the communities needs, builders would be required to build apartments in the following way: Every time they build 3 one-bedroom apartments, they should build 4 two-bedroom apartments and 1 three-bedroom apartment. Suppose a builder is planning to build a large apartment complex containing 30–40 apartments. How many apartments should be built to meet this regulation?

Suppose one built 32/40 apartments (choose one). How many one-bedroom, two-bedroom, and three-bedroom apartments would the apartment building contain?

Suppose one built 14 one-bedroom apartments, 18 two-bedroom apartments, and 4 three-bedroom apartments. Would the requirement be satisfied?

Fig. 4.8
*Five problems used to elicit children's use of unitizing and norming.*

*The pizza problem*. Consider the following drawing of children and pizzas.

Who gets more pizza, the girls or the boys?

*The alien problem*. Consider the following picture of aliens and food pellets.

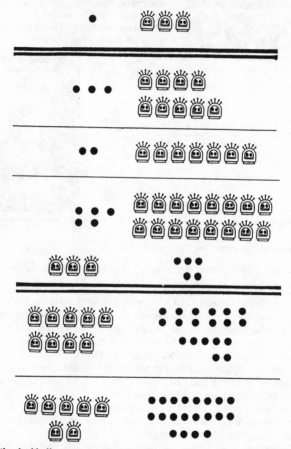

Above the double line, you can see some aliens and the number of food pellets they need to live for one day. Assume that all of the aliens eat about the same amount. Tell me if each group of aliens under the double line has the right number of food pellets, too many, or too few.

(Referring to the second picture:) Suppose I had 36 food pellets. How many aliens could I feed for one day?

Fig. 4.8
*Continued.*

though none of the children used symbols to represent his or her thinking, conceptual competency exceeded symbolic competency. Successful student strategies consistently involved a strong intuitive application of unitizing and norming.

*Balloons.* The successful processes used to solve the balloons problem follow. The number of students who used each of these processes is given in parentheses. Problem: Ellen, Jim, and Steve bought three helium-filled balloons and paid $2.00 for all three. They decided to go back and get enough balloons for everyone in their class. How much did they have to pay for twenty-four balloons?

1. For every third balloon, you pay $2.00. So you have eight packets (sets, groups, or bunches of eight) and 8 × 2 would be $16. (10)

2.
| | |
|---|---|
| 3 | $2.00 |
| 6 | 4.00 |
| 9 | 6.00 |
| 12 | 8.00 |
| 15 | 10.00 |
| 18 | 12.00 |
| 21 | 14.00 |
| 24 | 16.00 (2) |

3. $2.00/3 = .6666 . . . or .66 r2    .66 × 24 = $15.84 (3)

4. Three balloons were $2.00 and 2 divided by 3 is $2/3$, so I asked myself how many 24ths is $2/3$? The answer is 16. (1)

In subsequent questioning, the two students who used the building-up strategy (shown in 2) failed to show full recognition of the functional relationship in this situation. That is, they could not predict cost given the number of balloons. The tables they constructed began as convenient methods to keep track of a double counting process and were expanded as patterns were recognized. The students who used the two-step unit rate strategy (shown in 3) sacrificed accuracy; although correct, their thinking in terms of single units proved less powerful than thinking in terms of composite units. Without using the traditional symbolism, the last student

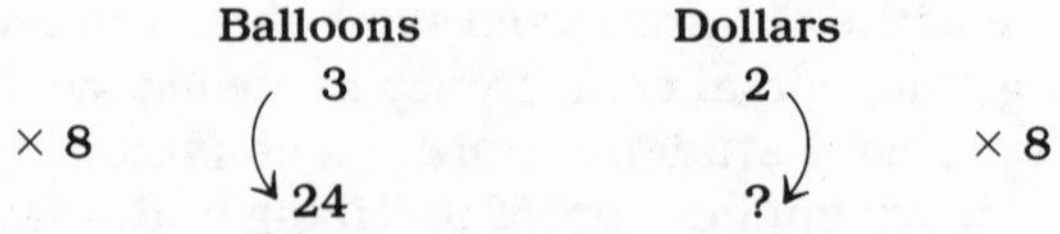

Balloons        Dollars

$\times 8 \quad \left( \begin{array}{c} 3 \\ \downarrow 24 \end{array} \right. \qquad \left. \begin{array}{c} 2 \\ ? \end{array} \right) \quad \times 8$

Fig. 4.9
*Solution to the balloon problem.*

engaged in proportional reasoning with full recognition of the structural relationships by equating rates. The first response indicates, however, that for the majority of the students who solved the problem, their ability to conceptualize the situation in terms of groups, sets, packets, or bunches— various class terms were used—enabled them to solve the problem. As shown in Figure 4.9, they solved a proportion consisting of within measure space ratios by applying the scalar operator between members of the first measure space to the second measure space.

*Subscription.* The subscription situation provided an interesting forum in which to investigate students' thinking because it provided a multitude of divisibility relationships, both within and between measure spaces, around which the student could base his or her solution. Problem: The student was shown a card in a magazine that offers three plans for subscribing to the magazine: (1) you may subscribe for a 6-month period, during which time you will receive three bills each for the amount of $4.00; (2) you may subscribe for 9 months and receive three bills of $6.00 each; or (3) you may subscribe for 12 months and pay three bills of $8.00 each. Do you get a cheaper rate if you buy the magazine for a longer period of time?

When the explicit rates as well as the comparable one-month and three-month rates are considered, this situation might be summarized as in Table 4.1. The thirteen students who were successful with this problem used interesting solution processes; their solutions follow. Again, the number of students who used each solution is given in parentheses.

  1. The price for two 6-month subscriptions is the cost of 12 months, and 9 is just the middle of that. (5)
  2. 4 is 2/3 of $6.00 and 6 is 2/3 of 9. (1)

TABLE 4.1. SUBSCRIPTION RATES

| $M_1$ Subscription Time in months | $M_2$ Cost in dollars | |
| --- | --- | --- |
| | Each payment | Total |
| 1 | 2/3 | 2 |
| 3 | 2 | 6 |
| 6 | 4 | 12 |
| 9 | 6 | 18 |
| 12 | 8 | 24 |

3. Three payments of $4.00 is $12.00; three payments of $6.00 is $18.00; three payments of $8.00 is $24.00. Divide each in half and you get 6, 9, and 12. (1)

4. The months go up by 3 and the dollars go up by 6, so it's the same any way you do it. (1)

5. In each case, if you figure out how much it costs a month, you get $2.00 a month. (1)

6. For the 6-month subscription, every 3 months would be $2.00; for the 9-month subscription, every 3 months is $2.00; the same for the 12-month subscription. (4)

Highlighting the cost of one payment tended to distract the students' attention from the fact that the total cost in dollars was twice the number of months of subscription. Therefore, only one student noticed the correspondence between the number of months and the cost of the subscription. In the fourth solution, the student confirmed that the rate must be the same merely by identifying patterns. The second solution identified and compared the scalar operators in both measure spaces. However, the majority of the solutions involved finding and comparing rates. It is interesting to note that of the eleven solutions using rates, only one student used a one-month unit. The other solutions used three-month or six-month units. It appears that the students made decisions about the most appropriate unit with which to reinterpret a situation, based on the specific conditions of the problem. In this case, most students chose the three-month unit as the most efficient unit because 3 was the greatest common factor of 6, 9, and 12. Also, most realized that they could deal with the cost of a single payment because every option involved three equal payments.

*Apartment.* Students' thinking about the apartment problem proved quite interesting when analyzed in terms of the formation of composite units. This problem requires the repeated formation of composite units, and then the decomposition of those units back into their constituent parts. Problem: In a certain town, the demand for rental units was analyzed and it was determined that, to meet the community's needs, builders would be required to build units in the following way: every time they build three single units, they should build four two-bedroom units and one three-bedroom unit. Suppose a builder is planning to build a large apartment complex containing between thirty and forty units. How many units should be built to meet this regulation? Suppose one built 32/40 (choose one). How many one-bedroom units, two-bedroom units, and three-bedroom units would the apartment building contain?

This problem may be analyzed as follows:

(Units)
1-bedroom apartments
2-bedroom apartments
3-bedroom apartments

(Units of units)
three units of 1-bedroom apartments
four units of 2-bedroom apartments
one unit of 3-bedroom apartments

(Units of units of units)
Those units of apartments together form one eight unit of apartments.

(Units of units of units of units)
How many 8 units are needed?
4 or 5 of that 8 unit

(Decomposition of composite units)
$40 = 5 (8) = 5 (1:3:4) = 5 + 15 + 20$

Fifteen out of twenty-four students were able to think through this situation. Eleven students used the composition and decomposition process just described. Laurie's protocol is typical of the solution strategies they used.

> *Laurie:* He should build fifteen one-bedroom apartments, twenty two-bedroom apartments, and five three-bedroom apartments.
> *Researcher:* Tell me how you thought about that.
> *Laurie:* Well, 3 and 4 and 1 make like one building and you can make five buildings. So there would be 5 of each kind of apartment.
> *Researcher:* What do you mean 5 of each kind?
> *Laurie:* There would be three one-bedrooms in each building, and five buildings, so that's fifteen one-bedrooms and like that for the other sizes.

The other four students did not create one eight unit, but instead, simultaneously added columns of three units, four units, and one units or the multiples of 3, 4, and 1 until they achieved a total of 32 or 40 apartments, and then noticed that they had 4 or 5 of each type. Kari's protocol illustrates this solution process.

> *Kari:* (Worked for a few minutes on paper.) 15 one-bedroom apartments, 20 two-bedroom apartments and 5 three-bedroom apartments.
> *Researcher:* Would you explain that answer? What were you doing here?
> *Kari:* (Showed the following work)

| | | | |
|---|---|---|---|
| 3 one-bedroom | 6 | 9 | 15 |
| 4 two-bedroom | 8 | 12 | 20 |
| 1 three-bedroom | 2 | 3 | 5 |
| | 16 | 24 | 40 |

First I doubled each number and added them up. That was 16, so I could get more. . . . So I did three more of each and I just added them up. Then I decided to try each one times 5 and that gave me 40.

In addition to the fifteen successful strategies, two more completed the composition of units (that is, decided that there were going to be 32 or 40 total apartments), but were unable to reverse the process and decompose into the numbers of one-, two-, and three-bedroom apartments.

An additional question further probed some students'

understanding of the situation: Suppose one built fourteen one-bedroom apartments, eighteen two-bedroom apartments, and four three-bedroom apartments. Would the regulation be met?

Predictably, the students who were able to decompose the units in the first question were able to decide that, in fact, the regulation would not be met. The two students who were not able to decompose were also not able to recognize that the total number of apartments was not a multiple of 8, nor were the numbers of apartment types divisible by a common factor that would yield the required ratio $3:4:1$.

*Pizza.* Sometimes, students' reasoning defied expected patterns. Divisibility relationships, integral ratios, and unit rates, for example, are thought to facilitate symbolic solutions of proportions (Karplus et al., 1983; Noelting, 1980a, 1980b), but for these students, who were not yet operating symbolically, the numerical relationships did not appear to be of major importance. Instead, these students used both ratios and rates as units and operated one upon the other, using a double-matching process. Their strategies for the pizza problem showed that their thinking did not involve either of the methods for setting up ratios predicted by dimensional analysis: between or within measure spaces. Problem: Student is shown the drawing of children and pizzas in Figure 4.8 and asked, "Who gets more pizza, the girls or the boys?"

Dimensional analysis would predict two ways in which to think about this situation. There are two measure spaces, one measuring the cardinality of the set of children, and the other measuring the cardinality of the set of pizzas, and one way to relate them is with a between strategy, as shown in Figure 4.10. The following rates would then be compared: $7/3$ and $3/1$. A second option is to think about each set of chil-

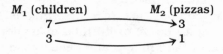

Fig. 4.10
*Solving the pizza problem using the functional relationship between measure spaces.*

$M_1$ (children)                    $M_2$ (pizzas)

$7$                                 $3$

$3$                                 $1$

Fig. 4.11
*Solving the pizza problem using a within measure space strategy.*

dren and pizzas separately, in which case quantities would be related using a within strategy, as shown in Figure 4.11. Consequently, the following ratios would be compared: 7/3 and 3/1. (Merely by coincidence, rates and ratios turn out to be the same for this particular problem.)

To compare, the expected method would be to first notice that the ratio-rate 3/1 is a unit rate and then determine what should be the corresponding ratio-rate for equivalence (Karplus et al., 1983). This process involves norming in terms of the unit rate, which, in this case, would produce 9/3 as the desired equivalent. At this point, the student could determine whether the ratios-rates are equivalent, but needs to invoke additional knowledge to determine the direction of the inequality. One might switch to a division interpretation of the fraction symbol or apply other informal knowledge concerning the size of fractions with like denominators to determine that $7/3 < 9/3$.

Eighteen of the twenty-four students interviewed in this study solved the problem, but not in the expected manner. They all used the same solution strategy explained by Kari.

> *Researcher:* Who gets more pizza, the girls or the boys?
> *Kari:* The girls.
> *Researcher:* Why do you think the girls do, Kari?
> *Kari:* Because the boys have to share three people with one pizza and if the girls did the same thing . . . if they had three people to this pizza (pointed to the first pizza and covered three girls) and three people to this pizza (pointed to the second pizza and covered three more girls), then one person would have a whole pizza to herself. So some people could go over to that one to get some more.

This process might be analyzed into the following steps:

1. Distinguish four kinds of units.

girls                    boys
girl pizzas              boy pizza

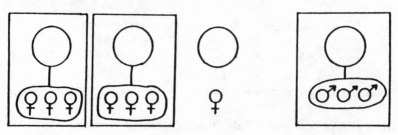

Fig. 4.12.
*The girl-to-girl pizza relationship reinterpreted in terms of the boy-to-boy pizza relationship.*

2. Create units of units.

| 7 girls | 3 boys |
| 3 girl pizzas | 1 boy pizza |

3. Create units of units. Here, thinking involves a between strategy and the rate becomes the new unit.

$$7:3 \qquad 3:1$$

4. Choose one rate as the standard unit and reinterpret the other in terms of that unit. In this case, the rate $3:1$ became the standard unit. How many $3:1$ units are in $7:3$? As shown in Figure 4.12, $7:3$ was reinterpreted in terms of the composite unit $3:1$.

5. Because $1:1$ yields more pizza per person than $3:1$, the girls get more pizza. Symbolically, this process might be represented in the following way:

$$7:3 = 2(3:1) + (1:1) = (6:2) + (1:1)$$

$$1:1 > 3:1$$

The fact that $3:1$ is a unit rate may or may not have been critical in its selection as the norming unit. However, students used that rate in a far different manner than would have been predicted. Both of its components were compared simultaneously with the components of the other rate.

*Aliens.* All twenty-four students solved the aliens problem in the same way; that is, using the double-matching strategy. Table 4.2 describes the process they used. Figure 4.13 shows the reinterpretation of the $9:19$ in terms of the complex unit $3:5$. Answer: $9:19 = 3(3:5) + 0:4 = 9:15 + 0:4$

TABLE 4.2.  HOW MANY $(3:5)$s ARE IN $(9:19)$?

| *Group 1 of aliens* | *Group 2 of aliens* |
|---|---|
| *Group 1 food pellets* | *Group 2 food pellets* |
| $3A_1$ | $9A_2$ |
| $5P_1$ | $19P_2$ |
| $3:5$ | $9:19$ |

(The $4P_2$ are extras). Again, the divisibility relationship between $3A_1$ and $9A_2$ probably led to the use of $3:5$ as the norming unit.

The purpose of the follow-up question to the alien problem was to see if a student who solved the problem by norming with the ratio-rate as a unit could reverse his or her thinking. Not all students were asked the questions, but of those who were, two students solved the aliens problem using a building-up process as follows:

> *Researcher:* Suppose I had thirty-six food pellets. How many aliens could I feed for one day? (Prior information told the student that three aliens could live on five food pellets for one day.)
>
> *Chris:* Well, 3 aliens need 5 pieces of food, so about 3 times it would be . . . 9 could use 15 (wrote 9 and 15.) . . . and timesed again, another 9 would get 15 (wrote 9 under the first 9, 15 under the first 15). So, we're up to 30. That means we could use 5 more for 3 more aliens . . . and there would be 1 left over (wrote 5 and 3 in the appropriate columns and added both columns). So we have enough food for 21 aliens.

Fig. 4.13
*The ratio 3:5 used as a norming unit.*

Symbolically, Chris' thinking could be represented in the following way:

$$3(3:5) + 3(3:5) + 1(3:5) + (0:1) = 9:15 + 9:15 + 3:5 + 0:1 = 21:36$$

## Discussion

Three-fourths of the students interviewed naturally formed ratios of otherwise unrelated sets and engaged in the process of norming, or reinterpreting one ratio in terms of the other. The presentation of the situation in the form of pictures probably facilitated the double-matching process. Naturally, some presentations make the semantic structure of a problem more transparent than others and are more likely to optimize student performance. Also, Case's (1978, 1980) argument, that easing the load on the working memory facilitates reasoning, may be applicable in this case. For whatever reason, students' ability to think about a ratio as an invariant composite unit and work simultaneously with both its components in a double-matching process (covariance) illustrated the kind of understanding we would like them to have about the meaning of ratio and proportion. See Rizzuti (1991) for a discussion of covariance as strong intuitive interpretation of function. Their thinking may be interpreted as a kind of presymbolic quantitative proportional reasoning. Through the use of presentations and representations that make the norming process transparent, it is likely that students' understanding of the notions of ratio and proportion and their connections would be enhanced. Modeling of this nature may be a key to understanding important relationships in the domain.

Although the students did not use symbolic representations, their thinking showed that they were quantifying relationships. Even more significant, they were performing arithmetic operations on those quantities. The extension of this quantification and arithmetic into symbols appeared imminent. In particular, student thinking was easily translatable into the representation $a:b$ and its appropriate arithmetic operations:

$$(a:b) \pm (c:d) = (a \pm c:b \pm d)$$
$$c(a:b) = (ca:cb)$$
$$(a:b)/(c:d) = M(c:d) + (a - Mc:b - Md)$$

where $M$ is the minimum of the integers in the quotients $a/c$ and $b/d$.

Students representations of the balloon and subscription problems showed that most students not only think in terms of composite units, but also make decisions about which unit is most appropriate. The most simplistic solution to both problems would have been to use a one unit as the norming unit, but in most cases, students saw an advantage to using a composite unit: a three unit of balloons and a three-unit or six-unit period of subscription. Therefore, levels of sophistication could also be distinguished in children's solutions to these problems based on whether or not they formed composite units and reinterpreted the problems in terms of those units, as well as on which units they chose; some units were more efficient than others. This observation would dictate against teaching students to use the unit ratio strategy (e.g., find the rate for one and multiply to get the rate for many) as a standard approach to proportion problems (cf. Herron and Wheatley, 1978; Goodstein and Boelke, 1980; Gold, 1978, 1980).

Although future research should pursue these ideas, it is apparent that students' preinstructional reasoning provides some basis on which to build an understanding of intensive quantities.

## UNITIZING AND NORMING AS A MECHANISM
## FOR MATHEMATICAL GROWTH

Both theoretical analyses and empirical results suggest that the unitizing and norming process may be an important mechanism by which more advanced reasoning evolves. Understanding ratio and proportion depends on one's ability to view a relationship as a single quantity and then to operate with it. Taking a look at the larger scheme of things, however, reveals that understanding the relative nature of quantities in a ratio may be just another level of sophistication built upon an already-complex foundation of unitizing processes. In Table 4.3, some of the processes used by children to make sense of ratio and proportion are placed in counterpoint to some of the unitizing processes employed in early childhood and for addition and subtraction, as well as those children use when thinking about multiplication and division. Although

TABLE 4.3.   THE INCREASING COMPLEXITY OF THE UNITIZING PROCESS AS REVEALED BY CHILDREN'S THINKING IN THE ADDITIVE AND MULTIPLICATIVE CONCEPTUAL FIELDS

| | + or − | × or ÷ | Ratio |
|---|---|---|---|
| Modeling and Counting | Counting objects 1, 2, 3, 4, 5 Subitizing: The visual formation of composite units △△△ △△ 5 | Segmented counting: forming experential composites How many 3s in 12: 1, 2, 3; 4, 5, 6; 7, 8, 9; 10, 11, 12 | Counting triggered by pictorial presentation; forming experential ratios: matching three aliens and five food pellets |
| Composing | Constructing a numerical composite: ✋ Ⴈ Ⴈ Ⴈ 5 6 7 8 | Early multiplication scheme: repetitively forming and counting three units: 1 − 2 − 3   1 1 − 2 − 3   2   4 × 3 1 − 2 − 3   3 1 − 2 − 3   4 | Repetitively matching and counting three units and five units 1 − 2 − 3   1 − 2 − 3 − 4 − 5   1 1 − 2 − 3   1 − 2 − 3 − 4 − 5   2 1 − 2 − 3   1 − 2 − 3 − 4 − 5   3 |
| Abstracting | Constructing an abstract numerical composite: thinking of one five-unit without considering its constituent one units | Forming iterable units: 3 is stripped of its composite quality: 3 is 1 6 is 2 9 is 3        double counting 12 is 4 | Iterable ratios: 6:10 and 9:15 represent the same relationship as 3:5 |
| Relating | PPW: the whole is a numerical composite whose parts are also composites | PPW, the distributive property: eight four units can be achieved by adding five four units and three four units; parts are multiples of the same composite units | PPW, a multiple of the whole is composed of multiples of different composite units: 5(8-units) = 5(3-units) + 5(4-units) + 5(1-units) |

more extensive research is warranted before we can posit any kind of theory, the table attempts to highlight some interesting parallels.

Moving from left to right between the columns and moving from top to bottom between the rows traces a route of increasing sophistication in children's thinking and increasing complexity in the types of units required. Thus, those cells in the matrix lying farther to the right and nearer the bottom represent more demanding cognitive activity.

Early elementary school children count objects and eventually are able to subitize or form a composite unit in the presence of the individual objects that compose it. Likewise, Steffe (1990) found that segmented counting helped older children experience the three units in twelve. It is particularly significant that each of the individual counting words, *one, two, three, four, . . . eleven, twelve,* were used. Middle school children's preinstructional, presymbolic efforts at making sense in ratios and proportions recapitulated the methods they invented and applied to early addition and multiplication concepts. Intuitive counting strategies emerged when pictorial presentations facilitated the coordination of all the quantities involved. In addition to reaffirming the importance of counting and modeling as important sense-making activities, the simultaneous matching process students employed suggests a valuable activity for facilitating understanding of a ratio. "Experiencing" a ratio through a concrete activity such as counting and pairing two sets of objects may be an important prerequisite to abstracting the concept of ratio, an important first step in a learning cycle.

Research on children's thinking in early addition and subtraction concepts (Carpenter and Moser, 1983) and in multiplication (Steffe, this volume) reveals that modeling and counting activities give way to composing units. The purposeful grouping of single items to make a single entity marks this milestone. In counting several such entities, what is significant is that a single counting word is attached to composite unit, even though its constituent unit elements are evident, as in the case of saying *five* in the presence of the five digits. Similarly, in an early multiplication scheme, children repetitively form three units instead of using all the individual counting words. In ratio, children showed the same ability to use composite units, two at a time. Three aliens and five food pellets were matched and the child counted "one day"; another three aliens were matched with another five food pellets and the child counted "two days," and so on.

The next apparent level of sophistication in cognitive activity is "abstracting." At this level, a child is able to operate with the composite unit in a manner oblivious to its constituent parts. Steffe (this volume) calls such units *iterable units* and claims that a double-counting process, such as "3 is 1; 6 is 2; 9 is 3,"", or the coordination of two counting schemes, such as "3, 6, 9, 12, 13, 14," is evidence that units are iterable. In ratio, units were clearly *not* iterable for the children interviewed. Their operations with the ratios of aliens to food pellets, for example, were highly dependent on their counting each group of aliens and food pellets before matching them. We can speculate that the child for whom ratio is an iterable unit might be able to name members in the equivalence class of 3:5 without counting groups of aliens and food pellets.

After constructing abstract numerical composites, even greater sophistication in thinking is enabled when those composites can be meaningfully related to others. In addition and subtraction, in the part-part-whole relationship, the whole is a numerical composite whose parts are also composites. In multiplication, the distributive property relates multiples of the same composite units. In ratio, another level of complexity is added as a child needs to cognitively coordinate multiples of different composite units. It is possible that some of the children were not able to complete the apartment problem because of the extreme difficulty entailed in cognitively coordinating multiples of several different composite units.

## CONCLUSION

These parallels suggest that the unitizing and norming process may be viewed as a spiraling cognitive activity in which increasingly complex unit formations are taken through the operations of modeling and counting, composing, abstracting, and relating. Research efforts should further investigate children's thinking in the more complex mathematical domains and seek to clarify possible connections with cognitive activities in less complex domains. The identification of such mechanisms capable of enhancing mathematical thinking is a compelling enterprise if we hope to facilitate higher order thinking in our classrooms.

## ACKNOWLEDGMENT

An earlier version of this chapter was presented at the 1990 Annual Meeting of the American Educational Research Association, April 16–20, Boston.

## REFERENCES

Behr, M., G. Harel, T. Post, and R. Lesh. 1990. On the operator concept of rational numbers: Towards a semantic analysis. Paper presented at the annual meeting of the American Educational Research Association, Boston.

Callanan, M. A., and E. M. Markman. 1982. Principles of organization in young children's natural language hierarchies. *Child Development* 53: 1093–1101.

Carpenter, T. P., and E. Fennema. 1988. Research and cognitively guided instruction. In *Integrating research on teaching and learning mathematics*, ed. E. Fennema, T. P. Carpenter, and S. J. Lamon, 1–17. Madison: Wisconsin Center for Education Research.

Carpenter, T. P., and J. M. Moser. 1983. Acquisition of addition and subtraction concepts. In *Acquisition of mathematics concepts and processes*, ed. R. Lesh and M. Landau, 7–44. Orlando, FL: Academic Press.

Case, R. 1978. Implications of developmental psychology for the design of effective instruction. In *Cognitive psychology and instruction*, ed. A. M. Lesgold, J. W. Pelligrino, S. D. Fokkema, and R. Glaser, 441–463. New York: Plenum Press.

———. 1980. Intellectual development and instruction: A neo-Piagetian view. In *1980 AETS yearbook: The psychology of teaching for thinking and creativity* ed. A. E. Lawson, 59–102. Columbus, OH: ERIC Clearinghouse for Science, Mathematics, and Environmental Education.

Fennema, E., T. P. Carpenter, and S. J. Lamon. (eds.) 1988. *Integrating research on teaching and learning mathematics.* Madison: Wisconsin Center for Education Research.

Freudenthal, H. 1983. *Didactical phenomenology of mathematical structures.* Dordrecht: D. Reidel.

Gold, A. P. 1978. Cumulative learning versus cognitive development: A comparison of two different theoretical bases for planning remedial instruction in arithmetic. Unpublished doctoral dissertation, University of California, Berkeley.

————. 1980. A developmentally based approach to the teaching of proportionality. In *Proceedings of the fourth international conference of psychology of mathematics education,* ed. R. Karplus. Berkeley, CA: Lawrence Hall of Science.

Goodstein, M. P., and W. W. Boelke. 1980. A prechemistry course on proportional calculation. Columbus, OH: ERIC/SMEAC.

Harel, G., M. Behr, T. Post, and R. Lesh. 1990. A scheme to represent the multiplicative conceptual field. Unpublished manuscript.

Hart, K. 1981. *Children's understanding of mathematics, 11–16.* London: Murray.

————. 1984. Ratio: Children's strategies and errors. *A report of the strategies and errors in secondary mathematics project.* London: NFER-Nelson.

Herron, J. D., and G. H. Wheatley. 1978. A unit factor method for solving proportion problems. *The Mathematics Teacher* 71: 18–21.

Hiebert, J., and D. Wearne. 1985. A model of students' decimal computation procedures. *Cognition and Instruction* 2: 175–205.

Karplus, R., S. Pulos, and E. Stage. 1983. Proportional reasoning of early adolescents. In *Acquisition of mathematics concepts and processes,* ed. R. Lesh and M. Landau, 45–90. Orlando, FL: Academic Press.

Kieren, T. 1983. Axioms and intuition in mathematical knowledge building. In *Proceedings of the fifth annual meeting of the North American chapter of the international group for the psychology of mathematics education,* ed. J. C. Bergeron and N. Herscovics, 67–73. Montreal.

Lamon, S. J. 1989. Ratio and proportion: Preinstructional cognitions. Unpublished doctoral dissertation, University of Wisconsin, Madison.

Lesh, R., T. R. Post, and M. Behr. 1988. Proportional reasoning. In *Number concepts and operations in the middle grades,* ed. J. Hiebert and M. Behr, 93–118. Reston, VA: National Council of Teachers of Mathematics.

Lybeck, L. 1978. *Studies of mathematics in teaching of science in Goteborg.* Goteborg: Institute of Education, No 72.

Mack, N. K. 1990. Learning fractions with understanding: Building on informal knowledge. *Journal for Research in Mathematics Education* 21(1): 16–32.

Markman, E. M. 1979. Classes and collections: Conceptual organization and numerical abilities. *Cognitive Psychology* 11: 395–411.

Noelting, G. 1980a. The development of proportional reasoning and the ratio concept. Part 1—Differentiation of stages. *Educational Studies in Mathematics* 11: 217–253.

———. 1980b. The development of proportional reasoning and the ratio concept. Part 2—Problem structure at successive stages; problem-solving strategies and the mechanism of adaptive restructuring. *Educational Studies in Mathematics* 11: 331–363.

Post, T. R., M. J. Behr, R. Lesh, and I. Wachsmuth. 1985. Selected results from the rational number project. In *Proceedings of the ninth international conference for the psychology of mathematics education.* Vol. 1. *Individual contributions,* ed. L. Streefland, 342–351. Noordwijkerhout, The Netherlands.

Post, T. R., G. Harel, M. Behr, and R. Lesh. 1988. Intermediate teachers' knowledge of rational number concepts. In *Integrating research on teaching and learning mathematics,* ed. E. Fennema, T. P. Carpenter, and S. J. Lamon, 194–217. Madison: Wisconsin Center for Education Research.

Post, T. R., I. Wachsmuth, R. Lesh, and M. Behr. 1985. Order and equivalence of rational numbers: A cognitive analysis. *Journal for Research in Mathematics Education* 16: 18–36.

Pothier, Y. 1981. Partitioning: Construction of rational numbers in young children. Unpublished doctoral thesis. University of Alberta, Edmonton.

Rizzuti, Jan. 1991. High school students' uses of multiple representations in the conceptualization of linear and exponential functions. Unpublished doctoral dissertation, Cornell University, Ithaca, NY.

Rupley, W. H. 1981. The effects of numerical characteristics on the difficulty of proportional problems. Unpublished doctoral dissertation, University of California, Berkeley.

Schwartz, J. 1988. Intensive quantity and referent transforming arithmetic operations. In *Number concepts and operations in the middle grades,* ed. J. Hiebert and M. Behr, 41–52. Reston, VA: National Council of Teachers of Mathematics.

Steffe, L. P. 1969. Proportionality. Unpublished manuscript. University of Georgia, Athens.

Steffe, L. P. and P. Cobb. 1988. Construction of arithmetical meanings and strategies. New York: Springer-Verlag.

Steffe, L. P. and R. B. Parr. 1968. The development of the concepts of ratio and fraction in the fourth, fifth, and sixth years of the elementary school. Technical Report No. 49. Madison: Wisconsin Research and Development Center for Cognitive Learning.

Streefland, L. 1983. The long term learning process for ratio. In Proceedings of the seventh conference for the psychology of mathematics education, ed. R. Hershkowitz, 182–187. Rehovot, Israel.

———. 1984. Search for the roots of ratio: Some thoughts on the long term learning process (towards . . . a theory). Part I— Reflections on a teaching experiment. Educational Studies in Mathematics 15: 327–348.

———. 1985. Search for the roots of ratio: Some thoughts on the long term learning process (towards . . . a theory). Part II—The outline of the long term learning process. Educational Studies in Mathematics 16: 75–94.

Tourniaire, F. 1986. Proportions in elementary school. Educational Studies in Mathematics 17: 401–412.

——— and S. Pulos. 1985. Porportional reasoning: A review of the literature. Educational Studies in Mathematics 16: 181–204.

Treffers, A., and F. Goffree. 1985. Rational analysis of realistic mathematics education—The Wiskobas Program. In Proceedings of the ninth international conference for the psychology of mathematics education, ed. L. Streefland, 97–121. Noordwijkerhout, The Netherlands.

Van den Brink, J., and L. Streefland. 1979. Young children (6–8)— Ratio and proportion. Educational Studies in Mathematics 10: 403–420.

Vergnaud, G. 1983. Multiplicative structures. In Acquisition of mathematics concepts and processes, ed. R. Lesh and M. Landau, 127–174. New York: Academic Press.

# 5 Units of Quantity: A Conceptual Basis Common to Additive and Multiplicative Structures

*Merlyn J. Behr*
*Guershon Harel*
*Thomas Post*
*Richard Lesh*

The issue of connections between and among mathematical knowledge structures is mentioned frequently in current writing concerning the knowing, learning, and teaching of mathematics. Less frequently given, however, are clear statements of what it means for there to be a connection or an explicit example of connected mathematical concepts. This chapter uses the notion that mathematical knowledge is composed of conceptual units and hypothesizes the type of knowledge structures that knowers of certain mathematical concepts have and learners of these concepts need to develop to be able to make connections between selected mathematical concepts. Through an analysis of mathematical structures in the additive and multiplicative conceptual fields, we make hypotheses about connections between concepts within the additive conceptual field and also across the additive and the multiplicative conceptual fields. The ultimate hypothesis of the work is that exploiting these connections in a curricular context will facilitate children's ability to expand their knowledge about additive structures and extend this knowledge to multiplicative structures.

## INTRODUCTION

Traditionally we have taught children the arithmetic of numbers with the ultimate objective of their being able to use this arithmetic to model and solve real-world problems. The order of this approach has been first to teach the arithmetic of numbers, essentially divorced from a social context, and then to teach problem solving by making attachments between numbers and operations on numbers with measure-

121

ments of and operations on quantities to model relationships between quantities. What is proposed by the arithmetic of quantity is that arithmetic should grow out of social contexts. That is, by beginning with the observation of how quantities behave and by attaching numbers to attributes of quantity through the process of measurement, the arithmetic of quantities and ultimately the arithmetic of numbers should be suggested. An important distinction to be made here is that arithmetic of numbers is apparently based on the assumption that all the numbers represent quantities of the same unit of one, a quantity of singleton units. The arithmetic of quantity involves composite conceptual units of various composition. The arithmetic of quantity requires special attention to measure units and different types of composite units. The arithmetic of numbers apparently assumes only singleton units.

Research in the Soviet Union has given some attention to children's ability to deal with different types of quantity units. Davydov (1982), describing experimental work conducted in the Soviet Union, found that first through third graders in a traditional program master procedures for addition and subtraction of single and multidigit numbers and can easily determine that 3 + 4 is equal to 7, for example. Many of the children were able to describe contexts for which the number sentence was a mathematical model. These same children were immediately perplexed when asked what possible sense they could make out of the unexpected sentence 3 + 4 = 5. In a teaching experiment he found some first grade children were successful with a problem that could be modeled with this sentence when represented with physical objects, three containers of the same size filled to the brim with water and four containers half the size also filled with water, and the question posed was of how many containers of the larger size would be needed to contain the water. This situation is easily represented as follows: 3 units of size 2 + 4 units of size 1 = how many units of size 2? Through a reformation of units, 4 units of size 1 can be thought of as 2 units of size 2 (Behr, Harel, Post, & Lesh, 1992a); then, the preceding can be rewritten as 3 units of size 2 + 2 units of size 2 = 5 units of size 2.

This discussion illustrates that a hidden assumption in the arithmetic of numbers is that all quantities are represented in terms of units of 1. This hidden assumption has a negative impact on the elementary and middle school curriculum. The current curriculum on whole number arithmetic gives problem situations in which children deal with quantities essentially expressed only in singleton units, rather than providing problem situations in which they represent quantity in various unit types (Steffe, 1988). Even in so-called multistep problems such as "John has four bags with six marbles in each and three bags with four in each. How many bags with two marbles in a bag can he make?" in which some of the given problem quantities are in composite units, the traditional approach is to change to units of one in the solution of the problem. The traditional approach is to change the six-marble and four-marble units to twenty-four and twelve one-marble units, add to find that there are thirty-six one-marble units, and then divide by 2 to find that there are eighteen two-marble units. An alternate approach is to reformulate the six-marble and four-marble units to two-marble units, a unit common to the two given and the unit referred to in the problem question. Representation and solution of the so-called multistep problems do not require a conversion to units of 1 in every case. Indeed there are problem situations in which it is more efficient to find a common unit other than a unit of 1, and in some cases, using a unit of 1 is actually contrary to the constraints of the problem situation. Freudenthal (1983) and more recently Lamon (1989) refer to the process of conceptualizing a situation in terms of a common unit other than 1 as *norming*.

Galperin and Georgiev (1969) suggest that the concept of unit has a special place in the formulation of elementary mathematical notions and, indeed, indicate that *all* mathematical concepts assume the notion of a unit. The work of Steffe and his colleagues (Steffe, Cobb, and von Glasersfeld, 1988) strongly point out the relationship between formation of units and the development of concepts of number, addition, and subtraction of whole numbers. More recently, they have suggested that concepts of multiplication and division and rational numbers depend on the formation of certain units.

We assert that giving children situations of whole number arithmetic that involve a variety of unit types and units of units and experience in representing and manipulating quantities that can be represented in these unit types will provide a more adequate foundation for learning and understanding whole number arithmetic and a cognitive bridge to learning and understanding rational number concepts and operations. Therefore the thrust of our analysis is to exhibit the units structure of these mathematical constructs.

## Why an Analysis of Elementary Mathematics Constructs?

There are at least two possible approaches to research and theory development on children's knowing and learning of mathematical constructs, such as concepts of whole and rational numbers and operations on these numbers. One approach, which is pursued by numerous researchers, is to work with children in ascertaining and teaching experiments on the assumption that this will lead to an understanding of children's mathematical knowing and learning and ultimately to an understanding of school mathematics and its organization for teaching. This approach depends on the development of mathematical tasks to which the children react individually, in small groups, or in other types of instructional settings. We maintain that the cognitive structures for which a child's response gives evidence are substantially task dependent. Therefore, any assumption that this research exposes a full range of children's knowledge of mathematics must presuppose that the developer of the tasks knows the full extent of the intricacies of the mathematical constructs under investigation. We believe that the psychological and mathematical intricacies of the mathematics is not that deeply understood.

A second perspective takes the position that much more needs to be known about the mathematical constructs, even for so-called elementary mathematics. This perspective calls for deep, careful, and detailed analysis of mathematical constructs both to exhibit their mathematical structure and to hypothesize about the cognitive structures necessary for understanding them. Such analysis would lead to a theory about mathematical knowing and learning that could guide cognitive research.

It is not the thesis of this argument that one perspective is right and the other is not. The position that we take is that the two perspectives interact. Information obtained from research along the first perspective must be used in the analysis of mathematical constructs to give psychological hypotheses about the knowing and learning of the constructs a base in reality. On the other hand, a deep analysis of the mathematical content and hypotheses about cognitive structures needed to construct mathematical knowledge gives a basis for deeper and broader inquiries into children's mathematical knowledge construction.

We are involved in an analysis of selected mathematical constructs through which we are contributing to this deeper understanding of the mathematics and to the formation of hypotheses about psychological aspects of the mathematics. To conduct the analysis and communicate findings, we have developed two nonstandard representational systems. Each of these representational systems is highly generalized and abstract. One is a generalized or *generic manipulative aid.* Representations in this system appear in the form of drawings. The other system is a *generalized notation for mathematics of quantity.* Representations in this system are symbolic. An analysis proceeds through interaction between the two systems and the final product is parallel representations in the two systems. The generic manipulative aid representation gives psychological integrity to the analysis, and the symbolic notation for generalized mathematics of quantity gives a corresponding mathematical integrity. We want to make it clear that the representational systems are not intended to be used to communicate mathematics to children, although appropriate instantiations of these abstract systems might be so used. The notational systems we have developed and communicate in this chapter were for theoretical analysis and communication within the research community. We do not advocate that these notational systems be used with children, but on the other hand, we do not argue against the possibility that particular instantiations of the notational systems might be used with children. The development of such instantiations goes beyond the scope of this chapter. We do argue, however, that the generic manipulative and unit analyses do provide a template for the construction of such instantiations to provide appropriate manipulative experiences for children.

We have found the standard mathematical symbolism

inadequate to represent the mathematics in terms of various unit types. Therefore, we created the generalized mathematics of quantity notation. This notation is general in the sense that it represents general unit types, specific instances of the system being units such as inch, centimeter, apple, pizza, and so forth. In our analysis, we have relied on manipulative aid representations to convey possible psychological and cognitive aspects of mathematics of quantity. Here again we have concern that any standard manipulative aid carries certain inherent features that may facilitate or inhibit the representation. Objects in standard manipulative aids have a particular shape, color, and other potentially conflicting or facilitating attributes. Therefore, any analysis we might conduct on the basis of such a manipulative aid might be questioned as to whether its findings are unique to the manipulative aid. For this reason we have created the generic manipulative aid in which objects might be considered place holders for objects from any standard manipulative aid. The only attribute we wish to deliberately assign to the manipulative aid is whether continuous or discrete quantity is represented. A brief description of the two systems will allow us to demonstrate its usefulness in analyzing some aspects of the additive and multiplicative fields; however, readers wishing a more in-depth discussion should refer to Behr, Harel, Post, and Lesh (1992a, 1992b).

Thus, this chapter has three purposes. One is to give some details about the two representational systems. A second is to demonstrate application of the analysis to selected additive and multiplicative structures. The third is to demonstrate through the analysis some cognitive structures that are common to the two conceptual fields; throughout the chapter, we address the question of how attention to unit types in whole number situations will facilitate learning and understanding of rational number concepts.

## Description and Notation of Units

*Notation for the Generic Manipulative Aid.* In the generic manipulative aid, symbols will appear in the following forms: 0, *, or #; (0), [*] ((000) (000)), (0000); or (000) with one or two shaded. Symbols in the first group are used to denote, or serve as place holders for, discrete *objects* such as an apple, an orange, or a stone; the second group of symbols are used to denote *units* of quantity. To indicate that a collection of objects or units is to be considered a unit, we enclose one or more symbols for objects or units within usual grouping symbols ( ), [ ], and { }. The usual device of bold face is used to designate fractional parts of units. Singleton units conceptualized as a one-apple unit, a one-orange unit, or a one-stone unit, we denote as (0), (*), or (#). Combinations of grouping symbols are used to represent complex units, such as composite units of units and units of units of units, and units of intensive quantity (ratios of units from the same or different measure spaces). We use the ( ) grouping symbols most, reserving the [ ] and { } symbols for when there is a need to distinguish among units from different measure spaces (Vergnaud, 1988) or to identify some special types of units of quantity such as units of intensive quantity. The concept of measure space as used here can be clarified with an example. In the two quantities represented by (000) and (00000) the placeholder, 0, would have the same replacement set in both for a given instantiation. On the other hand, in (000) and [000], the 0 could have different replacement sets.

*Notation for the Generalized Mathematics of Quantity.* In the generalized mathematics of quantity notation, the size of a unit will be notated as a number hyphen unit and enclosed in grouping symbols, (2-unit). The number of units that one has will be denoted as $n$(b-unit)s or 3(2-unit)s. Thus, we have introduced a notation for composite units (Steffe, 1988). Using this notation, symbols for units can be embedded within other symbols for units. For example, 1(3(2-unit)s-unit) denotes a composite unit conceptualized as a (3-unit) derived from uniting 3(2-unit)s. The notation 3(2-unit)s denotes three units. Three two-apple packages would represent an instantiation of this, which we would denote as 3(2-apple unit)s. To further illustrate these notational systems we give the following examples. For further details and discussion of the notational systems, the interested reader is referred to (Behr et al., 1992a, 1992b).

1. From a single object, 0, we can conceptualize a *singleton unit*, denoted in the generic manipulative aid as: (0)

and in the generalized mathematics of quantity as 1(1-unit), or we can conceptualize several singleton units (denoted in the respective systems) as

(0) (0) (0), 3(1-unit)s.

2. From several objects, 0 0 0, we can conceptualize a *composite unit:*

(0 0 0),   1(3-unit).

3. From several singleton units, (0) (0) (0) (0), or several composite (3-unit)s, (0 0 0) (0 0 0) (0 0 0) (0 0 0), we can conceptualize a *units of-units:*

((0) (0) (0) (0)),   1(4(1-unit)s-unit),
((0 0 0) (0 0 0) (0 0 0) (0 0 0)),   1(4(3-unit)s-unit).

4. From several composite units-of-units ((00)(00)(00)(00)) ((00)(00)(00)(00)) ((00)(00)(00)(00)), we can conceptualize a composite *unit-of-units-of-units:*

(((00)(00)(00)(00)) ((00)(00)(00)(00)) ((00)(00)(00)(00))),
1(3(4(2-unit)s-unit)s-unit).

5. From a composite unit such as a (4-unit), (0 0 0 0), and two measure spaces () and [], we can conceptualize a unit of *intensive quantity,* 1(4-unit) per [1-unit]

$$[(0\ 0\ 0\ 0)], \quad \frac{1(4\text{-unit})}{[1\text{-unit}]} \ .$$

(This illustrates the notation for an intensive quantity greater than 1 per 1. Fractional intensive quantities are illustrated later.)

6. From four singleton units such as a 4(1-unit)s, (0) (0) (0) (0), and two measure spaces () and [], we can conceptualize another unit type for *intensive quantity,* 4(1-unit)s per [1-unit]

$$[(0)\ (0)\ (0)\ (0)], \quad \frac{4(1\text{-units})}{[1\text{-unit}]} \ .$$

In both notational systems the number of embeddings of paired left and right parentheses indicates the depth of the embedding of units.

We next turn our attention to the issue of representing fractional quantities with the two notational systems. Fractional parts of units also are quantities that can be united as several different unit types. For example 2/4 can be conceptualized in terms of several unit types, among them are 2/4(4-unit), 1(2/4(4-unit)-unit), and 1[2/4-unit]. We now illustrate how the notational systems (summarized in Table 5.1) capture these subtleties.

1. From a composite unit, for example a (4-unit), made of singleton objects, (0 0 0 0), or units-of-units, ((0) (0) (0) (0)) or ((0 0 0) (0 0 0) (0 0 0) (0 0 0)), we can conceptualize one-fourth of a unit:

   (**0** 0 0 0),  $\frac{1}{4}$(4-unit),
   ((**0**) (0) (0) (0)),  $\frac{1}{4}$(4(1-unit)s-unit),
   ((**0 0 0**) (0 0 0) (0 0 0) (0 0 0)),  $\frac{1}{4}$(4(3-unit)s-unit),

or we can conceptualize 2/4 of a composite unit or as a unit of units:

   (**0 0** 0 0),  $\frac{2}{4}$(4-unit)
   ((**0**) (**0**) (0) (0)),  $\frac{2}{4}$(4(1-unit)s-unit)
   ((**0 0 0**) (**0 0 0**) (0 0 0) (0 0 0)),  $\frac{2}{4}$(4(3-unit)s-unit),

2. The two objects or units that form the fractional part of a unit can be taken as a unit. Thus from 2/4 of a (4-unit) we can conceptualize a *fractional unit of units*:

   ((**0 0**) 0 0),  1($\frac{2}{4}$(4-unit)-unit)
   (((**0**) (**0**)) (0) (0)),  1($\frac{2}{4}$(4(1-unit)s-unit)-unit)
   (((**0 0 0**)(**0 0 0**))(0 0 0)(0 0 0)),  1($\frac{2}{4}$(4(3-unit)s-unit)-unit).

3. From an intensive quantity or measure unit of the type 1(4-unit) per [1-unit], [(0 0 0 0)], we can conceptualize two-fourths of the measure unit:

   [(**0 0** 0 0)],  2/4[1-unit];

TABLE 5.1.   NOTATION SYSTEMS USED IN THE ANALYSIS

| Type of unit | Pictoral representation | Mathematics of quantity representation |
|---|---|---|
| A singleton object | 0, *, # | |
| A singleton unit | (0) | (1-unit) or 1(1-unit) |
| Three singleton units | (0) (0) (0) | 3(1-unit)s |
| A composite unit of 3(1-unit)s | (0 0 0) | 1(3-unit) |
| Four composite (3-unit)s | (0 0 0) (0 0 0) (0 0 0) (0 0 0) | 4(3-unit)s |
| A composite unit of 4 (3-unit)s, a unit-of-units | ((0 0 0) (0 0 0) (0 0 0) (0 0 0)) | (4(3-unit)s-unit) |
| An intensive unit of 4 (1-unit)s per 1[1-unit], a measure unit | [(0) (0) (0) (0)] | $\frac{4(1\text{-unit})s}{[1\text{-unit}]}$ |
| A composite unit of 2 intensive units, a unit-of-units | [[(0) (0) (0) (0)] [(0) (0) (0) (0)]] | $\frac{4(1\text{-unit})s}{[1\text{-unit}]}$ |
| An intensive unit of 1(4-unit) per [1-unit], a measure unit | [(0 0 0 0)] | $\frac{4(1\text{-unit})}{[1\text{-unit}]}$ |
| To designate 1/4 of a (4-unit) | ((**O**) (0) (0) (0)) or (**O** 0 0 0), or ((**O 0 0**) (0 0 0) (0 0 0) (0 0 0)) | 1/4(4-unit), 1/4(4(3-unit)s-unit) |
| To designate 2/4 of a (4-unit) | ((**O**) (**O**) (0) (0)) or (**O O** 0 0), or ((**O 0 0**) (**O 0 0**) (0 0 0) (0 0 0)) | 2/4(4-unit), 2/4(4(3-unit)s-unit) |
| To designate 1/4 of a 1(4-unit) per [1-unit] measure unit | [(**O** 0 0 0)] | 1/4[1-unit] |
| To designate 2/4 of a 1(4-unit) per [1-unit] measure unit | [(**O O** 0 0)] | 2/4[1-unit] |
| To designate a uni-tized 2/4 of a 1(4-unit) per [1-unit] measure unit | [((**O O**) 0 0)] | 1(2/4[1-unit]-unit) |

(continued)

TABLE 5.1. (*Continued*)

| Type of unit | Pictoral representation | Mathematics of quantity representation |
|---|---|---|
| To designate one 1/4-unit of a 4(1-unit)s/[1-unit] measure unit | [(**O**) (0) (0) (0)] | 1[1/4-unit] |
| To designate two 1/4-units of a 4(1-unit)s/[1-unit] measure unit | [(**O**) (**O**) (0) (0)] | 2[1/4-unit]s |
| To designate a unit-ized two 1/4-units of a 4(1-unit)s per [1-unit] measure unit | [((**O**) (**O**)) (0) (0)] | 1[2/4-unit] |

4. From the 2/4[1-unit] quantity we can conceptualize a unitization to a composite unit:

$$[(\mathbf{O}\ \mathbf{O})\ 0\ 0)], \quad 1[\tfrac{2}{4}[1\text{-unit}]\text{-unit}].$$

5. The [2/4[1-unit]-unit] can also be conceptualized as 1[2/4-unit].

6. From an intensive quantity of the type 4(1-unit)s per [1-unit] we can conceptualize a 1/4-measure unit:

$$[(\mathbf{O})\ (0)\ (0)\ (0)], \quad 1[\tfrac{1}{4}\text{-unit}]$$

or 2 1/4-measure units:

$$[(\mathbf{O})\ (\mathbf{O})\ (0)\ (0)], \quad 2[\tfrac{1}{4}\text{-unit}]s$$

and so forth.

*Problem Example: A Connection within the Additive Conceptual Field.* We next present a problem from the additive domain represented in units other than 1 and indicate a solution of this in terms of the two notational systems in figures 5.1 and 5.2, respectively. Some pilot data indicating

1. ((000000) (000000) (000000)) + [[0000] [0000] [0000] [0000] [0000]]

2. ((000000) (000000) (000000)) + ((0000) (0000) (0000) (0000) (0000))

3. (((00)(00)(00)) ((00)(00)(00)) ((00)(00)(00))) + [(0000) (0000) (0000) (0000) (0000)]

4. ((00)(00)(00)(00)(00)(00)(00)(00)(00)) + ((0000) (0000) (0000) (0000) (0000))

1. A representation of the problem. The marbles Joe had are expressed as a composite unit: 1(3(6-unit)-unit); those Tom gave him as 1[5[4-unit]s-unit].

2. The 1[5[4-unit]s-unit] is expressed as 1(5(4-unit)s-unit) to indicate that the marbles Tom gave to Joe now are Joe's marbles; that is, both sets of marbles now belong to the same measure space—Joe's marbles.

3. Each (6-unit) is reunitized to 3(2-unit)s so the 1(3(6-unit)s-unit) is reunitized to 1(3(3(2-unit)s-unit)s-unit).

4. The 1(3(3(2-unit)s-unit) is re-unitized to 1(9(2-unit)s-unit).

| | |
|---|---|
| 5. ((00)(00)(00)(00)(00)(00)(00)(00)(00)) + (((00)(00)) ((00)(00)) ((00)(00)) ((00)(00)) ((00)(00))) | **5. Each (4-unit) is reunitized to 2(2-unit)s so that the 1(5(4-unit)s-unit) is reunitized to 1(5(2(2-unit)s-unit)s-unit).** |
| 6. ((00)(00)(00)(00)(00)(00)(00)(00)(00)) + ((00)(00)(00)(00)(00)(00)(00)(00)(00)(00)) | **6. The 1(5(2(2-unit)s-unit)s-unit) is reunitized to 1(10(2-unit)s-unit).** |
| 7. (00)(00)(00)(00)(00)(00)(00)(00)(00) + (00)(00)(00)(00)(00)(00)(00)(00)(00)(00) | **7. The 1(9(2-unit)s-unit) is reunitized to 9(2-unit)s and the 1(10(2-unit)s-unit) is reunitized to 10(2-unit)s.** |
| 8. (00)(00)(00)(00)(00)(00)(00)(00)(00)(00) (00)(00)(00)(00)(00)(00)(00)(00)(00) | **8. The 9(2-unit)s and 10(2-unit)s are combined to 19(2-unit)s.** |

**Figure 5.1.**
***A representation and solution of addition Problem 1 using the generic manipulative aid representation.***

how children solve the problem is also given. The problem is one of several we generated according to the classification of addition and subtraction problems given by Riley, Greeno, and Heller (1983).

Problem 1: Joe had three bags, each containing six marbles. Tom gave him five more bags, each containing four marbles. If Joe wanted to keep them in bags of two marbles each, how many bags would he have? A manipulative aid representation of the solution of this problem is given in Figure 5.1.

Here is a portion of an interview with an 11-year-old girl who was presented the problem orally and also in written form and given marbles and plastic "baggies" with which to represent and solve the problem.

> (After reading the problem) OK, so he started with three bags with six marbles in each (makes up bags), and then Tom gave him five bags with four in each (makes new bags in another location and then pushes them together). So now we have to put all of these into bags of two. (Begins with the bags of six. Takes two out and puts into another bag, takes out another two and puts into another bag, leaves two in the bag. This is repeated with remaining bags). OK, so now he has one, two three, . . . (counts bags) nineteen bags of two marbles.

We follow this with a mathematics of quantity representation in Figure 5.2. In the symbolic representation of the problem we represent the problem quantities as follows: Joe's three bags with six marbles as 1(3(6-unit)s-unit) and Tom's five

1.  1(3(6-unit)s-unit) + 1[5(4-unit)s-unit]
2.      = 1(3(6-unit)s-unit) + 1(5(4-unit)s-unit)                    (2)
3.      = 1(3(3(2-unit)s-unit)s-unit) + 1(5(4-unit)s-unit)          (3)
4.      = 1(9(2-unit)s-unit) + 1(5(4-unit)s-unit)                   (4)
5.      = 1(9(2-unit)s-unit) + 1(5(2(2-unit)s-unit)s-unit)          (5)
6.      = 1(9(2-unit)s-unit) + 1(10(2-unit)s-unit)                  (6)
7.      = 9(2-unit)s + 10(2-unit)s
8.      = 19(2-unit)s

Fig. 5.2.

*A mathematics of quantity representation of addition Problem 1, which corresponds to the manipulative aid representation given in Figure 5.1.*

bags of four marbles that he gave to Joe as 1[5[4-unit]s-unit]].
Using () to denote units and the unit of units for Joes' marbles and [] to denote the units and unit of units for Tom's
marbles suggests that two measure spaces are involved in
the original problem representation. When Joe actually takes
possession of Tom's marbles, all of the units are denoted
with ().

   We next illustrate a connection between a mathematics
of quantity representation of a similar problem situation and
the arithmetic of number within the whole number domain.
Consider the Problem 2.

   Problem 2: Joan has five bags containing four marbles
and one bag with two marbles. Roger has two bags with four
marbles and six bags with two marbles. If they put their marbles together how many bags and of what size would there be
if they used the *fewest number of bags*? (We use only bags
that hold either two or four marbles.) The constraint of using
the fewest number of bags forces a representation in terms of
the larger size bags. The solution of this problem using our
generalized mathematics of quantity notation could be represented as follows:

$$(5(4\text{-unit})s + 1(2\text{-unit})) + (2(4\text{-unit})s + 6(2\text{-unit})s)$$
$$= 7(4\text{-unit})s + 7(2\text{-unit})s$$
$$= 7(4\text{-unit})s + 6(2\text{-unit})s + 1(2\text{-unit})$$
$$= 7(4\text{-unit})s + 3(4\text{-unit})s + 1(2\text{-unit})$$
$$= 10(4\text{-unit})s + 1(2\text{-unit}).$$

The similarity between this solution and the procedure
known to be used by some children to solve a problem such
as 28 + 35 (Hiebert and Wearne, 1992) as illustrated in the
following suggests foundational function of a problem situation and problem solution based on units of quantity other
than singleton units. In this case, the relationship between
such a problem and its solution illustrates a connection with
the multidigit whole number arithmetic as follows.

   The addition of 28 + 35 is solved by some children by
saying "2 tens and 3 tens is 5 tens and 8 and 5 is 13, so that's
5 tens and 1 ten more is 6 tens and 3 is 63." The accepted
place value representation of the sum of 28 and 35 corresponds to representing a discrete quantity in terms of the
fewest number of bags, where bags come in the size of 1, 10,

100, . . . . Moreover, the conversion of 13 to 1 ten and 3 suggests a decomposition of 13 units of 1 to 1 composite unit of
10 and 3 units of 1.

## A Connection Between the Additive and Multiplicative Conceptual Fields

We will draw attention to the conceptual similarity between
the representation and solution of the whole number addition problem (Problem 1), a problem from the additive conceptual field, and addition of rational numbers, a problem
from the multiplicative conceptual field. A rational number
such as 3/4 has several unit-type interpretations (Behr et al.,
1992a): 3/4(1-unit), 1(3/4-unit), 1/4(3-unit), and 3(1/4-unit)s.
Using the interpretation like 3/4 as 3(1/4-unit)s, for 3/7 and
2/7, then finding the sum of these two fractions is conceptually the same as finding the sum of 3(2-units) and 2(2-
unit)s. Each involves the addition of two quantities where
both are represented in terms of the same unit of quantity, a
2-unit quantity in the whole number case and a 1/7-unit
quantity in the rational number case.

Using notation for mathematics of quantity these two
sums could be represented as follows:

$$3(\text{2-unit})\text{s} + 2(\text{2-unit})\text{s} = 5(\text{2-unit})\text{s, and}$$

$$3(\tfrac{1}{7}\text{-unit})\text{s} + 2(\tfrac{1}{7}\text{-unit})\text{s} = 5(\tfrac{1}{7}\text{-unit})\text{s.}$$

Each of these problems could be made more realistic for children by using units of quantity from more realistic contexts;
that is, a specific instantiation of the generalized mathematics of quantity notation. A package of two apples can be conceptualized as a 2-unit, a two-apple unit. Similarly 1/7 of an
apple can be conceptualized as a 1/7-unit, a 1/7-apple unit.
Using this terminology the two additions could be symbolized as follows:

$$3(\text{2-apple package})\text{s} + 2(\text{2-apple package})\text{s} = 5(\text{2-apple package})\text{s, and}$$

$$3(\text{1/7-apple piece})\text{s} + 2(\text{1/7-apple piece})\text{s} = 5(\text{1/7-apple piece})\text{s.}$$

These additions can be modeled with real apples or with ma-
nipulatives that represent apples or 1/7's of apples.

If two quantities are represented with different units, then to express the joining of these quantities in terms of a single unit requires that one or both of the two quantities be reunitized so that the quantities are expressed in the same unit. This basic principle is essential to give a closed symbolic expression for the sum of two fractional quantities that are represented as fractions with unlike denominators. Unfortunately, the first time that a child in a traditional mathematics program meets a situation in which this principle is required is when addition of unlike fractions is considered. At the point in the curriculum when addition of fractions is considered, the child should already have knowledge of this principle. Unfortunately, the child's number experience up to this point would have included very little experience with expressing quantities in a common unit, or even finding a common unit for a pair of quantities. Although the child's experience probably includes finding the common and the greatest common factors, which involves this concept, this skill is not introduced to children from a units-of-quantity perspective. Modification of the curriculum to give the child experience with situations involving norming is needed.

We now give an example of a situation that requires one or both of the quantities to be reunitized. We present the addition of 5(6-apple unit)s and 3(4-apple unit)s and of 5/6 and 3/4 to demonstrate the conceptual similarity between these two additions. Direct observation of 6-apple packages and 4-apple packages will help children see that a (6-apple unit) could be reunitized to 3(2-apple unit)s and that a (4-apple unit) could be reunitized to 2(2-apple unit)s. A very important observation in this is that a (2-apple unit) is a subunit of both of the original units. Or more precisely, a (2-apple unit) is the largest unit that satisfies these requirements: It is contained by both quantities as a subunit, and the measure of each of the original quantities with respect to this unit is a whole number. The principle of sameness to be abstracted and generalized to apply to fractional units of quantity as well is that quantities can be expressed in terms of a common subunit.

A manipulative representation of this addition would be very similar to the demonstration in Figure 5.1, so we omit it. A mathematics of quantity representation, using a particular instantiation of the generalized mathematics of quantity rep-

1. (O O O O O O O O O O O O)

2. a. ((OO) (OO) (OO) (OO) (OO) (OO))

   b. ([OOO]) ([OOO]) ([OOO]) ([OOO])

3. ([[OO][O] [O](OO]] [(OO)[O] [O](OO]])

4. ([[OO] (**O**] [**O**] [(OO] [**O**] [**O**] [(OO]])

5. [[[(**O**](**O**]] [(**O**](**O**](**O**]] [(**O**](**O**](**O**]] [(**O**](**O**](**O**]]]

1. Choose a unit that can be partitioned into six parts and reunitized into 6(1/6-unit)s and partitioned into four parts and reunitized into 4[1/4-unit]s. A (12-unit) is chosen.

2. a. Partition the (12-unit) into 6(1/6-unit)s and reunitize the (12-unit) to 1(6[1/6-unit)s-unit).

   b. Partition the (12-unit) into 4[1/4-unit]s and reunitize the (12-unit) to 1(4[1/4-unit]s-unit).

3. Superimpose the two unitized partitions.

4. Identify a unit that is a common subunit of a (1/6-unit) and of (1/4-unit) and for which the measure of both units is a whole number, a (1/12-unit).

5. Reunitize the 4[1/4-unit]s in terms of the common (1/12-unit) by unitizing each (1/4-unit) to 3(1/12-unit)s. That is, show 4[1/4-unit]s reunitized to 4[3(1/12-unit)s-unit]s. Distinguish 3[3(1/12-unit)s-unit]s.

6. Reunitize the 1(4(3(1/12-unit)s-unit) to 1(12(1/12-unit)s-unit). This results in the 3[3(1/12-unit)s-unit]s subunits being reunitized to the subunit 1[9(1/12-unit)s-unit].

((([0]|0)|0)|0)|0)|0)|0)|0)|0)|0)|0)|0)) (0|0)|0))

7. Reunitize the 6(1/6-unit)s in terms of the common (1/12-unit). That is, show 6(1/6-unit)s reunitized to 6(2(1/12-unit)s-unit)s. Distinguish 5(2(1/12-unit)s-unit)s.

((([0]|0)) ((0)|0)) ((0)|0)) ((0)|0)||(0)|0)))

8. Reunitize the 1(6(2(1/12-unit)s-unit)s to 1(12(1/12-unit)s-unit). This results in the 5[2(1/12-unit)s-unit]s subunits being reunitized to 1(10(1/12-unit)s-unit).

((0)|0)|0)|0)|0)|0)|0)|0)|0)|0)|0)) (0|0))

9. Count to determine the total number of 1(1/12-unit)s in the two distinguished subunits, 1(10(1/12-unit)s-unit) and 1(9(1/12-unit)s-unit). The number is 19(1/12-unit)s.

((([0]|0)|0)|0)|0)|0)|0)|0)|0)|0)|0)) (0|0)|0))

     ↑
1(1/12-unit) ... 9(1/12-unit)s

((([0]|0)|0)|0)|0)|0)|0)|0)|0)|0)|0)|0)) (0|0))

 ↑      ↑
10(1/12-unit)s ... 19(1/12-unit)s

Fig. 5.3.

*A manipulative aid representation of 3/4 + 5/6.*

resentation, the addition of the quantities of apples, proceeds as follows:

1. 5(6-apple unit)s + 3(4-apple unit)s
2.    = 5(6-apple unit)s + 3(2(2-apple unit)s-unit)s
3.    = 5(3(2-apple unit)s-unit)s + 3(2(2-apple unit)s-unit)s
4.    = 5(3(2-apple unit)s-unit)s + 1(6(2-apple unit)s-unit)
5.    = 1(15(2-apple unit)s-unit) + 1(6(2-apple unit)s-unit)
6.    = 1(15(2-apple unit)s + 6(2-apple unit)s-unit)
7.    = 21(2-apple unit)s
8.    = 21(2-apple unit)s.

In figure 5.3 we give a manipulative representation of finding the sum 3/4 + 5/6. This is followed by a mathematics of quantity representation in figure 5.4.

1. 5/6 + 3/4 = 5(1/6-unit)s + 3(1/4-unit)s      (1)
2.        = 5(1/6-unit)s + 3[3(1/12-unit)s-unit]s      (5)
3.        = 5(2(1/12-unit)s-units + 3[3(1/12-unit)s-unit]s      (6)
4.        = 5(2(1/12-unit)s-unit)s + 1[9(1/12-unit)s-unit]      (7)
5.        = 1(10(1/12-unit)s-unit) + 1[9(1/12-unit)s-unit]      (8)
6.        = 10(1/12-unit)s + 9(1/12-unit)s      (9)
7.        = 19(1/12-unit)s.      (9)

Fig. 5.4.

*A mathematics of quantity representation corresponding to the manipulative representation given in Figure 5.3.*

Manipulative experience as suggested in figure 5.3 would lay the conceptual foundation for an arithmetic of numbers algorithm substantiated by the mathematics of quantity representation in Figure 5.4 as follows:

1. 5/6 + 3/4 = 5(1/6) + 3(1/4)
2.        = 5(1/6) + 3(3(1/12))
3.        = 5(2(1/12)) + 3(3(1/12))
4.        = 10(1/12) + 9(1/12)
5.        = 19(1/12)
6.        = 19/12.

## PRODUCT OF UNITS

In this section we will consider an analysis of the type of experiences that we hypothesize children need during their work with multiplication of whole numbers (a) to understand cross product multiplication and (b) to provide a foundation for constructing meaning for finding the product of rational numbers expressed in fraction form. In one part of this analysis we will show the conceptual similarity between problems such as 5/6 × 4/7 from the rational number part of the multiplicative conceptual field and problems such as the following from the whole number part of the multiplicative conceptual field:

1. If Jane has three blouses and five skirts, how many outfits can she make?
2. If three ships are in operation for five days, how many ship-days would be used?
3. If Bob has three varieties of apples and five varieties of oranges, how many apple-orange pairs can he make?
4. If the length of a rectangle is 3 inches and width is 5 cm, how many inch × centimeter units are in the area of the rectangle?

Each of these problems involves the product of units of quantity. The solution to these problems involves a type of multiplication not easily interpreted as repeated addition (Freudenthal, 1983; Dienes, 1960). In the problem involving blouses, skirts, and outfits three measure spaces are involved. We could denote these units as (1-blouse unit), (1-skirt unit), and (1-outfit unit). In this particular case the product of these two unit types occurs with sufficient frequency in our daily lives that we give the name of outfit to the product unit. Nevertheless, it is important to recognize that the unit of outfit is a pair, a blouse × skirt pair.

To proceed with our analysis of problems that involve the product of measures, we need to add to our two notation schemes.

### *Extension: Notation Schemes*

If we have two composite units from two different measure spaces, an (a-unit) and a [b-unit], where a and b are any positive rational numbers, we will denote their indicated product as 1(a-unit) × 1[b-unit]. The *product-unit* that comes about in the process of carrying out this multiplication of units is a unit we denote by 1(((a-unit) × [b-unit])-unit). The parentheses that enclose (a-unit) × [b-unit] in the notation serve only to group (a-unit) × [b-unit] as a grammatical entity. It is to aid in recognizing that (a-unit) × [b-unit] is the object that provides the perceptual basis for forming a conceptual unit, which we denote as shown. Decomposition of the (a-unit) to (a(1-unit)s-unit) or the [b-unit] to [b[1-unit]s-unit] will aid in transforming the product unit into unit types that are compatible for reforming this cross product unit into a unit of product units to meet certain problem constraints. One sequence of reunitizations leads to the unit of product units (ab(1((1-unit) × [1-unit])-unit)s-unit). These reunitizations are the hypothesized cognitive structures involved in a cross-product multiplication.

We use the example of finding the number of outfits possible with three blouses and four skirts to attempt to provide some intuition for these ideas. To form the cross product of three blouses and four skirts, a great deal of conceptual unit formation and reformation needs to occur. First the three blouses and four skirts are unitized as a (3-blouse unit) and a [4-skirt unit]. Then the (3-blouse unit) is decomposed to (3(1-blouse unit)s-unit); next each of these (1-blouse unit)s is paired with the [4-skirt unit] to form 3 (1((1-blouse unit) × [4-skirt unit])-unit)s. After the [4-skirt unit] is decomposed to [4[1-skirt unit]s-unit], each of the (1-blouse unit)s is "distributed over" the uniting operation on [1-skirt unit]s in [4[1-skirt unit]s-unit]s to form three units of the form {4(1((1-blouse unit) × [1-skirt unit])-unit)-unit}s. This describes at least some of the cognitive constructions that come about when the 3-by-4 array of blouse-skirt pairs is constructed. We state these results abstractly in the form of product-unitization principles using the generalized mathematics of quantity notation and then illustrate with a (3-unit), that is (000), and a [4-unit], that is, [****].

### Reunitizing Principles for Product Units

1. The principles we give are hypothesized to represent knowledge structures needed to represent a cross product of

two quantities with a manipulative aid or with standard notation. Although the principles are essential "axioms" for the notational systems, they also, we believe, represent essential knowledge for meaningful understanding of cross-product multiplication. Therefore, they also represent knowledge structures[1] that should be developed as intuitive knowledge by children.

 a. If the ($a$-unit) is reunitized to $a$(1-unit)s, the product unit (1(($a$-unit) × [$b$-unit])-unit) becomes (1(($a$(1-unit)s-unit) × [b-unit])-unit).

 b. If the [$b$-unit] is distributed over the uniting operation on $a$(1-unit)s, (1(($a$(1-unit)s-unit) × [$b$-unit])-unit) becomes a(1((1-unit) × [$b$-unit])-unit)s.

 c. If the [$b$-unit] in $a$(1((1-unit) × [$b$-unit])-unit)s is reunitized to $b$[1-unit]s, the result is ($a$(1((1-unit) × $b$[1-unit]s-unit])-unit)s-unit).

 d. If the (1-unit) in ($a$(1((1-unit) × [$b$[1-unit]s-unit])-unit)s-unit) is distributed across the operation of uniting of $b$[1-units], the product-unit becomes ($ab$(1((1-unit) × [1-unit])-unit)s-unit).

2. If the [$b$-unit] is reunitized to $b$[1-unit]s, the product unit (1(($a$-unit) × [$b$-unit])-unit) becomes 1(($a$(1-unit) × [$b$[1-unit]s-unit])-unit). A sequence of reunitizations similar to 1a–1c gives the same result as in 1d.

3. If the ($a$-unit) is reunitized to $a$(1-unit)s and [$b$-unit] is reunitized to $b$[1-unit]s, the product unit (1(($a$-unit) × [$b$-unit])-unit) becomes (1(($a$(1-unit)s-unit) × [$b$[1-unit]s-unit])-unit). Again, a sequence of reunitizations similar to 1a–1c gives the same result as in 1d.

We will give illustrations of these general comments using a (3-unit), that is, (000), and a [4-unit], that is, [****].

1. If (000) is reunitized to ((0) (0) (0)) then, the product-unit {(0 0 0) × [* * * *]} becomes, {((0) (0) (0)) × [* * * *]}, {1((3(1-unit)s-unit) × [4-unit])-unit}. And from this, distribution of [* * * *] over the uniting operation in ((0) (0) (0)) gives a decomposition of the product-unit to the unit-of-product-units shown in 1a:

 a. {((0) × [* * * *]) ((0) × [* * * *]) ((0) × [* * * *])}, 1{3(1((1-unit) × [4-unit])-unit)s-unit}; and by reunitizing [* * * *] to [[*][*][*][*]] in 1a we can form the unit of product units shown in 1b.

b. {((0) × [[*][*][*][*]]) ((0) × [[*][*][*][*]]) ((0 × [[*][*][*][*]])}, 1{3(1((1-unit) × [4[1-unit]s-unit])-unit)s-unit}; and with distribution of cross multiplication by (0) over the uniting operation in [[*] [*] [*] [*]] in each ((0) × [[*][*][*][*]]) gives a decomposition of the product of units in 1b to the unit of units of product units shown in 1c.

c. {((((0) × [*]) ((0) × [*]) ((0) × [*]) ((0) × [*])))
   (((0) × [*]) ((0) × [*]) ((0) × [*]) ((0 × [*]))
   (((0) × [*]) ((0) × [*]) ((0) × [*]) ((0 × [*])))},

1{3(4(1((1-unit) × [1-unit])-unit)s-unit)s-unit};

and by forming a composite unit from the unit of units of product units in 1c we get the unit of product units shown in 1d.

1d. {((0) × [*]) ((0) × [*]) . . . ((0) × [*])},

1{12(1((1-unit) × [1-unit])-unit)s-unit}.

2. If [****] is decomposed into singleton units to [[*] [*] [*] [*]] then, the product-unit {(0 0 0) × [* * * *]} becomes, {(000) × [[*] [*] [*] [*]]}, {1((3-unit) × [4[1-unit]s-unit])-unit}. From this reunitization of the product-unit and a sequence of steps similar to 1a–1c, we come to the same result as in 1d.

3. If (000) is reunitized to ((0) (0) (0)) and [****] is re-unitized to [[*] [*] [*] [*]], then the product-unit {(0 0 0) × [* * * *]} becomes, {((0) (0) (0)) × [[*] [*] [*] [*]]}, {(3(1-unit)s-unit) × [4[1-unit]s-unit]-unit}. From this reunitization of the product unit and a sequence of steps similar to 1a–1c, we come to the same result as in 1d.

Figure 5.5 offers a continuous, geometric interpretation corresponding to the sequence of reunitizations of discrete quantity given in 1, using the indicated product 1(3-in. unit) × 1[4-cm unit].

In the discrete case, the notation (1-unit) × [1-unit] denotes the pair (not necessarily ordered) of quantities (1(1-unit), 1[1-unit]). In the continuous case of the enclosed rectangular area, one can also think of (1-unit) × [1-unit] as denoting a pair of unit lengths arranged to form a segment of a right angle (i.e., like ⌐). Then juxtaposition of an appropriate number of these with the sides of the rectangle and with one another, partitions the rectangular region into rectangular units of area. This conception of the product of units of length measure can lead to an understanding of how the product of the measures of the lengths of the sides of a rectangle gives the measure of the area enclosed by the sides.

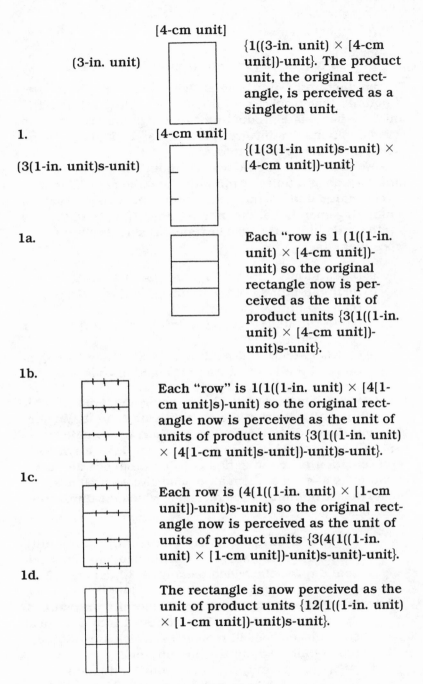

**[4-cm unit]**

(3-in. unit)

{1((3-in. unit) × [4-cm unit])-unit}. The product unit, the original rectangle, is perceived as a singleton unit.

**1.**

(3(1-in. unit)s-unit)

**[4-cm unit]**

{(1(3(1-in unit)s-unit) × [4-cm unit])-unit}

**1a.**

Each "row is 1 (1((1-in. unit) × [4-cm unit])-unit) so the original rectangle now is perceived as the unit of product units {3(1((1-in. unit) × [4-cm unit])-unit)s-unit}.

**1b.**

Each "row" is 1(1((1-in. unit) × [4[1-cm unit]s)-unit) so the original rectangle now is perceived as the unit of units of product units {3(1((1-in. unit) × [4[1-cm unit]s-unit])-unit)s-unit}.

**1c.**

Each row is (4(1((1-in. unit) × [1-cm unit])-unit)s-unit) so the original rectangle now is perceived as the unit of units of product units {3(4(1((1-in. unit) × [1-cm unit])-unit)s-unit)-unit}.

**1d.**

The rectangle is now perceived as the unit of product units {12(1((1-in. unit) × [1-cm unit])-unit)s-unit}.

Fig. 5.5
*Continuous notation.*

## More Principles for Reunitizing Product Units

The product unit reunitizing principles we have given cover situations that involve the product of a 1(a-unit) and 1[b-unit]. We next state product unit reunitization principles to cover situations that involve m(a-unit)s and n[b-unit]s, where m and n are nonnegative integers.

We will denote the indicated product of m(a-unit)s and n[b-unit]s as m(a-unit)s × n[b-unit]s and the product unit as (1((m(a-unit)s-unit) × [n[b-unit]s-unit])-unit). As a result of reunitizing operations, the product unit (1((m(a-unit)s-unit) × [n[b-unit]s-unit])-unit) can be transformed into several unit types including:

a. (m(1((a-unit) × [n[b-unit]s-unit])-unit)s-unit),
b. (n(1((m(a-unit)s-unit) × [b-unit])-unit)s-unit),
c. (mn(1((a-unit) × [b-unit])-unit)s-unit),
d. (x(n(v(1((a-unit) × [b-unit])-unit)s-unit)-unit), where 1 ≤ x, v and xv = m.
e. (t(u(v(1((a-unit) × [w[b-unit]s-unit])-unit)s-unit)s-unit)s-unit), where 1 ≤ t, u, v, m = tu, and n = vw.

We illustrate these reunitizations using the product unit ((1(30(2-in. unit)s-unit) × [6[5-cm unit]s-unit])-unit). We suggest that the reader imagine (maybe draw) a rectangle 30 (2-in. unit)s high and 6 [5-cm unit]s long. We will give a symbolic representation of reunitizations of this product unit corresponding to a–e, label them a'–e', and give a verbal description of the partitioning of the rectangle which the unitization describes.

a'. (30(1((2-in. unit) × [6[5-cm unit]s-unit])-unit)s-unit), the original rectangle is partitioned into 30 horizontal rows, embedded rectangles, which are 1(2-in. unit) high by 1[6[5-cm unit]s-unit] long.

b'. (6(1((30(2-in. unit)s-unit) × [5-cm unit])-unit)s-unit), the original rectangle is partitioned into 6 vertical columns, embedded rectangles, which are 1(30(2-in. unit)s-unit) high by 1[5-cm unit] long.

c'. (180(1((2-in. unit) × [5-cm unit])-unit)s-unit), the original rectangle is partitioned into 180 embedded rectangles, which are 1 (2-in. unit) high and 1 [5-cm unit] long.

d'. (10(6(3(1((2-in. unit) × [5-cm unit])-unit)s-unit)s-unit)s-unit), the original rectangle is partitioned into 10 horizontal rows of embedded rectangles with 6 rectangles in each row, which are 1(3(2-in. unit)s-unit) high by 1[5-cm unit] long.

e'. (10(3(3(1((2-in. unit) × [2[5-cm unit]s-unit])-unit)s-unit)s-unit)s-unit), the original rectangle is partitioned into 10 horizontal rows of embedded rectangles with 3 rectangles in each row, which are 1(3(2-in. unit)s-unit) high by 1[2[5-cm unit]s-unit] long.

It is interesting to note how the verbal description of the partition of the original rectangle has words that suggest the embeddedness of rectangles corresponding to each level of unit embeddedness suggested by the generalized mathematics of quantity representation: the height × length product defines an embedded rectangle that is embedded in a row of rectangles, which in turn are embedded in the original rectangle.

### Whole Number Problems

In this section we will give some problem examples that suggest the type of problems we believe children should have experience solving during their study of whole number arithmetic. These problems will enhance their understanding of product-of-units multiplication and also develop a foundation for understanding the operation of multiplication of rational numbers as a product of units. After we have given the solution of some whole number multiplication problems, we will give the solution to some multiplication of rational numbers to show the conceptual and procedural similarity in the representation and solution of the problems. Along with each problem we will give a representation of the solution in both the manipulative aid notation system and in the mathematics of quantity notation system. Extensive detail in the analysis hypothesizes the type of units formed in the conceptual process of solving a product of measures multiplication.

*Example Problem 1.* Jane has a large quantity of three different varieties of apples and four varieties of oranges. She is going to package the fruit so that she has one apple and

1. A representation of the problem: Pairs of 1 apple, 1 orange, that can be made from three varieties of apples and four varieties of oranges.

2. The indicated product is transformed to a product unit.

3. The composite units (01 02 03) and [*1 *2 *3 *4] are reunitized to units of units.

4. Distribution of cross-product multiplication by [[*1] [*2] [*3] [*4]] over the uniting operation in ((01) (02) (03)) is begun.

   a. Distribution of cross-product multiplication by (01) over the uniting operation in [[*1] [*2] [*3] [*4]] is begun.

   b. Distribution of cross-product multiplication by (01) over the uniting operation in [[*1] [*2] [*3] [*4]] continues. Product units (e.g., ((01) × [*1]) and ((01) × [*2])) are unitized into a unit of units, a unit of two product units.

   c. Distribution of multiplication over uniting continues, a unit of three product units is formed.

5. Distribution of cross-product multiplication by (01) over the uniting operation in [[*1] [*2] [*3] [*4]] is completed. A unit of four product units is formed.

---

1. (01 02 03) × [*1 *2 *3 *4]

2. {(01 02 03) × [*1 *2 *3 *4]}

3. {[(01) (02) (03)) × [[*1] [*2] [*3] [*4]]}

4. {[(01) × [[*1] [*2] [*3] [*4]])
   ((02) (03)) × [[*1] [*2] [*3] [*4]])}

   a. {( ((01) × [*1]) ((01) × [[*2] [*3] [*4]]) )[1]
   ( ((02) (03)) × [[*1] [*2] [*3] [*4]]) )}

   b. {( ((01) × [*1]) ((01) × [*2])  ((01) × [[*3] [*4]]) )
   ( ((02) (03)) × [[*1] [*2] [*3] [*4]]) )}

   c. {( ((01) × [*1]) ((01) × [*2]) ((01) × [*3])) ((01) × [*4]))
   ( ((02) (03)) × [[*1] [*2] [*3] [*4]]) )}

5. {( ((01) × [*1]) ((01) × [*2]) ((01) × [*3]) ((01) × [*4]) )
   ( ((02) (03)) × [[*1] [*2] [*3] [*4]]) )}

6. {[ ((01) × [*1]) ((01) × [*2]) ((01) × [*3]) ((01) × [*4]) )
((02) × [[*1] [*2] [*3] [*4]])
((03) × [[*1] [*2] [*3] [*4]])}

6. Attention is returned to distribution of cross-product multiplication by [[*1] [*2] [*3] [*4]] over the uniting operation in ((02) (03)) and is completed.

7. {[ ((01) × [*1]) ((01) × [*2]) ((01) × [*3]) ((01) × [*4]) )
( ((02) × [*1]) ((02) × [*2]) ((02) × [*3]) ((02) × [*4]) )
((03) × [[*1] [*2] [*3] [*4]])}[2]

7. Distribution of cross-product multiplication by (02) over the uniting operation in [[*1] [*2] [*3] [*4]] is completed. A second unit of four product units is formed.

8. {[ ((01) × [*1]) ((01) × [*2]) ((01) × [*3]) ((01) × [*4]) )
( ((02) × [*1]) ((02) × [*2]) ((02) × [*3]) ((02) × [*4]) )
( ((03) × [*1]) ((03) × [*2]) ((03) × [*3]) ((03) × [*4]) )}[2]

8. Distribution of cross-product multiplication by (03) over the uniting operation in [[*1] [*2] [*3] [*4]] is completed. A third unit of four product units is formed.

9. {((01) × [*1]) ((01) × [*2]) ((01) × [*3]) ((01) × [*4])
((02) × [*1]) ((02) × [*2]) ((02) × [*3]) ((02) × [*4])
((03) × [*1]) ((03) × [*2]) ((03) × [*3]) ((03) × [*4])}

9. The three units of four product units are united into one unit of twelve product units.

10. ((01) × [*1]) ((01) × [*2]) ((01) × [*3]) ((01) × [*4])
((02) × [*1]) ((02) × [*2]) ((02) × [*3]) ((02) × [*4])
((03) × [*1]) ((03) × [*2]) ((03) × [*3]) ((03) × [*4])

10. The one unit of twelve product units is reconceptualized as twelve product units.

Fig. 5.6.

*A manipulative aid representation of the solution to Example Problem 1.*

[1]We introduce a special notation. In the pair, left parentheses, space is to be considered a grouping symbol closed by the pair space, right parentheses. This special notation has been used only in instances where sums of quantities are algebraically grouped.
[2]Detail similar to steps 4a–4c could be shown here.

1. $1\{3\text{-unit}\} \times 1[4\text{-unit}] = 1\{1[(3\text{-unit}) \times [4\text{-unit}])\text{-unit}\}$      (1, 2)[1]

2. $= 1\{1[(3[1\text{-unit}\}s\text{-unit}) \times [4[1\text{-unit}|s\text{-unit}]\text{-unit}\}$      (2, 3)

3. $= 1\{1[ 1[1((1\text{-unit}) \times [4[1\text{-unit}|s\text{-unit}])\text{-unit}) +$
   $(1[(2[1\text{-unit}]s\text{-unit}) \times [4[1\text{-unit}]s\text{-unit}]) \text{-unit}]\ )\text{-unit}\}^{2}$      (3, 4)

a. $= 1\{( 1[ (1[((1\text{-unit}) \times [1\text{-unit})]) + 1[1((1\text{-unit}) \times [3[1\text{-unit}]s\text{-unit}])\text{-unit} )\ +$
   $1[1((2[1\text{-unit}]s\text{-unit}) \times [4[1\text{-unit}]s\text{-unit}])\text{-unit}\ )\text{-unit}\}$      (4, 4a)

b. $= 1\{1[ (1[((1\text{-unit}) \times [1\text{-unit}])\text{-unit}] + 1[ 1[1((1\text{-unit}) \times [1\text{-unit}])\text{-unit}] +$
   $1[(1\text{-unit}) \times [2[1\text{-unit}]s\text{-unit}])\text{-unit}]\ ) +$
   $1[1((2[1\text{-unit}]s\text{-unit}) \times [4[1\text{-unit}]s\text{-unit}])\text{-unit})\text{-unit}\}$      (4a, 4b)

c. $= 1\{1[ 1[( 1[((1\text{-unit}) \times [1\text{-unit}])\text{-unit}] + 1[1((1\text{-unit}) \times [1\text{-unit}])\text{-unit}] )\text{-unit}] +$
   $1[(1\text{-unit}) \times [2[1\text{-unit}]s\text{-unit}])\text{-unit}] +$
   $1[1((2[1\text{-unit}]s\text{-unit}) \times [4[1\text{-unit}]s\text{-unit}])\text{-unit})\text{-unit}\}$      (4b)

d. $= 1\{1[ 2[1((1\text{-unit}) \times [1\text{-unit}])\text{-unit}]s + 1[1((1\text{-unit}) \times [2[1\text{-unit}]s\text{-unit}]\text{-unit}]\text{-unit}) +$
   $1[1((2[1\text{-unit}]s\text{-unit}) \times [4[1\text{-unit}]s\text{-unit}])\text{-unit})\text{-unit}\}$      (4b, 4c)

e. $= 1\{1[ 2[1((1\text{-unit}) \times [1\text{-unit}])\text{-unit}]s + 1[( 1[ 1((1\text{-unit}) \times [1\text{-unit}])\text{-unit}] +$
   $1[(1\text{-unit}) \times [1\text{-unit}])\text{-unit}]\ )\text{-unit}] +$
   $1[1((2[1\text{-unit}]s\text{-unit}) \times [4[1\text{-unit}]s\text{-unit}])\text{-unit})\text{-unit}\}$      (4c)

f. $= 1\{1[ ( 2[1((1\text{-unit}) \times [1\text{-unit}])\text{-unit}]s + 1[1((1\text{-unit}) \times [1\text{-unit}])\text{-unit}]\ ) +$
   $1[(1\text{-unit}) \times [1\text{-unit}])\text{-unit}] +$
   $1[1((2[1\text{-unit}]s\text{-unit}) \times [4[1\text{-unit}]s\text{-unit}])\text{-unit})\text{-unit}\}$      (4c)

g. $= 1\{1[ 1[3[1((1\text{-unit}) \times [1\text{-unit}])\text{-unit}]s\text{-unit}] + 1( 1[1((1\text{-unit}) \times [1\text{-unit}])\text{-unit}]\ ) +$
   $1[1((2[1\text{-unit}]s\text{-unit}) \times [4[1\text{-unit}]s\text{-unit}])\text{-unit})\text{-unit}\}$      (4c, 5)

h. $= 1\{( 1[ 1[3[1((1\text{-unit}) \times [1\text{-unit}])\text{-unit}]s\text{-unit} + 1[1((1\text{-unit}) \times [1\text{-unit}])\text{-unit}]\ ) +$
   $1[1((2[1\text{-unit}]s\text{-unit}) \times [4[1\text{-unit}]s\text{-unit}])\text{-unit})\text{-unit}\}$      (5)

i. $= 1\{1[ 4[1((1\text{-unit}) \times [ 1\text{-unit}])\text{-unit}]s\text{-unit}] +$
   $1[1((2[1\text{-unit}]s\text{-unit}) \times [4[1\text{-unit}]s\text{-unit}])\text{-unit})\text{-unit}\}$      (5)

4.
$$= 1\{1( \ 4(1((1\text{-unit}) \times [1\text{-unit}])\text{-unit})s \ + \tag{5}$$
$$1(1((2(1\text{-unit}\}s\text{-unit}) \times [4[1\text{-unit}\}s\text{-unit}])\text{-unit}\} )\text{-unit}\}$$

5.
$$= 1\{1( \ 4(1((1\text{-unit}) \times [1\text{-unit}])\text{-unit})s \ + \ ( \ 1(1((1\text{-unit}) \times [4[1\text{-unit}\}s\text{-unit}])\text{-unit}])\text{-unit})^3 \ + \tag{5, 6}$$
$$1(1((1\text{-unit}) \times [4[1\text{-unit}\}s\text{-unit}])\text{-unit}\} )\text{-unit}\}$$

6.
$$= 1\{1( \ 4(1((1\text{-unit}) \times [1\text{-unit}])\text{-unit})s \ + \ (1 \ 4(1((1\text{-unit}) \times (1\text{-unit}))\text{-unit}) )\text{-unit})^3 \ + \tag{6}$$
$$1(1((1\text{-unit}) \times [4[1\text{-unit}\}s\text{-unit}])\text{-unit}\} )\text{-unit}\}$$

7.
$$= 1\{1( \ ( \ 4(1((1\text{-unit}) \times [1\text{-unit}])\text{-unit})s \ + \ 4(1((1\text{-unit}) \times [1\text{-unit}])\text{-unit})\text{-unit})s^3 \ + \tag{6, 7}$$
$$1(1((1\text{-unit}) \times [4[1\text{-unit}\}s\text{-unit}])\text{-unit}\} )\text{-unit}\}$$

8.
$$= 1\{1 \ 2(4(1((1\text{-unit}) \times [1\text{-unit}])\text{-unit})s\text{-unit}\}s \ + \tag{7, 8}$$
$$1(1((1\text{-unit}) \times [4[1\text{-unit}\}s\text{-unit}])\text{-unit}\} )\text{-unit}\}$$

9.
$$= 1\{1 \ 2(4(1((1\text{-unit}) \times [1\text{-unit}])\text{-unit})s\text{-unit}\} \ + \tag{7}$$
$$1(4[1((1\text{-unit}) \times [1\text{-unit}])\text{-unit})s\text{-unit}\} )\text{-unit}\}$$

10.
$$= 1\{3(4[1((1\text{-unit}) \times [1\text{-unit}])\text{-unit})s\text{-unit}\}\text{-unit}\} \tag{7, 8}$$

11.
$$= 1\{12(1((1\text{-unit}) \times [1\text{-unit}])\text{-unit})s\text{-unit}\} \tag{8, 9}$$

12.
$$= 12(1((1\text{-unit}) \times [1\text{-unit}])\text{-unit})s \tag{9, 10}$$

Fig. 5.7.

*A mathematics of quantity representation of the solution to Example Problem 1.*

[1]The numbers in parentheses key the steps in this demonstration to the steps in the demonstration in Figure 5.6.
[2]We introduce a special notation. In the pair, left parentheses, space, is to be considered a grouping symbol closed by the pair, space, right parentheses. This special notation has been used only in instances where sums of quantities are algebraically grouped.
[3]Detail similar to steps 3a–3i could be given here to show the step-by-step reunitization of this quantity to go from step 5 to step 7, and similarly to go from step 8 to step 10.

one orange per package and has every apple-orange variety pair represented. How many different kinds of packages will there be?

We analyze the problem as follows: The three varieties of apples and four varieties of oranges, respectively, constitute a (3-apple unit) and a (4-orange unit). We denote these respectively in the manipulative aid representation as (01 02 03) and [*1 *2 *3 *4]. The mathematics of quantity model for the problem in terms of our general mathematics of quantity representation is 1(3-unit) × 1[4-unit] = ?((1-unit) × [1-unit]-unit)s. Alternatively we could think of the three varieties of apples and four varieties of oranges as (3(1-apple unit)s-unit) and [4[1-orange unit]s-unit], respectively. Very little difference results, however, in carrying out the cross-product multiplication. One of the first steps in the first interpretation is to change the (3-apple unit) to a (3(1-apple unit)s-unit), and the [4-orange unit] to a [4[1-orange unit]s-unit].

The manipulative aid representation of the solution is given in Figure 5.6 and the mathematics of quantity representation is given in Figure 5.7. A possible conception of finding a cross product shown in these two representations is to hypothesize the units a child would construct by completing a two-dimensional array. The two composite units that represent the factors in the product would be conceptualized as orthogonal to each other, one along a vertical dimension and the other along a horizontal dimension.

The cross-product multiplication is then carried out by first reunitizing the two composite units that represent the vertical and horizontal dimensions to two units of units (i.e., the (3-apple unit) is reunitized to (3(1-apple unit)s-unit) and the [4-orange unit] to [4[1-orange unit]s-unit]). Next, as illustrated in Figure 5.6, the cross product multiplication by the "horizontal" unit—arbitrarily taken to be [4[1-orange unit]s-unit]—is distributed over the uniting operation in the "vertical" unit, giving units of the form (((1-apple unit) × [4[1-orange unit]s-unit])-unit). In Figure 5.6, this distribution is accomplished over several steps; in between steps of this application of distributivity, the distribution of the cross-product multiplication by a (1-apple unit) over the uniting operation in [4[1-orange unit]s-unit] is carried out giving a unit of the form (4(1((1-apple unit) × [1-orange unit])-unit)s-unit). Stated less formally, this cross-product multiplication is carried out by pairing the "first" (1-apple unit) with the [4[1-orange unit]s-unit] and then forming four ordered pairs—((1-apple unit), [1-orange unit]) pairs. The "second" (1-apple unit) is then paired with the [4[1-orange unit]s-unit], and again four ordered pairs are formed. This is continued until the (1-apple unit)s have all been used. This process is shown diagrammatically in Figure 5.8, which is given to suggest how this process might be accomplished in an interactive computer microworld.

We have used the analogy to a rectangular array only because that seems to be a useful model to use with children. Theoretically, the two dimensions need not be orthogonal nor do they need to be linear. Moreover, the analysis given here and in Figures 5.6 and 5.7 for discrete quantities would have a very similar parallel in continuous quantities, where the problem would be to form units of area from the linear units of the lengths of the sides.

*Example Problem 2.* A rectangle has a length of 6 inches and a height of 4 centimeters. What is the area of the rectangle?

We have deliberately avoided specifying the units in which the area should be expressed to suggest that this is an appropriate way to ask the question of a student. The problem of deciding what unit to use to express the area will force the student to consider what units are possible, based on the units in which the lengths of the sides are expressed, a consideration that concerns the relationship between units of area and the product of linear units. Many units are possible, among them (((1-cm unit) × [1-in. unit])-unit)s, (((1-cm unit) × [2-in. unit])-unit)s, (((2-cm unit) × [2-in. unit])-unit)s, (((2-cm unit) × [3-in. unit])-unit)s. In fact, one correct answer is that the area is (1((4-cm unit) × [6-in. unit])-unit), which could be appropriate in certain circumstances. The manipulative aid and mathematics of quantity models of the problem situation and manipulations of the models to answer the question of how many (((1-cm unit) × [1-in. unit])-unit)s are given Figures 5.9 and 5.10.

The demonstration in Figure 5.9 is intended to hypothesize the type of units a child would form in finding the area of a rectangle by using a 1-cm by 1-in. object, say, a block of this

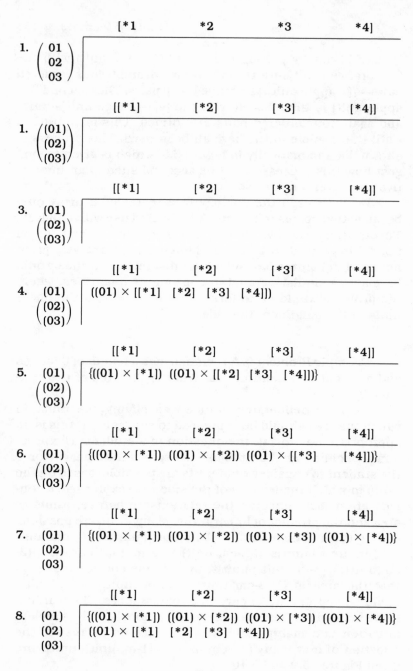

1.  $\begin{pmatrix} 01 \\ 02 \\ 03 \end{pmatrix}$  [*1          *2          *3          *4]

1.  $\begin{pmatrix} (01) \\ (02) \\ (03) \end{pmatrix}$  [[*1]        [*2]        [*3]        [*4]]

3.  $\begin{pmatrix} (01) \\ (02) \\ (03) \end{pmatrix}$  [[*1]        [*2]        [*3]        [*4]]

4.  $\begin{pmatrix} (01) \\ (02) \\ (03) \end{pmatrix}$  [[*1]        [*2]        [*3]        [*4]]
((01) × [[*1]   [*2]   [*3]   [*4]])

5.  $\begin{pmatrix} (01) \\ (02) \\ (03) \end{pmatrix}$  [[*1]        [*2]        [*3]        [*4]]
{((01) × [*1]) ((01) × [[*2]   [*3]   [*4]])}

6.  $\begin{pmatrix} (01) \\ (02) \\ (03) \end{pmatrix}$  [[*1]        [*2]        [*3]        [*4]]
{((01) × [*1]) ((01) × [*2]) ((01) × [[*3]     [*4]])}

7.  (01)   [[*1]        [*2]        [*3]        [*4]]
    (02)
    (03)
{((01) × [*1]) ((01) × [*2]) ((01) × [*3]) ((01) × [*4])}

8.  (01)   [[*1]        [*2]        [*3]        [*4]]
    (02)
    (03)
{((01) × [*1]) ((01) × [*2]) ((01) × [*3]) ((01) × [*4])}
((01) × [[*1]   [*2]   [*3]   [*4]])

Etc.

Fig. 5.8.

*A diagrammatic sequence to suggest how the cross-product multiplication analyzed in Figure 5.6 might appear in a computer interactive microworld.*

size, as follows. The child first places the object in, say, the upper lefthand corner of the rectangle and traces around it. This is then repeated to make a row of six 1-cm by 1-in. rectangles, giving a unit of product units, (6(1-cm unit) × [1-in. unit])s-unit). This is then repeated making a second, third, and fourth row, resulting in four such units of product units. Eventually the child may see the area as the product of the number of 1-cm by 1-in. rectangles in one row and the number of such rows. Whether this experience could lead a child to see the area of the rectangle as being related to the Cartesian product of the two length measures is clearly an open question. Because the solution is "driven" by the manipulative aid (a physical rectangular region and rectangular pieces) rather than by a predetermined plan and an anticipated solution on the part of the student, it is unlikely that these connections will be made. Having the student use the length and width of the 1-cm by 1-in. object to measure the respective sides of the rectangle and making a prediction about the number of copies of the rectangle that would be needed to cover the rectangle would likely be helpful.

The representation in Figure 5.10 corresponds to a method of determining the area by forming a grid in the rectangle. The process would be to first partition one of the sides into unit lengths—six 1-in. lengths or four 1-cm lengths. In Figure 5.10 the one 4-cm length is partitioned in four 1-cm lengths. Next lines are drawn parallel to the other side emanating from the partition points on the adjacent side. After this, the other side is partitioned into unit lengths (six 1-in. lengths in Figure 5.10) and lines are drawn parallel to the adjacent side emanating from each partition point on the first side. This grid seems to be closer to the manner in which children would complete an array to find the Cartesian product of two discrete units and might facilitate a child's ability to construct a connection between product-of-measure multiplication and the concept of using multiplication to determine area.

The manipulative aid demonstration in Figure 5.9 reflects what a child would do to determine the area of the rectangle by placing 1-cm by 1-in. blocks in rows. Steps 5a through 5f (Figure 5.9) corresponds to the child making a first row, and steps 7, 9, and 10 correspond to making subsequent rows. The action in this measurement process could

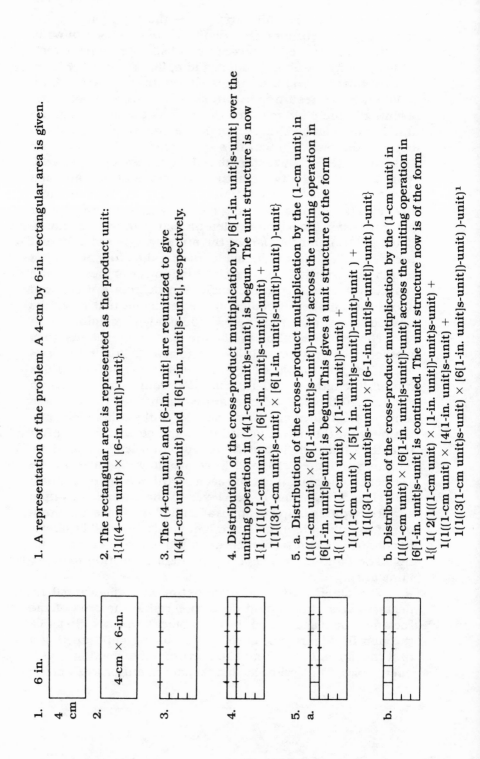

1. A representation of the problem. A 4-cm by 6-in. rectangular area is given.

2. The rectangular area is represented as the product unit:
1{1(([4-cm unit] × [6-in. unit])-unit}.

3. The (4-cm unit) and [6-in. unit] are reunitized to give
1{4(1-cm unit]s-unit) and 1[6[1-in. unit]s-unit], respectively.

4. Distribution of the cross-product multiplication by [6[1-in. unit]s-unit] over the uniting operation in (4{1-cm unit}s-unit) is begun. The unit structure is now
1{ (1(1([1-cm unit) × [6[1-in. unit]s-unit])-unit] +
1(1((3{1-cm unit]s-unit) × [6[1-in. unit]s-unit] )-unit}

5. a. Distribution of the cross-product multiplication by the (1-cm unit) in
(1{[1-cm unit] × [6[1-in. unit]s-unit])-unit] across the uniting operation in
[6[1-in. unit]s-unit] is begun. This gives a unit structure of the form
1{( 1( 1(1([1-cm unit) × [1-in. unit])-unit) +
1(1[1-cm unit) × [5[1 in. unit]s-unit])-unit)-unit ) +
1(1((3{1-cm unit]s-unit) × [6-1-in. unit]s-unit])-unit] )-unit}

b. Distribution of the cross-product multiplication by the (1-cm unit) in
(1{[1-cm unit] × [6[1-in. unit]s-unit])-unit] across the uniting operation in
[6[1-in. unit]s-unit] is continued. The unit structure now is of the form
1{( 1( 2(1([1-cm unit) × [1-in. unit])-unit]s-unit) +
1(1[1-cm unit) × [4[1-in. unit]s-unit] ) +
1(1((3{1-cm unit]s-unit) × [6[1-in. unit]s-unit])-unit] )-unit )-unit}[1]

. . .

. . .

**f.** The sixth, and final, step in the distribution of the cross-product multiplication by the (1-cm unit) in (1((1-cm unit) × [6[1-in. unit]s-unit])-unit) across the uniting operation in [6[1-in. unit]s-unit] is completed. The unit structure now is of the form

1{1[ 1(6(1((1-cm unit) × [1-in. unit])-unit]s-unit) +
1(1((3(1-cm unit]s-unit) × [6[1-in. unit]s-unit])-unit) )-unit}.

**6.** Attention is returned to continuation of distrubition of cross-product multiplication by [6[1-in. unit]s-unit] over the uniting operation in (4(1-cm unit]s-unit}. The unit structure is now of the form:

1{1[ 1(6(1((1-cm unit) × [1-in. unit])-unit]s-unit} +
1(1((1-cm unit) × [6[1-in. unit]s-unit])-unit} +
1(1((2(1-cm unit]s-unit) × [6[1-in. unit]s-unit])-unit) )-unit}.

**7.** A second time, the six steps in the distribution of the cross-product multiplication by the (1-cm unit) in (1((1-cm unit) × [6[1-in. unit]s-unit])-unit) across the uniting operation in [6[1-in. unit]s-unit] is completed. The unit structure now is of the form

1{1[ 2(6(1((1-cm unit) × [1-in. unit])-unit]s-units +
1(1((2(1-cm unit]s-unit) × [6[1-in. unit]s-unit])-unit) )-unit}.

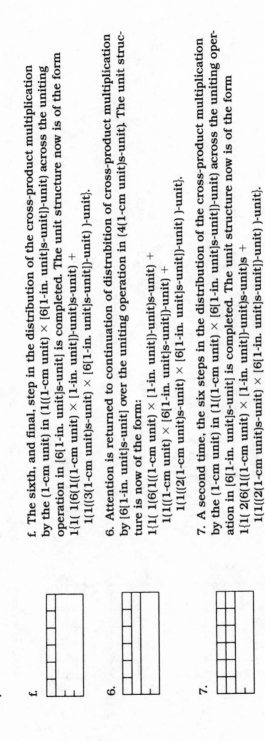

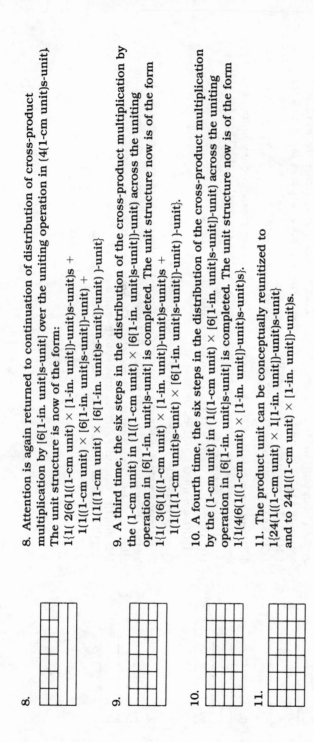

8. Attention is again returned to continuation of distribution of cross-product multiplication by [6[1-in. unit]s-unit] over the uniting operation in (4[1-cm unit]s-unit}. The unit structure is now of the form:

1{1( 2(6(1(((1-cm unit) × [1-in. unit])-unit]s-unit}s +
  1(1((1-cm unit) × [6[1-in. unit]s-unit])-unit) +
  1(1((1-cm unit) × [6[1-in. unit]s-unit])-unit) )-unit}

9. A third time, the six steps in the distribution of the cross-product multiplication by the (1-cm unit) in (1((1-cm unit) × [6[1-in. unit]s-unit])-unit) across the uniting operation in [6[1-in. unit]s-unit] is completed. The unit structure now is of the form

1{1( 3(6(1(((1-cm unit) × [1-in. unit])-unit]s-unit}s +
  1(1((1-cm unit]s-unit) × [6[1-in. unit]s-unit])-unit) )-unit}.

10. A fourth time, the six steps in the distribution of the cross-product multiplication by the (1-cm unit) in (1((1-cm unit) × [6[1-in. unit]s-unit])-unit) across the uniting operation in [6[1-in. unit]s-unit] is completed. The unit structure now is of the form

1{1(4[6(1((1-cm unit) × [1-in. unit])-unit]s-unit}s}.

11. The product unit can be conceptually reunitized to

1{24(1((1-cm unit) × 1[1-in. unit])-unit]s-unit}
and to 24(1((1-cm unit) × [1-in. unit])-unit]s.

[1]For completeness in detail, these steps should appear in a sequence. These details were omitted for considerations of space.

Fig. 5.9.

*A manipulative aid representation of Example Problem 2 to determine the area in (((1-cm unit) × [1-in. unit]-unit]s.*

be quite like the measurement of a linear quantity. The process could be perceived by the child as simply placing units end to end and counting the number of units. Therefore, for such an activity to lead to the formation of units as we hypothesize—single units based on single blocks, then doubleton units, triplets, and finally a single row of blocks—some teacher intervention might be necessary. When the first row of blocks is formed, the child's ability to conceptualize this as a unit might be facilitated by teacher intervention with questions such as these:

> How can you describe the blocks that you have here?
> (Looking for an answer such as "one row.")
> How many rectangles are in the row?
> What is the measure of this side of the rectangle?
> Does this row measure this side of the rectangle?

Such questions would draw the child's attention to a relationship between the number of rectangles in a row and the number of units of measure in one side of the rectangle. As the child repeats the process to form additional rows similar appropriate questions (How many rows? Does the number of rows equal the measure of the other side of the rectangle?) would be brought up for discussion. We see this activity as aiding a child in conceptualizing the area as a product: the number of rows times the number in each row. But this product is just a shortcut for counting the number of singleton rectangles. The children would need to make the connection that the number of rows corresponds to the measure of one side and the number in each row to the measure of the other dimension, and in this way the product of the number of rows times the number in each row would correspond to the product of the measures of the two sides. That this manipulative activity will aid children in conceptualizing the relationship between the one-dimensional measure of the sides with the two-dimensional measure of the area still seems in doubt. One troublesome circumstance in the sequence is that the measures of the sides of the rectangle do not enter the discussion until after the rows have been made. Attention might be drawn to this in a subsequent but similar activity. And, starting the sequence by asking children to predict the number in a row and the number of rows could create the appro-

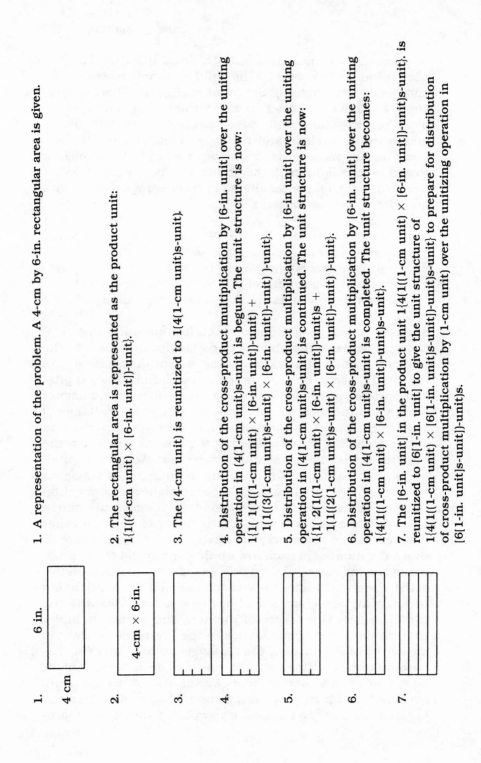

1. A representation of the problem. A 4-cm by 6-in. rectangular area is given.

1. 6 in.

4 cm

2. The rectangular area is represented as the product unit:
1{1((4-cm unit) × [6-in. unit])-unit}.

2. 4-cm × 6-in.

3. The (4-cm unit) is reunitized to 1(4(1-cm unit)s-unit).

3.

4. Distribution of the cross-product multiplication by [6-in. unit] over the uniting operation in (4(1-cm unit)s-unit) is begun. The unit structure is now:
1{1( 1(1((1-cm unit) × [6-in. unit])-unit) +
    1(1((3(1-cm unit)s-unit) × [6-in. unit])-unit) )-unit}.

4.

5. Distribution of the cross-product multiplication by [6-in unit] over the uniting operation in (4(1-cm unit)s-unit) is continued. The unit structure is now:
1{1( 2(1(1((1-cm unit) × [6-in. unit])-unit)s +
    1(1((2(1-cm unit)s-unit) × [6-in. unit])-unit) )-unit}.

5.

6. Distribution of the cross-product multiplication by [6-in. unit] over the uniting operation in (4(1-cm unit)s-unit) is completed. The unit structure becomes:
1{4(1((1-cm unit) × [6-in. unit])-unit)s-unit}.

6.

7. The [6-in. unit] in the product unit 1{4(1((1-cm unit) × [6-in. unit])-unit)s-unit}, is reunitized to [6[1-in. unit] to give the unit structure of
1{4(1((1-cm unit) × [6[1-in. unit])-unit)s-unit) to prepare for distribution of cross-product multiplication by (1-cm unit) over the unitizing operation in [6[1-in. unit]s-unit])-unit}.

7.

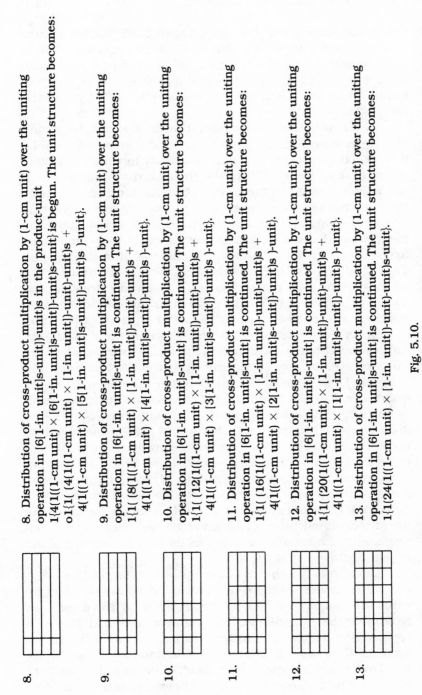

8.

9.

10.

11.

12.

13.

8. Distribution of cross-product multiplication by (1-cm unit) over the uniting operation in [6[1-in. unit]s-unit]) in the product-unit is begun. The unit structure becomes:

1{4(1[(1-cm unit) × [6[1-in. unit]s-unit])-unit]s-unit} +

o1{1( 4(1[(1-cm unit) × [1-in. unit])-unit]s +

4(1[(1-cm unit) × [5[1-in. unit]s-unit])-unit]s )-unit}.

9. Distribution of cross-product multiplication by (1-cm unit) over the uniting operation in [6[1-in. unit]s-unit] is continued. The unit structure becomes:

1{1( 8(1[(1-cm unit) × [1-in. unit])-unit]s +

4(1[(1-cm unit) × [4[1-in. unit]s-unit])-unit]s )-unit}.

10. Distribution of cross-product multiplication by (1-cm unit) over the uniting operation in [6[1-in. unit]s-unit] is continued. The unit structure becomes:

1{1( 12(1[(1-cm unit) × [1-in. unit])-unit]s +

4(1[(1-cm unit) × [3[1-in. unit]s-unit])-unit]s )-unit}.

11. Distribution of cross-product multiplication by (1-cm unit) over the uniting operation in [6[1-in. unit]s-unit] is continued. The unit structure becomes:

1{1( 16(1[(1-cm unit) × [1-in. unit])-unit]s +

4(1[(1-cm unit) × [2[1-in. unit]s-unit])-unit]s )-unit}.

12. Distribution of cross-product multiplication by (1-cm unit) over the uniting operation in [6[1-in. unit]s-unit] is continued. The unit structure becomes:

1{1( 20(1[(1-cm unit) × [1-in. unit])-unit]s +

4(1[(1-cm unit) × [1[1-in. unit]s-unit])-unit]s )-unit}.

13. Distribution of cross-product multiplication by (1-cm unit) over the uniting operation in [6[1-in. unit]s-unit] is continued. The unit structure becomes:

1{1(24(1[(1-cm unit) × [1-in. unit])-unit]s-unit}.

Fig. 5.10.

*A alternative manipulative, and mathematics of quantity, representation of Example Problem 2 to determine the area in (((1-cm unit) × [1-in. unit])-unit]s.*

priate cognitive disequilibrium (Piaget, 1985) to lead toward an accommodation of this relationship. The process of making a prediction would force the child to attempt to anticipate the arrangement of singleton rectangles into rows and columns within the constraints of the length and height of the rectangle.

The manipulative demonstration in Figure 5.10, which also deals with the problem of finding the area of a 4-cm by 6-in. rectangle, corresponds to an activity in which a child is led to first actively indicate the linear units of measure on two adjacent sides of the given rectangle and then to "grid" the rectangle to form 1-cm by 1-in. rectangles. Might this experience, along with teacher-child discussion, help the child to see that the formation of the 1-cm by 1-in. rectangles based on the units of the two dimensions of the rectangle? This "griding" also involves the formation of subunits of the given 4-cm by 6-in. rectangular unit. Each line drawn at the end of a 1-cm unit of length parallel to the 6-in. dimension suggests a 1-cm by 6-in. unit of area. When all possible units of this type are formed, each line drawn at the end of 1-in. units parallel to the 4-cm dimension forms four 1-cm by 1-in. units of area or one composite unit of units made up of a column of four 1-cm by 1-in. units. When all of these lines are drawn the area can be conceptualized as the composite unit of units of units made up of six composite unit of units (the six 4-cm by 1-in. columns). This would lead to the idea that the area can be conceptualized as the product of the number of 1-cm by 1-in. rectangles in one column times the number of columns. We suggest that this type of manipulative activity, along with appropriate teacher-student interaction, could help children conceptualize the relationship between the one-dimensional units of measure of the sides of the rectangle with the two-dimensional units of the measure of the region enclosed by these sides.

Of course, the choice of whether to first draw lines (Figure 5.10) parallel to the 6-in. dimension so that we end up with columns of 1-cm by 1-in. rectangles or to first draw lines parallel to the 4-cm dimension so that we would end up with rows of rectangles was arbitrary in the case of this demonstration. The same is true for the demonstration in Figure 5.9. For children, experiences need to be provided for both choices. And children need to be allowed to make their own choice as well.

Again, an aspect of this intervention would be to engage the child in a discussion about drawing grid lines as suggested and at various points in the sequence to have the student make a prediction about the number of 1-cm by 1-in. rectangles that would result. Helpful interventions leading in this direction could be to mark off (or suggest that the child mark off) 1-cm parts of the 4-cm side and ask how many 1-cm by 6-in. rectangles there would be. One could have the child draw (and, eventually, mentally picture) lines from the end points of the 1-cm segments. Then, suggesting that the child keep an image of the 1-cm by 6-in. rectangle, mark (or have the child mark) points to partition the 6-in. side into 1-in. parts and think about drawing lines emanating from each partition point and to predict the number of small rectangles (1-cm by 1-in.) that would result in each 1-cm by 6-in. rectangle, and finally to predict the number of small rectangles in all.

One should go even further and engage the child in situations where the question is about the number of 2-cm by 1-in., 1-cm by 2-in., 2-cm by 2-in., 2-cm by 3-in., . . . small rectangles in the whole rectangle, after partitioning the sides of the rectangle appropriately. One should even involve the child in situations in which the height is partitioned into four 1-cm units and the length into two 2-in. units plus two 2-in. units and have the child determine the types and numbers of each type of units that would result by taking the product of these measures of the height and length. A similar activity would be to partition the height into one 2-cm units plus two 1-cm units and the length as earlier.

This kind of activity would set the stage for further connection between area and multiplication of units to the point where the activity of partitioning the sides of a rectangle into appropriate lengths and determination of the units of area resulting from this partitioning would give a strong conceptual basis for multiplication of multidigit numbers, such as $34 \times 25$. How this might be done is suggested in Figure 5.11. Flexibility in partitioning the sides of the rectangle might lead the child to invent alternate strategies to perform computations as suggested in Figure 5.12.

## Multiplication of Rational Numbers

In this section, we will look more directly at how problems such as Example Problem 1 (Jane has a large quantity of

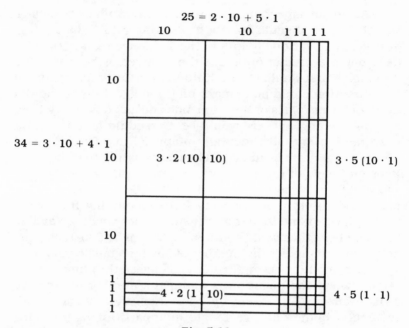

Fig. 5.11

*An area representation of 34 times 25 as a product of units of 10 and units of 1.*

three different varieties of apples and four varieties of oranges. She is going to package the fruit so that she has one apple and one orange per package and has every apple-orange variety pair represented. How many different kinds of packages will there be?) can be used to form a foundation for finding the product of rational numbers. There are alternative ways to look at Example Problem 1. One modification of this problem to move it in the direction of providing additional support for rational number multiplication is as follows: Jane has a large quantity of three different varieties of apples, two of them are red and the other a different color. How many packages among all of the packages will contain a red apple?

A representation of this problem might be:

{((**O**1) (**O**2) (O3)) × [[*1] [*2] [*3] [*4]]}, that is, as {(3(1-apple unit)s-unit) × 4[1-unit]s-unit])-unit}.

$$34 \cdot 25 = 30 \cdot 25 + 4 \cdot 25$$

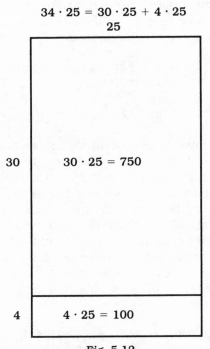

Fig. 5.12

*An alternate representation of 34 times 25 with an area model.*

Unitization of the red apple varieties would give

{((((**O**1) (**O**2)) (03)) × [[\*1] [\*2] [\*3] [\*4]]}, or
{(((2(1-red-apple unit)s-unit) + (1-apple unit))
   × [4[1-orange unit]s-unit]}.

And distribution of cross product multiplication by [[\*1] [\*2]
[\*3] [\*4]] over the first-level uniting operation in (((**O**1) (**O**2))
(03)) gives

{((((**O**1) (**O**2)) × [[\*1] [\*2] [\*3] [\*4]])
   (03) × [[\*1] [\*2] [\*3] [\*4]])}, or
{((2(1-red-apple unit)s-unit) × [4[1-orange unit]s-unit]
   + (1-apple unit)) × 4[1-orange unit]s-unit]}.

If a spatial arrangement was used to display the states in the
process of carrying out this cross product, as might be done

with an interactive computer microworld, the final result might look like the following. Each of the two units of product units corresponds to one of the terms in the distributed product.

$$((\mathbf{O}1) \times [*1]) \;((\mathbf{O}1) \times [*2]) \;((\mathbf{O}1) \times [*3]) \;((\mathbf{O}1) \times [*4])$$
$$((\mathbf{O}2) \times [*1]) \;((\mathbf{O}2) \times [*2]) \;((\mathbf{O}2) \times [*3]) \;((\mathbf{O}2) \times [*4])$$
$$\cdot \;\; \cdot \;\; \cdot$$
$$((O3) \times [*1]) \;((O3) \times [*2]) \;((O3) \times [*3]) \;((O3) \times [*4]).$$

From this display one can easily count, or multiply, to find that eight out of twelve packages have a red apple. Another modification, to move further in the direction of providing additional support for rational number multiplication, is to suppose as above, that two of the three varieties of apples are red and that three of the four varieties of oranges are California oranges. Then consider the question, How many package types have a red apple and a California orange, and how many apple-orange packages are there in all?

A representation of the problem in the generic manipulative aid might be

$$\{((\mathbf{O}1) \,(\mathbf{O}2) \,(O3)) \times [[[*]1] \,[*]2] \,[*]3] \,[*4]]\}, \text{ or}$$
$$\{(3(1\text{-apple unit})\text{s-unit}) \times [4[1\text{-unit}]\text{s-unit}])\text{-unit}\}.$$

Unitization of the red apple varieties and the California oranges gives

$$\{((((\mathbf{O}1) \,(\mathbf{O}2)) \;\;(O3)) \times [[[[*]1] \,[*]2] \,[*]3]] \;\;[*4]]\},$$

or the product-unit resulting from the cross-product multiplication of

$$((2(1\text{-red-apple unit})\text{s-unit}) + (1\text{-apple-unit})) \text{ and}$$
$$[[3[1\text{-California-orange unit}]\text{s-unit}] + [1\text{-orange unit}]].$$

Distribution of appropriate cross-product multiplications over unitizing operations and arrangement of the results into four quadrants corresponding with the four major terms of the distributed product would give the following:

$$((\mathbf{O}1) \times [[*]1]) \;((\mathbf{O}1) \times [[*]2]) \times ((\mathbf{O}1) \times [[*]3]) \cdot ((\mathbf{O}1) \times [*4])$$
$$((\mathbf{O}2) \times [[*]1]) \;((\mathbf{O}2) \times [[*]2]) \;((\mathbf{O}2) \times [[*]3]) \cdot ((\mathbf{O}2) \times [*4])$$
$$\cdot \;\; \cdot \;\; \cdot$$
$$((O3) \times [[*]1]) \;((O3) \times [[*]2]) \;((O3) \times [[*]3]) \cdot ((O3) \times [*4]).$$

The answer to the question can be obtained by interpreting the product units in the upper lefthand quadrant of the display and finding the number of product units in it.

Because the question can be answered through two separate computations, one can ask what cognitive advantages there are in looking at the problem in this form. Each of the factors—two apple varieties out of three and three orange varieties out of four—in the cross-product multiplication is a part-whole relationship; therefore, the product is a product of part-whole relationships. The product has this more general representation:

(2(1-unit)s out of 3(1-unit)s) × (3[1-unit]s out
   of 4[1-unit]s) =
?(((1-unit) × [1-unit])-unit)s out of ?(((1-unit)
   × [1-unit])-unit)s.

The problem takes on the character of a product of rational numbers when the unit of two red-apple varieties is considered to be 2/3 of all the varieties of apples, and the unit of three California orange varieties is considered to be 3/4 of all the varieties of oranges. The interpretation of the question then becomes 2/3 × 3/4 = 6/12.

An alternative interpretation to the problem can be given if we replace each of (1-red-apple unit) and (1-apple unit) by the general unit notation (1/3-unit) and each of [1-California-orange unit] and [1-orange unit] by a [1/4-unit]. Then the cross-product multiplication of

((2(1-red-apple unit)s-unit) + (1-apple unit)) by
[[3[1-California-orange unit]s-unit] + [1-orange unit]],

becomes the following cross-product unit:

{((2(1/3-unit)s-unit) + (1/3-unit)) × [[3[1/4-unit]s-unit]
   + [1/4-unit]]}.

Careful investigation of the units structures involved in carrying out appropriate distribution of cross-product multiplications over unitizing operations to reformulate this product unit into a composite unit of product units would lead to the interpretation that

$$2/3 \times 3/4 = \{(2(1/3\text{-unit})s\text{-unit}) \times (3[1/4\text{-unit}]s\text{-unit}]\text{-unit}\}$$
$$= \{6((1/3\text{-unit}) \times [1/4\text{-unit}])s\text{-unit}\}$$
$$= \{6(1/12\text{-unit})s\text{-unit}\}$$
$$= 6(1/12\text{-unit})s.$$

This leads to an algorithm for multiplication of rational numbers, represented as multiplies of unit fractions, which is illustrated by

$$2/3 \times 3/4 = 2(1/3) \times 3(1/4)$$
$$= (2 \times 3)(1/3 \times 1/4)$$
$$= 6(1/12)$$
$$= 6/12.$$

Thus, some basic understandings for multiplication of rational numbers are that any rational number $a/b$ can be represented as $a$ units of size $1/b$, $a(1/b)$, that the product of any two unit fractions $1/x$ and $1/b$ is $1/(x \cdot b)$, and that the product of two rational numbers is conceptually a product-of-units problem (Vergnaud, 1988).

In the remainder of this section we will look more directly at a discrete quantity manipulative aid model for multiplication of rational numbers. The types of conceptual units that, according to our hypothesis, are needed to understand the multiplication are similar to those shown in the detailed analysis given in Figures 5.6 and 5.7. A corresponding mathematics of quantity model will further emphasize the formation and reformation of these units. We start with a manipulative aid model based on discrete quantity to find the product of 1/6 and 1/4 in Figure 5.13.

The product of 1/6 and 1/4 was found in the context of finding the cross products of the two units implied by the fractions 1/6 and 1/4—a (6-unit) and a (4-unit). This procedure resulted in a representation of the product unit and the product of the fractions as the result of a single procedure. This is important in the following sense. For a child who carries out such a procedure to quantify the product of 1/6 and 1/4, it is necessary that he or she quantify the part structure of the product unit (the whole of which 1/6 times 1/4 is part). This part structure can and should be quantified by the child in numerous ways: as $6(1/6) \times 4(1/4)$, as $6 \times 4(1/6 \times 1/4)$, as $24(1/6 \times 1/4)$, as $24(1/24)$ and as 24/24. Once the

1. [0 0 0 0 0 0]

   (0 0 0 0)

1. Choose two units, one which can be re-unitized into 1/6s and one into 1/4s.

2. [[0] [0] [0] [0] [0] [0]]

   ((0) (0) (0) (0))

2. Reunitize to 6[1/6[6-unit]s-unit] and 4(1/4(4-unit)s-unit).

3. [[**0**] [0] [0] [0] [0] [0]]

   ((**0**) (0) (0) (0))

3. Distinguish 1[1/6[6-unit]-unit] and 1(1/4(4-unit)-unit).

4. [[**0**] [0] [0] [0] [0] [0]] × (**0**) (0) (0) (0))

4. A representation of the problem as an indicated product.

5. {[[**0**] [0] [0] [0] [0] [0] × (**0**) (0) (0) (0))}

5. The indicated product is transformed to a product unit.

6. {[([**0**] × (**0**)) ([**0**] × (0)) ([**0**] × (0)) ([**0**] × (0))
   ([0] × (**0**)) ([0] × (0)) ([0] × (0)) ([0] × (0))
   ([0] × (**0**)) ([0] × (0)) ([0] × (0)) ([0] × (0))
   ([0] × (**0**)) ([0] × (0)) ([0] × (0)) ([0] × (0))
   ([0] × (**0**)) ([0] × (0)) ([0] × (0)) ([0] × (0))
   ([0] × (**0**)) ([0] × (0)) ([0] × (0)) ([0] × (0))}[1]

6. A representation of the product unit as a unit of units of product units after carrying out the cross-product multiplication. The special product unit ([**0**] × (**0**)), that is, (([1/6[6-unit]-unit] × ((1/4(4-unit)-unit), can be seen to be one product unit out of a total of twenty-four product units in the composite unit of product units, {24(((1/6-unit) × [1/4-unit])-unit)s-unit}.

[1]Detail similar to that shown in Figure 5.6 could be provided here. The type of units suggested in finding the product in Figure 5.6 would also be formed in finding the Cartesian product here.

Fig. 5.13.
*A manipulative display to show that the product of a [1/6[6-unit]-unit] and a (1/4(4-unit)-unit) is a (1/24(24-unit)-unit).*

1. Choose two units, one that can be reunitized into 1/6s, a [6-unit], and one that can be reunitized into 1/4s, a (4-unit).

2. The [6-unit] is reunitized to 6[1/6[6-unit]-unit]s and the (4-unit) to 4(1/4(4-unit)-unit)s.

3. Distinguish 4[1/6[6-unit]-unit]s and 3(1/4(4-unit)-unit)s.

4. Reunitize, the unit structures now are [4[1/6[6-unit]-unit]s + 2[1/6[6-unit]-unit]s-unit], and ([3(1/4(4-unit)-unit)s + 1(1/4(4-unit)-unit)-unit).

5. The problem is represented as an indicated product.

6. The problem is represented as a product unit.

7. The cross-product multiplication by 1[1/6-unit] is distributed over the uniting operation on 3(1/4-unit)s and 1(1/4-unit). This gives {1[3([[1/6-unit] × (1/4-unit))-unit]s + 1[[1/6-unit] × (1/4-unit)-unit]-unit}.

8. The cross-product multiplication of a second 1[1/6-unit] is distributed over the uniting operation on 3(1/4-unit)s and 1(1/4-unit). This gives {2[3([[1/6-unit] × (1/4-unit))-unit]s + 1[[1/6-unit] × (1/4-unit))-unit}.

9. The cross-product multiplication by a third 1[1/6-unit] is distributed over the uniting operation on 3(1/4-unit)s and 1(1/4-unit). This gives {3[3([[1/6-unit] × (1/4-unit))-unit]s + 1[[1/6-unit] × (1/4-unit))-unit}.

10. The cross-product multiplication by a fourth 1[1/6-unit] is distributed over the uniting operation on 3(1/4-unit)s and 1(1/4-unit). This gives {4[3([[1/6-unit] × (1/4-unit))-unit]s + 1[[1/6-unit] × (1/4-unit))-unit}.

1. [0 0 0 0 0 0]
   (0 0 0 0)

2. [[0] [0] [0] [0] [0] [0]]
   ((0) (0) (0) (0))

3. [[■] [■] [■] [*] [*]]
   ((#) (#) (0))

4. [[[■] [■] [■] [■]] [*] [*]]
   (((#) (#) (#)) (0))

5. [[[■] [■] [■]] [*] [*]] × ((((#) (#) (#)) (0))

6. {[[[■] [■] [■]] [*] [*]] × ((((#) (#) (#)) (0))}

7. ([■] × (#)) ([■] × (#)) ([■] × (#))   ([■] × (0))

8. ([■] × (#)) ([■] × (#)) ([■] × (#))   ([■] × (0))
   ([■] × (#)) ([■] × (#)) ([■] × (#))   ([■] × (0))

9. ([■] × (#)) ([■] × (#)) ([■] × (#))   ([■] × (0))
   ([■] × (#)) ([■] × (#)) ([■] × (#))   ([■] × (0))
   ([■] × (#)) ([■] × (#)) ([■] × (#))   ([■] × (0))

10. ([■] × (#)) ([■] × (#)) ([■] × (#))   ([■] × (0))
    ([■] × (#)) ([■] × (#)) ([■] × (#))   ([■] × (0))
    ([■] × (#)) ([■] × (#)) ([■] × (#))   ([■] × (0))
    ([■] × (#)) ([■] × (#)) ([■] × (#))   ([■] × (0))

11. The cross-product of a fifth 1[1/6-unit] is distributed over the uniting operation on 3(1/4-unit)s and 1(1/4-unit). This gives {5[3([[1/6-unit] × (1/4-unit)-unit]s + 1([[1/6-unit] × (1/4-unit)-unit]-unit}.

12. The cross-product multiplication by the sixth 1[1/6-unit] is distributed over the uniting operation on 3(1/4-unit)s and 1(1/4-unit). This gives {6[3([[1/6-unit] × (1/4-unit)-unit]s + 1([[1/6-unit] × (1/4-unit)-unit)-unit}.

13. The cross-product of the 6[1/6-unit]s and 4(1/4-unit)s is interpreted as

4 · 3[[1/6-unit] × (1/4-unit)-unit]s +
4 · 1[[1/6-unit] × (1/4-unit)-unit]s +
2 · 3[[1/6-unit] × (1/4-unit)-unit]s +
2 · 1[[1/6-unit] × (1/4-unit)-unit]s.

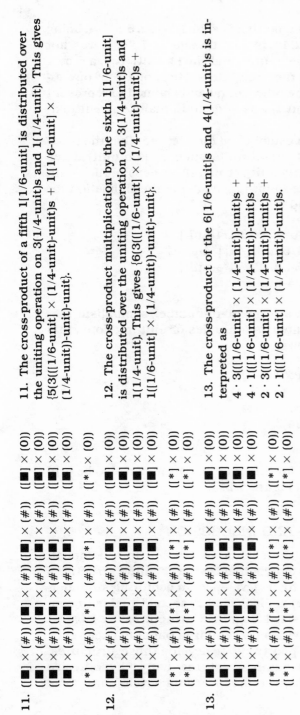

14. Reference to the manipulative display in step 13, now makes possible the inference that the product 4[1/6-unit]s × 3(1/4-unit)s = 4 · 3[[1/6-unit] × (1/4-unit)-unit]s or 12[([1/6-unit] × (1/4-unit)-unit]s. That each [([1/6-unit] × (1/4-unit)-unit]s to unit is (1/24[[6-unit] × (4-unit)-unit] can be established by comparing any one of the [([1/6-unit] × (1/4-unit)-unit]s to the [([6-unit] × [4-unit)-unit] that appears in the form 6 · 4[([1/6-unit] × (1/4-unit)-unit]s. The final inference that 4[1/6[6-unit]-unit]s × 3(1/4(4-unit)-unit]s = 12(1/24[([6-unit] × (4-unit)-unit]-unit]s, and thus that 4[1/6-unit]s × 3[1/4-unit]s = 12(1/24-unit]s, is now possible.

Fig. 5.14.

*A manipulative display of the product of 5/6 times 3/4 using a quantity interpretation of 5/6 and of 2/4.*

whole product unit is quantified as 24(1/24), 1/6 × 1/4, being one of twenty-four 1/24s, is quantifiable as 1/24. It was not that the product of the (6-unit) and the (4-unit) (e.g., a 6 by 4 array) was given and then the intersection of 1/6 of one dimension and 1/4 of the other found and claimed to represent the product of 1/6 and 1/4 as is done in many current textbooks.

Once it has been established for a learner that the product of any two unit-fraction units such as (1/$a$-unit) times [1/$b$-unit] equals 1{1/$ab$-unit}, it might be possible for children to work with the following symbolic multiplication for the product of 4/6 and 3/4:

1.  4/6 × 3/4 = 4(1/6-unit) × 3[1/4-unit]
2.          = 4 · 3(1/6-unit) × [1/4-unit]
3.          = 12(1/6-unit) × [1/4-unit]
4.          = 12{1/24-unit}.

Or more simply, with arithmetic-of-number symbolism having fractions represented as multiples of unit fractions, this would appear as follows:

1.  4/6 × 3/4 = 4(1/6) × 3(1/4)
2.          = 4 · 3((1/6) × (1/4))
3.          = 12(1/6) × (1/4))
4.          = 12(1/24)
5.          = 12/24.

Manipulative support for this multiplication can also be provided as suggested in Figure 5.14.

## CONCLUSIONS AND DISCUSSION

This analytical work represents an attempt to accomplish two goals. One goal is to hypothesize the cognitive structures that develop, or need to be developed, in acquiring an understanding of the concepts discussed. A second goal is to consider these hypothesized cognitive structures to suggest kinds of learning activities that children ought to experience so that they have an opportunity to develop those structures. A conjecture that these unit structures that we hypothesize correspond to mental structures learners develop has some support from cognitive science (S. Ohlsson, personal communication, August 7, 1991).

We wish to emphasize that the notational systems we have developed and communicated in the chapter were developed for theoretical analysis and communication within the research community. We neither advocate that the notational systems be used with children nor disavow the possibility that particular instantiations of the notational systems might be used with children. However, the development of these instantiations goes beyond the scope of this chapter. We do argue, however, that the generic manipulative aid unit analyses do provide a template for the construction of such instantiations to provide appropriate manipulative experiences for children.

Whether or not the cognitive structures we hypothesize develop during, or are necessary for, the learning and understanding of the concepts under discussion remains to be determined by research. That the analysis given in this chapter has validity is supported by work in progress by Simon and Blume (1991). In their research with preservice teachers, they have investigated the development of the understanding of the area of a rectangular region as a multiplicative relationship between the lengths of the sides. Their work supports, first of all, that insights into how the area of a rectangle is related to the product of the length–measures of the two sides is essentially nonexistent prior to instructional intervention. Through some insightful instructional moves on the part of these researchers, preservice teachers developed their understanding of this relationship to the point where their insight into area was essentially that of the cross product of units of width and units of length. The units used were long and short sticks. This understanding was displayed in the context of the following problem: Two people work together to measure the size of a rectangular region, one measures the length and the other the width. They each use a stick to measure with. The sticks, however, are of different lengths. Louisa says, "The length is four of my sticks." Ruiz says, "The width is five of my sticks." What have they found out about the area of the rectangular region?

Students worked on this problem in groups of three. A response by Tonie is particularly appropriate to the analysis presented in this paper: "I think that if you had enough sticks to build an entire rectangle, they would fall naturally into miniature rectangles to fill it. If you had all the sticks . . . it would go four across, four across, four across, and then it would naturally form rectangles inside the rectangle." It seems apparent that this student had in mind that successive right-angle pairs of sticks juxtaposed to the edges of the rectangle or previously placed pairs would enclose rectangular regions. Further protocol analysis would be needed to determine the closeness of the match between this student's cognitive structures and those hypothesized in Figures 5.9 or 5.10.

Connection between mathematical concepts is a notion that gets considerable attention in current discussions about mathematics learning (e.g., *Curriculum and evaluation standards for school mathematics*, 1989). Clear specifications of what constitutes a connection appears less frequently. We argue that attention to units of quantity does point to such connections and suggests both the cognitive and mathematical connection. Using a problem involving addition of problem quantities given in "unlike" units (other than units of 1) we were able to show connections between such a problem structure and addition of multidigit numbers and also between such a problem structure and addition of rational numbers. In the section Multiplication of Rational Numbers, we showed how interpretation of the same problem in whole number units and in fractional number units provides a connection between a whole number problem structure and a rational number problem structure. Exemplification of these connections in instructional situations, we hypothesize, will enrich children's understanding of problem and computational procedures and facilitate the extension of children's knowledge about whole number situations to rational number situations.

## ACKNOWLEDGMENTS

The development of this chapter was in part supported with funds from the National Science Foundation under Grant No. DPE 84-70077 (The Rational Number Project). Any opinions, findings, or conclusions expressed are those of the authors and do not necessarily reflect the view of the National Science Foundation.

## NOTE

1. Although this notation *represents* knowledge structures, a more in-depth investigation of the mental operations and actions that constitute those knowledge structures is given by other researchers in this volume, e.g., Confrey, Steffe, and Thompson.

## REFERENCES

Behr, M. J., G. Harel, T. Post, and R. Lesh. 1992a. Rational number, ratio, and proportion. In *Handbook of research on mathematics teaching and learning*, ed. D. Grouws. New York: Macmillan.

———. 1992b. Rational numbers: Toward a Semantic Analysis— Emphasis on the operator construct. In *Rational numbers: An integration of research*, ed. T. Carpenter and E. Fennema. Hillsdale, NJ: Lawrence Erlbaum.

*Curriculum and evaluation standards for school mathematics.* 1989. Reston, VA. National Council of Teachers of Mathematics.

Davydov, V. V. 1982. The psychological characteristics of the formation of elementary mathematical operations in children. In *Addition and subtraction: A cognitive perspective*, ed. T. P. Carpenter, J. Moser, and T. A. Romberg. Hillsdale, NJ: Lawrence Erlbaum Associates.

Dienes, Z. 1960. *Building up mathematics*. London: Hutchinson Educational.

Freudenthal, H. 1983. *Didactical phenomenology of mathematical structures*, 133–209. Boston: D. Reidel.

Galperin, P. Y., and L. S. Georgiev. 1969. The formation of elementary mathematical notions. In *Soviet studies in the psychology of learning and teaching mathematics*, ed. J. Kilpatrick and I. Wirszup. Chicago: The University of Chicago.

Hiebert, J., and D. Wearne. 1992. Links between teaching and learning place value with understanding in first grade. *Journal for Research in Mathematics Education.* 23(2): 98–122.

Lamon, S. J. 1989. Ratio and proportion: Preinstructional cognitions. Unpublished doctoral dissertation, University of Wisconsin, Madison.

Piaget, J. 1985. *The equilibration of cognitive structures*, trans. T. Brown and K. J. Thampy. Chicago: University of Chicago Press, originally published in 1975.

Riley, M., J. G. Greeno, and J. I. Heller. 1983. Development of children's problem-solving ability in arithmetic. In *The develop-*

*ment of mathematical thinking*, ed. H. Ginsburg, 153–196. New York: Academic Press.

Simon, M., and G. Blume. 1991. Building and understanding multiplicative structures: A study of prospective elementary teachers. Unpublished manuscript, Pennsylvania State University, University Park.

Steffe, L. 1988. Children's construction of number sequences and multiplying schemes. In *Research agenda for mathematics education: Number concepts and operations in the middle grades*, ed. J. Heibert and M. Behr, 119–140. Reston, VA: The National Council of Teachers of Mathematics.

———, P. Cobb, and E. von Glasersfeld. 1988. *Construction of arithmetical meanings and strategies*. New York: Springer-Verlag.

Vergnaud, G. 1988. Multiplicative structures. In *Number concepts and operations in the middle grades*, ed. H. Hiebert and M. Behr, 141–161. Reston, VA: National Council of Teachers of Mathematics.

# III
# RATIO AND RATE

# 6  The Development of the Concept of Speed and Its Relationship to Concepts of Rate

## *Patrick W. Thompson*

Can we measure the speed of a car in miles per century? *No, because you would die or the car would rust away before a century.* Suppose we traveled from here to Peoria at 30 miles per hour and returned at 70 miles per hour. What would be our average speed for this trip? *50 miles per hour—because 100 divided by 2 is 50.* What does it mean when we say we averaged 45 miles per hour on a trip? *That we mostly went 45 miles per hour.*

## INTRODUCTION

Hidden in these questions and several fifth grader's responses are a myriad of issues regarding not only a concept of speed, but also measurement, proportionality, ratio, rate, fraction, and function. But, do we see primitive conceptions of speed and primitive conceptions of average in these children's responses, or do we simply see the effects of schooling? I propose that it is all three, but predominantly the last. Also, I will demonstrate that when we remove certain insidious practices of schooling, such as insisting that students be able to calculate any expression they encounter in solving a problem, and focus students' attention on their conceptions of situations, students can advance far beyond what we typically expect of them.

The setting for the interviews from which the preceding excerpts were taken is a project to investigate the development of students' quantitative reasoning and the relationship of quantitative reasoning to reasoning algebraically in applied situations. The premise behind this investigation is that when people reason mathematically about situations, they are reasoning about *things* and *relationships*. The

179

"things" reasoned about are not objects of direct experience and they are not abstract mathematical entities. They are objects derived from experience—objects that have been constituted conceptually to have qualities that we call *mathematical*.

Abstract mathematical objects are the constructions of relatively few people who have reflectively abstracted qualities and relationships constructed through experience into formal systems of relationships and operations. If the person doing the reasoning about a situation is also a person who has reflectively abstracted systems of relationships and operations having to do with conceiving the situation, reconstituting them as *mathematical* objects (Harel and Kaput, 1991), then, from that person's point of view, the situation is composed of mathematical objects. This project aims to capture the multiple reconstitutions that take place in individuals as they progress toward the construction of mathematical objects such as ratio and rate. Vergnaud (1983, 1988, this volume) and Greer (1988, this volume) have given valuable explications of broad characteristics of situations that lend themselves to being conceptualized in terms of ratios and rates. It is also valuable to have an idea of how individual people come to comprehend such situations and how such comprehensions become possible.

One crucial part of sound mathematical development is students' construction of powerful and generative concepts of rate. What that means is a source of controversy. I will discuss some controversies surrounding the concept of rate in the next section as part of my explication of quantitative reasoning. In the third section of this chapter, I will discuss a teaching experiment done to investigate individual children's construction of the concept of speed as a rate and investigate the components of a conception of average rate.

## Quantitative Reasoning

To characterize quantitative reasoning I must first make clear what I mean by *quantity* and a *quantitative operation*. I will draw heavily on Piaget's constructs of internalization, interiorization, mental operation, and scheme.

The notion of goal-oriented action (praxis) is foundational to understanding Piaget's notions of internalization, interiorization, mental operation, and scheme. It may prove helpful if I point out that, in Piaget's usage, an action is not the same as an observable behavior. An action is an activity of the mind that might be expressed in behavior, but it need not be expressed in anything detectable by an observer (Powers, 1973, 1978). It is important to keep this distinction in mind when giving meaning to terms like *action*, *image*, and *operation*.

*Constructs from Piaget's Theory.* Actions are tied more or less to experience. At the lowest level are reflexes, at the highest level is intelligent thought. Internalization is the process of reconstructing actions to enable mental imagery of situations involving them (Lewin, 1991). Piaget (1967, pp. 294–296) distinguished among three types of images:

1. An "internalized act of imitation . . . the motor response required to bring action to bear on an object . . . a *schema* of action."

2. "In place of merely representing the object itself, independently of its transformations, this image expresses a phase or an outcome of the action performed on the object. . . . [but] the image cannot keep pace with the actions because, unlike operations, such actions are not coordinated one with the other."

3. "[An image] that is dynamic and mobile in character . . . entirely concerned with the transformations of the object. . . . [The image] is no longer a necessary aid to thought, for the actions which it represents are henceforth independent of their physical realization and consist only of transformations grouped in free, transitive and reversible combination."

*Internalization* refers to an assimilation, an initial "image having" (Kieren and Pirie, 1991). *Interiorization* refers to the progressive reconstruction and organization of actions to enable them to be carried out in thought, as mental operations. Later in this chapter I will say that a child "internalized" the text of a situation. By that I will mean that he or she read the text and constructed an image— probably unarticulated—of the situation and its elements, that contains actions of moving, filling, comparing, and so on.

As intimated in Piaget's quotations concerning images, a mental operation is a system of coordinated actions that can be implemented symbolically, independent of images in which the operation's actions originated. Mental operations are always implemented in an image, but the image need not be one tied historically to the origins of the operation (von Glasersfeld, 1991). Mathematicians speak frequently of such things as, say, partitions with the understanding that the actions taken to make one have definite requirements, characteristics, and products, yet if they conjure an image it has no special status relative to the generality of the operation of partitioning.

What makes an operation repeatable? This was the problem Piaget addressed with his notion of scheme. His characterization of a scheme, "whatever is repeatable and generalizable in an action" (Piaget, 1971a, p. 42), hardly seems helpful. Cobb and von Glasersfeld (1983) provide a useful elaboration. A scheme is an organization of actions that has three characteristics: an internal state that is necessary for the activation of actions composing it, the actions themselves, and an imagistic anticipation of the result of acting. The imagistic anticipation need not be iconic. It could just as well be symbolic, kinesthetic, or any other re-presentation of experience (von Glasersfeld, 1980)—although the specific characters of its actions and anticipations may affect the generality with which a scheme is applied (Steffe, this volume).

Piaget stated his view on the source of schemes quite clearly: Schemes emerge from assimilations of experience to ways of knowing.

> Assimilation thus understood is a very general function presenting itself in three nondissociable forms: (1) functional or reproductive assimilation, consisting of repeating an action and of consolidating it by this repetition; (2) recognitive assimilation, consisting of discriminating the assimilable objects in a given scheme; and (3) generalizing assimilation, consisting of extending the field of this scheme. . . . It is therefore assimilation which is the source of schemes . . . assimilation is the operation of integration of which the scheme is the result. Moreover it is worth stating that in any action the driving force or energy is naturally

of an affective nature (need and satisfaction) whereas the structure is of a cognitive nature. To assimilate an object to a scheme is therefore simultaneously to tend to satisfy a need and to confer on the action a cognitive structure. (Piaget, 1977, pp. 70–71)

The third form of assimilation, "generalizing assimilation," provides one way of understanding the important notion of transfer, and it provides a foundation for understanding how people meet occasions requiring reflection or distinction. We recognize situations by the fact we have assimilated them to a scheme. When features of that situation emerge in our understanding that do not fit what we would normally predict, we introduce a distinction, and the original scheme is accommodated by differentiating between conditions and subsequent implications of assimilation.[1] One way to understand the idea of generalizing assimilation is to consider that all situations as constituted in someone's comprehension are the products of actions and operations of thought, and that when there is a large commonality between the operations of thought activated on different occasions, the person constituting the situations will experience a feeling that the later comprehension is somehow similar to the earlier one.[2] Later on in this chapter I will discuss an attempt to orient a student so that she would construct a scheme for speed that would be powerful enough that she would recognize (what we take as) more general rate situations as being largely the same as situations involving speed. My attempts were to get this child to engage in generalizing assimilations.

Reflective abstraction is perhaps the most subtle, important, and least understood of Piaget's constructs (Steffe, 1991a; von Glasersfeld, 1991). It is the motor of interiorization—the process whereby actions become organized, coordinated, and symbolized (Bickhard, 1991; Thompson, 1985a, 1991). As von Humbolt put it, "In order to reflect, the mind must stand still for a moment in its progressive activity, must grasp as a unit what was just presented, and thus posit it as object against itself" (1907, quoted in von Glasersfeld, 1991).

Reflective abstraction has two aspects in Piaget's theory. The first is the reconstruction of actions so as their activation is progressively less dependent on immediate experience. The second is an assimilation to higher levels of thought—to schemes of operations. The first is closely allied with what we normally think of as learning, whereas the second is closely allied with what we normally think of as comprehending (Piaget, 1980).

*Quantity.* Quantities are conceptual entities. They exist in people's conceptions of situations. A person is thinking of a quantity when he or she conceives a quality of an object in such a way that this conception entails the quality's measurability. A quantity is schematic: It is composed of an object, a quality of the object, an appropriate unit or dimension,[3] and a process by which to assign a numerical value to the quality. Variations in people's conceptions of a quantity occur in correspondence with variations in level of development of components within their schemes. Also, I want to make clear that objects are constructions a person takes as given and that qualities of an object are imbued by the subject conceiving it—as Piaget pointed out repeatedly. To a young child watching a passing car, the car probably is an object, and it may have the quality *motion.* However, for this young child the car probably does not have the quality *distance moved in an amount of time.*

My characterization of quantity differs somewhat from others. Schwartz (1988), Shalin (1987), and Nesher (1988) characterize quantities as ordered pairs of the form (*number, unit*). To characterize quantities as ordered pairs may be useful formally, but it provides no insight into what people understand when they reason quantitatively about situations, and it severely confounds notions of number and quantity (Thompson, 1989, in press). Steffe (1991b), in extending the work of Piaget (1965), characterizes quantity as the outcome of unitizing or segmenting operations. I take the operations of unitizing and segmenting as foundational to a person's creation of quantities, but I have found it productive to use *quantity* more broadly than Steffe does. By characterizing the idea of quantity schematically we are able to capture important structural characteristics of people's reasoning when they reason about complex situations that involve a myriad of multiply related quantities (Thompson, 1989, in press).

*Quantification.* Quantification is a process by which one assigns numerical values to qualities. That is, quantification is a process of direct or indirect measurement. One does not need to actually carry out a quantification process for a quantity to conceive it. Rather, the only prerequisite for a conception of a quantity is to have a process in mind. Of course, one's grasp of a process may change as one reconceives the quality and the process in relation to one another.

Piaget (1965) made a valuable distinction between two dramatically different kinds of quantification: gross quantification and extensive quantification. *Gross quantification* refers to a conception of a quality in ways that objects having it can be ordered by some experiential criteria (e.g., "appears bigger than"). *Extensive quantification* refers to a conception of a quality as being composed of numerical elements that arise by operations of unitizing or segmenting (see also Steffe, 1991b; Steffe, von Glasersfeld, Richards, & Cobb, 1983). Both kinds of quantification are essential for conceiving situations quantitatively; persons limited to gross quantification are blocked from conceiving situations mathematically.

### Quantitative Operation

A quantitative operation is a *mental operation* by which one conceives a new quantity in relation to one or more already-conceived quantities. Examples of quantitative operations are combine two quantities additively, compare two quantities additively, combine two quantities multiplicatively, compare two quantities multiplicatively, instantiate a rate, generalize a ratio, combine two rates additively, and compose two rates or two ratios. It is important to distinguish between constituting a quantity by way of a quantitative operation and evaluating the constituted quantity. One can conceive of the difference between one's height and a friend's height without giving the slightest consideration to evaluating it.[4]

A quantitative operation creates a quantity in relation to the quantities operated upon to make it. For instance, comparing two quantities additively creates a difference; comparing two quantities multiplicatively creates a ratio. A quantitative operation creates a structure—the created quantity in relation to the quantities operated upon to make it.

Quantitative operations originate in actions: The quantitative operation of combining two quantities additively originates in the actions of putting together to make a whole and separating a whole to make parts; the quantitative operation of comparing two quantities additively originates in the action of matching two quantities with the goal of determining excess or deficit; the quantitative operation of comparing two quantities multiplicatively originates in matching and subdividing with the goal of sharing. As one interiorizes actions, making mental operations, these operations in the making imbue one with the ability to comprehend situations representationally and enable one to draw inferences about numerical relationships that are not present in the situation itself.

Not every action is interiorized as a mental operation. The problem-classification scheme developed by Carpenter and Moser (1983), for instance, can be thought of as corresponding to classes of action schemata that become internalized as mental images of action and then interiorized as the mental operations of additive combination and additive comparison.

*Quantitative Operations vs. Numerical Operations.* To understand the notion of quantitative reasoning, it is crucial to understand the distinction between quantitative operations that create quantities and numerical operations used to evaluate quantities. The following problem was used in a teaching experiment on complexity and additive structures with six fifth graders (Thompson, in press). "Team 1 played a basketball game against Opponent 1. Team 2 played a basketball game against Opponent 2. The captains of Team 1 and Team 2 argued about which team won by more. The captain of Team 2 won the argument by 8 points. Team 1 scored 79 points. Opponent 1 scored 48 points. Team 2 scored 73 points. How many points did Opponent 2 score?"

Five children, after varying degrees of effort and intervention, subtracted 48 from 79 (getting 31), then added 8 to 31 (getting 39). None of the five children could say what "39" stood for (i.e., the difference between Team 2's score and Opponent 2's score). It was evident in the interviews that these children understood that the situation was about comparing two differences, but it also seemed evident that to them, having *added* to get 39, 39 could not be the value of a difference. These children had not distinguished among the numerical operation actually used to evaluate a quantity (in this case addition), the quantitative operation used to create the quantity (in this case additive comparison), the kind of quantity being evaluated (in this case a difference), and the operation used to evaluate the quantity under stereotypical conditions (in this case subtraction). They had over-identified the quantitative operation of additive comparison with the numerical operation of subtraction.

The distinction between quantitative and numerical operations is largely tacit in mathematics education, and the two are often confounded, as in the following quote. "A partitive model for 12 ÷ 3 would be 12 miles divided by 3 hours equals 4 miles per hour. A quotitive interpretation would be 12 miles divided by 3 miles per hour equals 4 hours. Finally, as the inverse of cross product multiplication, a model could be 12 outfits divided by 3 skirts equals 4 blouses" (Hiebert and Behr, 1988, p. 5). Here is a confusion: division is a numerical operation, it is not a quantitative operation. We do not "divide" one object by another. If 12 ÷ 3 interpreted as quotitive division means "how many composite units of 3 are contained in the composite unit of 12" (Hiebert and Behr, 1988, p. 5), then "12 miles divided by 4 miles per hour" means we are asking how many units of 4 miles/hour are contained in 12 miles? A decision to divide need not be based always on considerations of partition or quotition. It could also be made relationally—as when one conceives of a situation multiplicatively and the information being sought pertains to an initial condition, such as "How long must one travel at 4 miles/hour to go 12 miles?" or "How many blouses does Sally have if she has three skirts and twelve skirt-blouse pairs?"

We may, in the final analysis, shortchange our students by holding numerical operations on high, proposing models for them so that they become meaningful. I have difficulty seeing how the "models" approach differs from some ways of teaching formal, axiomatic systems: Instantiate the axioms, thereby creating a model of the formal system. One proposes a model to make the formal system more tangible, but the ultimate aim is to teach a formal system that has no special relationship with the proposed model. In some ways this turns the idea of learning on its head.

A quantitative operation is nonnumerical; it has to do

with the *comprehension* of a situation. Numerical operations are used to evaluate a quantity. It is understandable, however, that the two are easily confounded. First, in simple situations being conceived by an adult who has both, the two are so highly related that they appear indistinguishable. It is in more complex situations that distinctions must be made, and often people make them so rapidly that they make them unconsciously. Second, we do not have a conventional notation for quantitative operations independent of the arithmetic operations of evaluation. Therefore, arithmetic notation has come to serve a double function. It serves as a formulaic notation for prescribing evaluation, and it reminds the person using it of the conceptual operations that led to his or her inferences of appropriate arithmetic. The double use of arithmetic operations provides power and efficiency for persons who can make these subtle distinctions while using the notation. It is a source of confusion to many students and teachers who have not constructed distinctions between quantitative relationship and numerical operations. In Thompson (1989) I describe an extension of a notational system for quantitative operations and quantitative relationships originally devised by Shalin (Greeno, 1987; Shalin, 1987) that can be used to represent nonnumerical comprehensions of situations.

*Comprehension of Quantitative Situations.* A person comprehends a situation quantitatively by conceiving of it in terms of quantities and quantitative operations. Each quantitative operation creates a relationship: The quantities operated upon with the quantitative operation in relation to the result of operating. For example, the quantities "girls in this class" and "boys in this class" being compared multiplicatively produces the quantity "ratio of girls to boys." Those three quantities in relation to one another constitute a quantitative relationship. Comprehensions of complex situations are built by constructing *networks* of quantitative relationships.

The process of constructing a comprehension of a specific situation is a dialectic among reflectively abstracting features of the situation to schemes of operations, expressing mental operations in action schemata, and reflecting results of activated schemata back to schemes of operations. Variations among people's comprehensions of a specific situation correspond to variations in the operational constitution of these schemes and to the classes of actions represented by their mental operations.

## Ratio vs. Rate

The fact that we have two terms *ratio* and *rate* would suggest that we have two ideas different enough to warrant different names. Yet, there is no conventional distinction between the two, and there is widespread confusion about such distinctions. The confusion is not limited to school classrooms; confusion is evident even in the mathematics education research literature. Ohlsson (1988) wrote the following as an analysis of ratio:

> If I get 8 miles per gallon out of my car during the first leg of a journey but for some reason get only 4 miles per gallon during the second leg, then the correct description of my car's performance over the entire trip is *not* (4 + 8) = 12 miles per gallon. Similarly, if one classroom has a ratio of 2 girls per 3 boys and another classroom has 4 girls per 3 boys, the combined class does not have $^2/_3 + {}^4/_3 = {}^6/_3$, or 6 girls per 3 boys. . . . The correct analysis of the examples is, I believe, the following. The fuel consumption during the first leg of the journey in the first example was 8 miles per gallon. We represent this with the vector (8,1). The fuel consumption during the second leg of the journey was (4,1). Adding these vectors . . . gives (12,2), which is equivalent to (i.e., has the same slope as) the vector (6,1). Vector theory predicts that the fuel consumption during the entire trip was 6 miles per gallon, which is correct. (Ohlsson, 1988, p. 81; emphasis in original)

Claims of correctness notwithstanding, Ohlsson's analysis of his car's mileage is correct only if he traveled $8x$ miles in the first leg of his journey and $4x$ miles in the second leg. Otherwise, it is incorrect. He evidently took a *rate* of consumption (8 miles per gallon) for actual consumption (went $8x$ miles and used $x$ gallons). Without information about the relative number of miles traveled or relative number of gallons used in each leg of his journey, we can say nothing about his car's *rate* of fuel consumption.

Perhaps the lack of conventional distinction between ratio and rate is the reason that the two terms are used often without definition. Lesh, Post, and Behr noted that "there is disagreement about the essential characteristics that distinguish, for example rates from ratios. . . . In fact, it is common to find a given author changing terminology from one publication to another" (1988, p. 108). The most frequent distinctions given between ratio and rate are

1. A ratio is a comparison between quantities of like nature (e.g., pounds vs. pounds) and a rate is a comparison of quantities of unlike nature (e.g., distance vs. time; Vergnaud, 1983, 1988).

2. A ratio is a numerical expression of how much there is of one quantity in relation to another quantity; a rate is a ratio between a quantity and a period of time (Ohlsson, 1988).

3. A ratio is a binary relation that involves ordered pairs of quantities. A rate is an intensive quantity—a relationship between one quantity and one unit of another quantity (Kaput, Luke, Poholsky, and Sayer, 1986; Lesh et al., 1988; Schwartz, 1988).

Although there is an evident controversy about distinctions between ratio and rate, each of these distinctions seems to have at least some validity. My explanation for this controversy is that these distinctions have been based largely upon situations per se instead of the mental operations by which people constitute situations. When we shift our focus to the operations by which people constitute *rate* and *ratio* situations, it becomes clear that situations are neither one nor the other. Instead, how one might classify a situation depends upon the operations by which one comprehends it. In Thompson (1989) I illustrate how an "objective" situation can be conceived in fundamentally different ways depending on the quantitative operations available to and used by the person conceiving it. When we take the perspective that ratios and rates are the products of mental operations, classification schemes for separating situations into "rate" and "ratio" categories are no longer of great importance.

*A Ratio Is the Result of Comparing Two Quantities Multiplicatively.* This definition is in accord with most given in

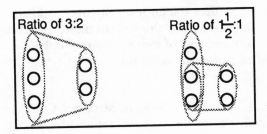

Fig. 6.1

the literature. One slight difference is that it does not specify how the result of a multiplicative comparison is denoted or expressed. For example, a collection of three objects can be compared multiplicatively against a collection of two objects in either of two ways: a comparison of the two collections per se, or a comparison of one as measured by the other (Figure 6.1). The first comparison is of the two collections as wholes. The second comparison is of one quantity measured in units of the other. Both are expressions of a multiplicative comparison of the two quantities. The second is propitious for concepts of fraction and may be more sophisticated than the first. It needs to be noted about the second comparison that even though the result is expressed in the same way as what is often called a *unit rate*, the comparison described is between two specific, nonvarying quantities, and hence is a ratio comparison.

One conception of ratio that is propitious for constructing rates is the conception of a ratio with a fixed value being the result of comparing two quantities having indeterminate values. I use the term *ratio* even in this last case, since there are two specific quantities (e.g., a car's distance and that same car's time of travel, thought of as fixed but unknown) present in thought when the person conceives the comparison.

When we focus on the mental operation of multiplicative comparison, it is evident that it makes no difference if the quantities are of the same dimension or not. What matters is that the two quantities are being compared multiplicatively. If the quantities being compared are measured in the same unit, then the comparison happens to be a direct comparison of qualities. If the quantities are measured in different units, then segmentations (measures) of their qualities are being compared. In either case, the salient mental operation is multiplicative comparison of two specific quantities, and the result of the comparison is a ratio.[5]

*A Rate Is a Reflectively Abstracted Constant Ratio.* A rate is a reflectively abstracted constant ratio, in the same sense that an integer is a reflectively abstracted constant numerical difference (Thompson, 1985a, 1985b; Thompson and Dreyfus, 1988). A specific numerical difference, as a mental structure, involves a minuend, a subtrahend, and the result of subtracting. An integer, as a reflectively abstracted numerical difference, symbolizes that structure as a whole, but gives prominence to the constancy of the result—leaving minuend and subtrahend variable under the constraint that they differ by a given amount.

Similarly, a specific ratio in relation to the quantities compared to make it is a mental structure. A rate, as a reflectively abstracted constant ratio, symbolizes that structure as a whole, but gives prominence to the constancy of the result of the multiplicative comparison.[6]

Once a specific situation is conceived in a way that involves a rate, it is implicit in the way the concept of rate is constructed that the values of the compared quantities vary in constant ratio. Hence, a specific conceived rate is (from my point of view) a linear function that can be instantiated with the value of an appropriately conceived structure. To say that an object travels at 50 miles/hour quantifies the object's motion, but it says nothing about a distance traveled nor about a duration traveled at that speed (Schwartz, 1988). However, conceiving speed of travel in relation to an amount of time traveled produces a specific value for the distance traveled.[7]

When one conceives of two quantities in multiplicative comparison and conceives of the compared quantities as being compared in their *independent, static* states, one has made a ratio. As soon as one reconceives the situation as being that the ratio applies generally outside of the phenomenal bounds in which it was originally conceived, then one has generalized that ratio to a rate (i.e., reflected it to the level of mental operations). The wording "as soon as one reconceives" is important. It is possible, perhaps likely, that people first conceive a multiplicative comparison in terms of a ratio and reconceive that ratio as a rate. One occasion for "reconceiving" a ratio as a rate in schools is when students are asked to "assume (something) continues at the same rate" when the initial situation is described in a way that there are two quantities to be compared multiplicatively.[8]

In responding to the question of how people come to conceive of rates I rely on Piaget's notion of reflective abstraction. The first sense of reflected abstraction is that a class of actions are reconstructed and symbolized at the level of mental operations. The second sense of reflected abstraction is that as a situational conception is constructed, the figurative aspects of the conception are "reflected" to the level of mental operations. In the first sense, we would say a person has learned (e.g., learned speed as a rate). In the second sense we would say a person has comprehended (e.g., conceived an object's motion as a rate).

One activity that provides occasions for students to think in ways propitious for constructing rate as a reflection of constant ratio is the use of "building-up" strategies in the solution of proportional reasoning tasks (Hart, 1978; Kaput & West, this volume). If a child is trying to find, say, how many apples there are in a basket where the ratio of pears to apples is 3 : 4 and there are twenty-four pears, and the child thinks "three pears to four apples, six pears to eight apples, . . . , twenty-four pears to thirty-two apples" (a succession of equal ratios), then this provides an occasion for the child to abstract the relationship "three apples for every four pears" (an iterable ratio relating collections of apples and pears as the amounts of either might vary), and eventually "there will be ¾ of an apple or part thereof for every pear or proportional part thereof" (an accumulation of apples and pears that carries the image that the values of both can vary, but only in constant ratio to the other). The former conception—accumulations made by iterating a ratio—I call an *internalized* ratio, whereas the latter conception—total accumulations in constant ratio—I call an *interiorized* ratio, or a *rate*.[9]

## Motion vs. Speed

Piaget (1970), in investigating the development of children's concepts of movement and speed, made a clear distinction between motion as a phenomenon—the experience or observation of movement—and speed as a quantity. Objects move; speed is a quantification of motion. Thus, the quantity normally identified as speed is more accurately thought of as motion together with a quantification of it.

Piaget explained the development of children's concepts of speed in terms of the emergence of the general mental operations entailed in proportional reasoning, and he used the language of centration (focus of attention), decentration (coordinations of centrations), and regulations (construction of mental operations that balance conflicting results of centrations, such as a conflict between noticing that one object moved farther than another, which suggests it went faster, and noticing that it took more time than the other, which suggests it went slower). He concluded that the concept of speed is constructed as a proportional correspondence between distance moved and time of movement—"the elaboration first of concrete and later of formal metrical operations" (p. 259).

Piaget described the emergence of a concept of speed (quantified motion) as a process wherein children first reconceive motion as entailing changes in position over changes in time, then coordinate the two dimensions of distance and time as changing in proportion to one another (Piaget, 1970, pp. 279–280). He intimated a distinction between ratio and rate, but he did not elaborate on it (p. 280).

The emergence of conceptions of speed investigated by Piaget will inform our discussions of ratio and rate, but only to a point. Conceptions of ratio and rate were not explicit issues in Piaget's study, so we must not lean too heavily on his analyses. Moreover, in his investigation of speed, as in most investigations he directed, Piaget was not interested in the actual operations used by individual children to solve specific problems, which is of considerable interest to researchers today. Rather, he was interested in the broad characteristics of mental operations held by the "epistemic subject"—a knower in general. He had no great interest in the "psychological subject"—an individual knower (Piaget, 1971a). Today we realize the importance of understanding individual children in order to understand children in general. Also, we know now that we can see wider variations in operativity of a concept when it is required for comprehension of a situation that is more complex than situations dominated by the concept itself. That is, Piaget's investigation may not have captured concepts of speed richly enough to suggest how students must understand speed so that they may understand, for example, the concept of average speed.

## A TEACHING EXPERIMENT

As a preliminary step to investigating students' construction of the quantities speed, average speed, and the construction of rates in general, I simply spoke with children and adults about what they meant when, in specific situations, they concluded that something had a particular average speed. The explanations of average speed given by adults and school students who could solve the problems posed were quite consistent: An average speed is the constant speed at which a different trip would be taken so that it would entail traveling the same distance in the same amount of time as the original trip. The explanations centered around the same person repeating the trip at a constant speed or a different person following the same path as the original. In either case, a second trip always was involved.

I wrote a computer microworld (described later) that captured the sense of two trips happening sequentially or simultaneously and presented components of speed: distance traveled, time spent traveling, and rate of travel. This microworld then served as a metaphor within which instruction on speed took place. The microworld as metaphor was later used canonically in that instruction on rates per se always referred back to the metaphor established by the microworld.

The logic of a teaching experiment is to use instruction as the primary site for probing students' comprehensions and gaining insights into their constructions (Cobb & Steffe, 1983; Hunting, 1983; Steffe, 1991c; Thompson, 1982). Instruction at the outset of a teaching experiment is highly bounded by conceptual analyses done prior to instruction. However, instruction is modified continually according to limitations in students' thinking as evidenced in their struggles and according to avenues that arise which promise greater insight into powerful ways of thinking. The hallmark of a teaching experiment is that it is opportunistic; one must continually rely on serendipity.

### Subject

One 10-year-old fifth grader, JJ, took part in the teaching experiment. JJ was a quiet, reflective student; she was not especially strong mathematically according to conventional indicators—Iowa Test of Basic Skills percentile scores for JJ were Concepts, 84; Problem Solving, 87; and Computation, 83. JJ participated in two earlier teaching experiments, one on complexity and additive structures (Thompson, in press), the other on area and volume.

### Computer Microworld

The computer microworld (called Over and Back) presents two animals, a turtle and a rabbit, who run along a number line (Figure 6.2). Both can be assigned speeds at which to run. The turtle's speed can be assigned two values: one for the turtle's speed while running "over," the other for the speed at which it will run "back." The rabbit's speed can be assigned only one value, which applies to both its run over and its run back. Each animal can be made to run separately from the other or they can be made to run simultaneously (as in a race). A timer shows elapsed time as either of the animals runs. One can press the Pause button to interrupt a race; when "paused," the distances traveled by either or both animals is displayed on the screen (Figure 6.3). Any assigned value can be changed during a pause; the animals will renew their race with speeds having the reassigned values.

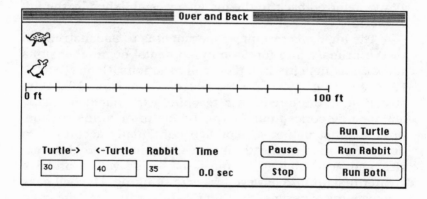

Fig. 6.2

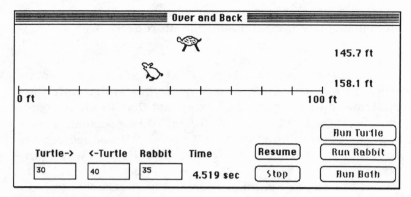

Fig. 6.3

### Overview of Instruction

Instruction occurred over eight sessions, each lasting approximately 55 minutes, between April 25 and May 14, 1990. The instructional format was that of a prolonged clinical interview. My remarks were given usually as questions, asking for clarification, or given to orient JJ's attention. Direct instruction was given only to establish conventions, to explain how the program worked, and to present directions for the performance of tasks.

Tasks for a session were based on JJ's progress during that session and in previous sessions. Sometimes tasks were modified during a session when opportunities arose for generalization or clarification. JJ had a computer at home on which to run Over and Back, so homework often involved her use of the program.

### Development of Concepts of Speed, Average Speed, and Rate

My discussion of the teaching experiment will be given in three parts, which correspond to a division of the teaching experiment's activities from my perspective. I saw the teaching experiment pass through three phases. Phase I focused on probing and extending JJ's concept of speed. Phase II focused on extending JJ's concept of speed to include what we normally take as an understanding of average speed. Phase III focused on giving JJ the opportunity to extend her

concept of speed to a more general concept of rate by generalizing assimilations and subsequent reflection.


*Concept of Speed.* I began the first session by demonstrating Over and Back to JJ and trying to determine her
understanding of speed. She explained that 40 ft/sec means
that "every second he runs 40 feet." JJ's explanation suggested a good comprehension of speed; it soon became evident that I had given it greater significance than was appropriate.

My intention was to get as quickly as possible to problems involving average speed. With this in mind, I asked JJ to
determine in advance how much time it would take for the
rabbit to go over and back at speeds of 50 ft/sec and 40 ft/sec,
and how long it would take the turtle to go over and back
when it would go over at 20 ft/sec and back at 40 ft/sec. She
answered each question successfully. I then asked JJ to give
the rabbit a speed so that it and the turtle would tie, where
the turtle went over at 20 ft/sec and back at 40 ft/sec. JJ said,
"30 ft/sec for the rabbit. [Why?] Because its 20 and 40 and in
between is 30." It was at this moment that the teaching experiment began in earnest.

JJ's method for determining an amount of time needed
to go a given distance at a given speed emanated from a primary notion of speed as one speed-length in one unit of time
(e.g., one length of 40 feet in 1 second). She conceived the
total distance as being measured in units of speed-length. JJ
could even reason proportionally in making a determination
of time: the rabbit would go 100 feet at 30 ft/sec in 3 1/3
seconds, "because there are three 30s in 100, and 10 is 1/3 of
30, so 3 1/3 seconds" (Session 1). However, in JJ's conception, an amount of time was *made* by traveling, and its determination was *produced* by comparing a total distance with a
speed-length. In other words, for JJ it was the case that *speed
was a distance* (how far in one second) and *time was a ratio*
(how many speed-lengths in some distance).

JJ's early conception of speed and time constrained the
ways in which she could constitute speed, distance, and time
in relation to one another. On four occasions in Session 1,
when asked to give the rabbit a speed that would make it and
the turtle tie, JJ resorted to a guess-and-check strategy of

"pick a speed, figure a time, adjust up or down" even when she knew that the rabbit and turtle needed to use the same amount of time, and she knew the amount of time in which the rabbit needed to complete its trip. JJ's actions were strongly suggestive of trying to produce a desired measure of a given length by adjusting and readjusting the length of the unit according to by how much she missed the desired measure.

The notion that speed is a distance was predominant in JJ's thinking. For example, she claimed that the turtle would always be 10 feet behind the rabbit when the turtle moved at 20 ft/sec and the rabbit moved at 30 ft/sec. A change in her conception began to appear after she played with the Pause button (see Figure 6.3) while running the turtle and rabbit. She eventually reconciled her prediction and her observation that the turtle was rarely 10 feet behind the rabbit by reconstituting the comparison as "the distance [between them] will increase by 10 [each second] as they are going." JJ's conception of *distance* was changing, but not her conception of speed. Distance could accumulate, but the accumulation was still measured in units of speed-length.

JJ's homework after Session 1 had two parts. In the first part she was to formulate predictions for how much time the turtle or rabbit would take to go over and back at various speeds, and then test her predictions using Over and Back. The second part was to formulate predictions of who would win when the two raced with various speeds, again testing her predictions using Over and Back. JJ formulated her time predictions with ease, using her measuring technique. On one task, which asked JJ to formulate a time prediction for when the turtle was to go over at 20 ft/sec and back at 50 ft/sec, JJ understood the task as "make them tie." She said that she tried to "figure it out, but got confused" and resorted to a guess-and-test method. I took this as our starting point for Session 2.

In Session 2 I attempted to orient JJ's reasoning toward rethinking her conception of time. She had previously conceived time of travel as being produced by moving in increments of speed-length, so I designed situations that, from my perspective, she might assimilate by that scheme, but in which she needed to work from the basis of a given time. The situations all presented a fixed distance (200 feet total) and times that were integral divisors of 200. JJ seemed to assimilate them as I predicted, looking for rate-lengths that when multiplied by the given number of seconds would produce 200 feet, or when multiplied by half the given time would produce 100 feet. My intention was that JJ eventually form an image of two segments, one for distance and one for time, where the segments were partitioned proportionally according to units of time.

Late in Session 2 JJ began to envision a segmented, total amount of time in relation to a to-be-traversed distance, and to envision that a segmentation of time imposed a segmentation on distance (Excerpt 1, paragraphs 1–5). That is, JJ began to conceive of distance and time as measured attributes of completed motion, but she was still more aware of distance than of time.

Excerpt 1 (Session 2)
1. *PT:* Okay, how about if we give a little tricky one. Make it [rabbit] go over and back in 6 seconds. (JJ sits up and looks at screen. Pause.) What are you saying to yourself?
2. *JJ:* Well, if it were 6 seconds there and back, it would have to go 3 seconds there and 3 seconds back. And, well, I was trying to figure out . . . (pause).
3. *PT:* Excuse me just a second. (Gets up and shuts the door.) It's too noisy out there. Okay. You were saying, 3 seconds over and 3 seconds back?
4. *JJ:* Yeah, and, well, you could, if you could figure out how many times 3 went into 100, that would, well . . .
5. *PT:* Okay, if we figured out how many times 3 would go into 100?
6. *JJ:* Well . . . (pause).
7. *PT:* . . . You know what you want to do?
8. *JJ:* Yeah.
9. *PT:* So what is it we need to do to see how many times 3 goes into 100.
10. *JJ:* Divide 100 by 3. (*Long* pause.)
11. *PT:* What is it that you've got in your hand?
12. *JJ:* (Has calculator in hand. Looks at it and laughs. Calculates $100 \div 3$.) 33.3.

JJ's comment ("How many times 3 goes into 100," Excerpt 1, paragraph 4) was ambiguous. It is not evident whether she was thinking of "how many 3s in 100" or of cutting 100 feet into 3 pieces. This was clarified in Session 3 (she meant the latter). Also, JJ's application of division in the previous excerpt was not generalizable by her to nonintegral values of time. Excerpt 2 presents the discussion following immediately after that in Excerpt 1.

> Excerpt 2 (Session 2)
>
> 1. *PT:* (PT and JJ have been discussing what all the 3s after a decimal place might stand for and whether JJ is confident of her answer. She is, "mostly.") Mostly. Well let's give it a shot. (PT enters 33.333 in the "Rabbit's Speed" box.) 6 seconds on the nose. You're right. How about making it go over and back in 7 seconds? (Pause.) What do you need to do?
> 2. *JJ:* Well, it has to go across one way in 3 and one-half seconds. And if it went 30 feet per second it would go 3 and one-third of a second. So, it would have to be more than 30.
> 3. *PT:* Well, last time you didn't do that. You didn't . . . estimate "has to be more than 30 or less than 50," or something like that. Last time you said you needed to know how many times 3 goes into 100. Why is this different, or is it different?
> 4. *JJ:* It . . . well . . . (*pause*). Um . . .
> 5. *PT:* Is this different from the last time?
> 6. *JJ:* Not really. It's just that I'm not, I'm not dividing it. But I'm sort of guessing it and then checking it.
> 7. *PT:* Um, would it make sense to ask the same kind of question you asked last time? (*Pause.*) It's going to go from the beginning to the end in 3 and one-half seconds, right?
> 8. *JJ:* Uh-huh.
> 9. *PT:* And so what question would you ask if you were to ask the same kind of question as you did last time?
> 10. *JJ:* How many times does 3 and one-half go into 100, or how many times does 7 go into 200? (JJ has a quizzical look on her face. Bell rings, ending Session 2.)

JJ had formed a strong association between the operation of cutting up a segment and the numerical operation of division, which would explain her remark that the situations involving 3 seconds and 3 1/2 seconds were not really different, but in the latter case she was not dividing (Excerpt 2, paragraphs 5–6). I suspected then, and still do, that JJ's discomfort in the latter case about using division to evaluate speed stemmed from this association. My hunch was that her discomfort would cease if she came to understand the operation of "cutting up" as one of making a proportional correspondence between two quantities. With this image of "cutting up," division is no less appropriate when cutting up segments into 3 1/2 parts than when cutting segments into 3 parts.

JJ's homework for that day was to determine speeds to give the rabbit so that it would make a complete trip in given amounts of time (5, 10, 8, 6, 7, 6.5, 7.5, and 8.3 seconds). During the next session JJ said that, in doing her homework, she figured some out with her calculator and did some by "guess and check," using Over and Back. She used "guess and check" for 6.5 seconds and 7.5 seconds, but actually calculated a speed for 8.3 seconds. Excerpt 3 presents JJ's justification for over and back in 6 seconds. In it I raised the question of what she was finding when she divided 100 by 3.

> Excerpt 3 (Session 3)
> 1. *PT*: In 8 seconds, 25; 6 seconds, 33.3. Can you tell me how you got 33.3? You don't have to tell me blow-by-blow details.
> 2. *JJ*: Well, I . . . (pause)
> 3. *PT*: What was your method?
> 4. *JJ*: Well, I, I, um, I, well, half of 6 is 3. And so I, um, took 3 into 100.
> 5. *PT*: Okay, and what were you finding out when you divided the 100 by 3.
> 6. *JJ*: Um, how many, how many, well, how many threes were in 100.
> 7. *PT*: How many threes were in 100? Is that because the rabbit is jumping in threes? (Points to the number line on screen.)
> 8. *JJ*: Um,     well . . . (pause) . . . the . . . okay,     um, well . . . if the rabbit has to go over and back in 6 seconds and so, and that would be 200, he has to go 200 feet in 6 seconds. So, if I made it so he can go 100 feet in 3 seconds. And I divided 100 by 3 . . . to figure out . . . how many, well I can't remember what I did.
> 9. *PT*: Can you, can you sort of explain, perhaps using . . . pointing to the number line, what it was that you were finding?
> 10. *JJ*: Well . . .
> 11. *PT*: Would that help?
> 12. *JJ*: Okay, I wanted . . . well to see if you could divide

this (pointing to number line) into three different parts that are equal.

13. *PT:* Uh, I see. Now I get it, okay. So you're trying to divide this into three parts that are the same size (points to number line indicating three equal sections), that would mean that each part he goes . . . takes how long to go in each part?

14. *JJ:* 3.

15. *PT:* 3 seconds for each part? (Pointing to three sections on the number line)

16. *JJ:* Oh, 1 second for each part.

Excerpt 3 shows that distance still predominated in JJ's reasoning, but that she now took both the total distance *and total time* as given and that she constituted completed motion as segmented distance in relation to a segmentation of time (3 units of time, so 3 segments of distance). This is further suggested in Excerpt 4, where, for the first time, JJ argued for the "sensibleness" of segmenting distance so that it (the segmentation) corresponded to a nonintegral value for amount of time traveled.

Excerpt 4 (Session 3)
(JJ and PT have discussed how to determine half of 8.3 seconds. JJ figured it would be "4.1 and one-half" and guessed that she should use "4.15" with her calculator.)

1. *PT:* Allright. Now, if you didn't already have this written down on the paper, then how would you figure out how fast it needs to go to go from 0 to 100 in 4.15 seconds? (Points to 0 feet and 100 feet on the number line.)

2. *JJ:* Well, you could divide 100 by 4.15.

3. *PT:* Now, I followed you when you said you would divide 100 by 3 because you're thinking of the rabbit going in 3 . . . 3 pieces, one second for each piece, but what do you have in mind when you say you would divide by 4.15?

4. *JJ:* Well, you want to figure out how many feet the rabbit can go in, per second, and . . . and it would have, it would have to be equal up to 4.15 seconds when you get to the end of the 100 feet.

I interpreted JJ's remark, "it would have to be equal up to 4.15 seconds when you get to the end of 100 feet," as an expression of an image that time accumulated (in seconds) as the rabbit moved. This, together with her frequently expressed conception that distance accumulated as the rabbit moved, suggested that she had come to conceive of completed motion as being constituted by a segmented, total distance in relation to a segmented, total amount of time. As such, it was important to see if JJ was actually thinking primarily in terms of segments of time, so I focused upon the significance she gave to 0.15 seconds in relation to distance traveled by the rabbit.

Excerpt 5 (Session 3)
    (JJ has evaluated 100 + 4.15 on her calculator.)
1. PT: Okay, 24.096385. . . . What, what did that number stand for?
2. JJ: How many, how many feet the rabbit would go per second.
3. PT: That's how far he's going to go each second. Okay? So now can you tell me precisely how far he will go in 4 seconds?
4. JJ: 96 feet.
5. PT: 96 feet? Exactly?
6. JJ: Um.
7. PT: You nodded your head yes. Is he going exactly 24 feet per second?
8. JJ: No.
9. PT: So then how would you tell me exactly where he's going to be?
10. JJ: Um, you would multiply this number by 4.
11. PT: Okay, why don't you do that.
12. JJ: (Uses calculator.) Ninety-six point three eight five five four (96.38554).
13. PT: And that number is a number of what?
14. JJ: Of how far he will be after 4 seconds.
15. PT: In miles, inches?
16. JJ: Feet.
17.     (PT runs rabbit, pausing it at 4.1 seconds. Rabbit is at 98.8 feet.)
18. PT: Oops, missed it. At the end of 4 seconds it will be near the end. What about that little bit left over, that extra 4 feet . . . a little less than 4 feet? How long will it take him to go the rest of the way?
19. JJ: Point one five (.15).

After the discussion presented in Excerpt 5, I asked JJ to give the rabbit a speed to tie with the turtle when the turtle goes over at 20 ft/sec and returns at 40 ft/sec. JJ calculated the total time for the turtle (7.5 sec), and then divided 200 by 7.5. This was the first occasion where JJ understood that the turtle's time would determine the rabbit's speed if they were to tie. JJ had evidently abstracted time from her intuitions of the covariation of distance and time. Amount of time was no longer tantamount to a number of speed-lengths.

To further verify that JJ had reconstituted speed by way of quantitative operations, rather than merely having abstracted a pattern in the numerical operations she used to answer questions, I asked JJ to consider a major variation in the kinds of problems she had been solving: The turtle goes over at some speed, comes back at 70 ft/sec, and the rabbit goes over and back at 30 ft/sec. Give the turtle an "over speed" that will make it and the rabbit tie (Excerpt 6).

Excerpt 6 (Session 3)

1. *PT:* This one is a little bit different. It says the turtle is supposed to go over at some speed that you find out, come back at 70 feet per second, and the rabbit is going to go at 30 feet per second. The distance is 100 feet, and they're supposed to tie, but we don't know the turtle's over speed. So how, how could you figure that out?

2. *JJ:* Well, first you could divide 30 into 200 to figure out how many seconds it would take for the rabbit to go . . .

3. *PT:* All right.

4. *JJ:* . . . there and back.

5. *PT:* Okay.

6. *JJ:* And then, um, you would divide 70 into 100 to see how many seconds or how many times it would take, how many times the turtle could go in, how many seconds the turtle could travel per feet at 70 feet per second.

7. *PT:* All right.

8. *JJ:* So you would divide . . . so you would divide 70 into 100. And then you would subtract, could subtract that number by the, how many seconds the rabbit took to go there and back, and then figure out how many seconds the turtle would have to go, and how many feet the turtle would have to go.

9. *PT:* How many feet?

10. *JJ:* Or, how many, well, how many feet per second.

The significance of JJ's reasoning as presented in Excerpt 6 is that she presented a plan for finding the values of various quantities. Every calculation identified by JJ was selected to evaluate a quantity (e.g., divide 70 into 100 *to see how many seconds it would take*; subtract turtle's "back time" from rabbit's total time *to see how many seconds it would take for the turtle to go over only*). I take JJ's solution to this problem as the first solid indication that she had constituted "turtle and rabbit" situations through a scheme of mental operations, for only through mental operations could she have supplied quantities that were unmentioned in my presentation but were related structurally to those that were mentioned. JJ evidently had an operative image of the situation. That is, her quantitative operations were part of her image of the situation, which means that she constituted the situation with them. This gave her both something to reason *about* and something to reason *with* in determining appropriate numerical operations for evaluating the various quantities that were present in her image.

JJ's homework after Session 3 was to solve more "turtle and rabbit tying" problems. In the next session, to further test the conclusion that JJ was using arithmetical operations representationally rather than constitutionally, I made up a situation similar to that presented in Excerpt 6. Excerpt 7 presents the first part of the discussion.

> Excerpt 7 (Session 4)
> > (Text of problem reads, "Tomorrow you are going to be given the turtle's "over" and "back" speeds and the length of the race track. Tell what arithmetic you would do to determine a speed to give the rabbit so that it and the turtle tie")
>
> 1. *PT:* Okay, so let me write *Turtle over* for the number that the turtle goes over. *Turtle back* for the number the turtle goes back, okay? (Writes *TO* for turtle over and *TB* for turtle back.) (See Figure 6.4).
> 2. *JJ:* And . . .
> 3. *PT:* We'll call them *TO* and *TB*.
> 4. *JJ:* Okay. Well . . . okay, you divide the number that the turtle goes over . . .
> 5. *PT:* Okay.
> 6. *JJ:* . . . in, into 100.
> 7. *PT:* And that number stands for what? *TO* stands for what?

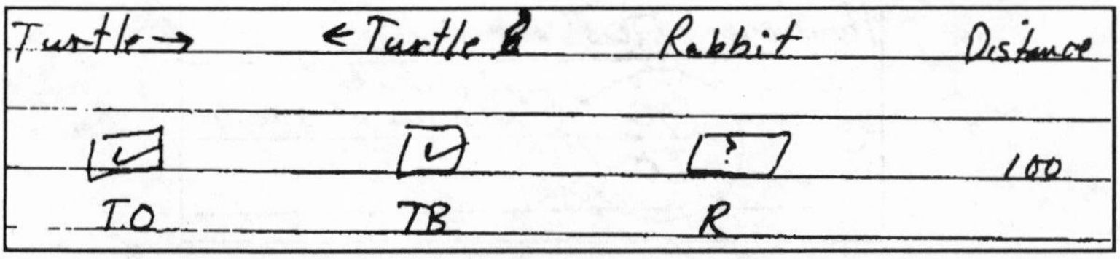

Fig. 6.4

8. *JJ:* How many feet per second he went over.
9. *PT:* Okay. So 100, and then you divide that by the speed that the turtle goes over by.
10. *JJ:* And, um, and then you, and you should get how many times, you should get how many times that goes into 100 and so you could change that into seconds.
11. *PT:* And what is, what does this stand for?
12. *JJ:* Uh, that stands for how many . . .
13. *PT:* So, I'll write that down, how many . . . (Writes *How many.*)
14. *JJ:* Um, how many feet, well, how many feet per second the turtle went over in 100 feet.
15. *PT:* Okay, what do you get when you do that division?
16. *JJ:* A number.
17. *PT:* And what does that number stand for?
18. *JJ:* It stands for how many times that number goes into 100.
19. *PT:* All right.
20. *JJ:* But then, you can . . . you can convert it into seconds.
21. *PT:* Convert it into seconds? How do you convert it into seconds?
22. *JJ:* Well change it to seconds.
23. *PT:* You just say it's the number of seconds?
24. *JJ:* Yeah. Well, like if you take 25 into 100, it goes 4 times, so you can say that's 4 seconds.
25. *PT:* Oh, I see. Okay. All right. So that's how many, how many of those numbers are in 100 and change that to a number of seconds. (Writes an expression.) I'm going to write division like this . . . like it's a fraction. (records what JJ said. See Figure 6.5.) Is that right?
26. *JJ:* Um-hum (yes).
27. *PT:* Okay.

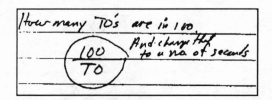

Fig. 6.5

28. *JJ:* And then, you do the same thing for the turtle to go back.
29. *PT:* Well, what, what do you mean the same thing?
30. *JJ:* Well . . .
31. *PT:* Divide 100 by *TO*?
32. *JJ:* No.
33. *PT:* By the turtle's over speed?
34. *JJ:* The turtle back speed.
35. *PT:* Oh, okay. Divide 100 by the turtle back speed and what does that stand for?
36. *JJ:* How many, um, of the thing, how many whatever, well how many of the numbers you have in . . .
37. *PT:* Okay, so that's . . .
38. *JJ:* How many . . .
39. *PT:* What number?
40. *JJ:* How many *TBs* you have in 100.
41. *PT:* So how many *TBs* there are in 100. (See Figure 6.6.) Okay.
42. *JJ:* And then you change that to seconds, too.
43. *PT:* Change that to seconds. All right.
44. *JJ:* And then, you add those two things of seconds, those two. (Pointing to the paper.)
45.    (PT and JJ discuss how she might write the addition of the two seconds. JJ writes the two "seconds" expressions in vertical format. She labels the expression with "The number you get will be how many seconds the turtle travels over and back.")

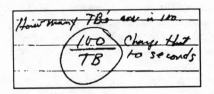

Fig. 6.6

46. *PT:* Okay. So, now you've got the number of seconds the turtle travels over and back, and . . . what was it that we were trying to do? I forgot.
47. *JJ:* To figure out what the number is of the rabbit going over and back.
48. *PT:* To do what?
49. *JJ:* So they tie.
50. *PT:* . . . Now, what would you do to find out the rabbit's speed over and back so they tie?
51. *JJ:* Uh . . . you would divide the number you get into 100, or . . . well . . . you could divide that into . . . well . . . you could either divide that into 200 with that number you get, or divide that by 2 and then divide it into 100.

Two things are significant about the conversation in Excerpt 7. First, JJ did not need actual values to reason about what arithmetic was appropriate. She inferred appropriate arithmetic according to what quantities she wished to evaluate and according to relationships of those quantities to others, and she did all of this with indeterminate values.

Second, JJ referred to the result of dividing the value of distance traveled by the value of speed as something that needed to be *converted into* a number of seconds (Excerpt 7, paragraphs 15–24). That JJ needed to "convert it into seconds" provides insight into her emerging comprehension of speed, distance, and time taken together as a quantitative structure. In Session 1 JJ needed to actually imagine measuring a given distance with a given speed-length to determine an amount of time. In Excerpt 7 we see that she had interiorized this notion to a level of thought, so that she understood that *any* rate-length would determine a *partition* of the total distance into units of rate-length and parts thereof and that this partition would correspond to a partition of the total time into units of seconds and proportional parts thereof (figure 6.7).

I asked one further question, following the discussion in Excerpt 7, as a final assessment of the degree to which JJ had established speed as a proportional correspondence between between distance and time: What would happen if you changed the distance from 100 to 200 feet and kept the turtle's speeds the same? JJ responded that "It would be pretty much the same . . . the time would just double." I then asked

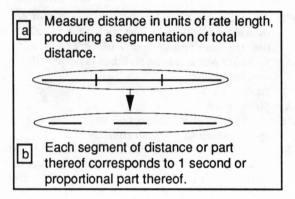

a | Measure distance in units of rate length, producing a segmentation of total distance.

b | Each segment of distance or part thereof corresponds to 1 second or proportional part thereof.

Fig. 6.7

JJ to suppose she wanted to keep the time the same, to which she responded "You would make it so it would be more feet per second. [How much more?] Doubled." JJ's explanation was very imagistic: "If you just divide it in half (pointing to the Over and Back screen) it would be just like it used to be." The evident obviousness to JJ that the time would double if the speeds were to remain the same, and that all the speeds would have to double if the time were to remain the same, suggests strongly that she had interiorized "turtle and rabbit" situations as a quantitative structure—that she had developed an image of turtle and rabbit situations that she constituted by way of quantitative operations.

*Average Speed.* To further investigate the generality of JJ's mental operations I introduced the general notion of "average speed," characterizing it as one person making a trip at (possibly) varying speeds, a second person starting at the same time and traveling the same itinerary at a constant speed, and they tie. The second person's speed is the first person's average speed. (I also recast it as the first person repeating his trip at a constant speed so that he takes the same amount of time.) My hypothesis was that, if JJ had actually constructed a speed as a scheme of operations, then the idea of average speed should be easily generalized to other situations of average rate (e.g., average price).[10] This is shown in Excerpt 8.

Excerpt 8 (Session 5)

1. *JJ:* (Reads text.) John traveled 35 miles per hour for 100 miles and 45 miles per hour for 50 miles, what was John's average speed on this trip?

2. *PT:* Okay, now don't, don't think that, uh, [I'm asking you to give] an answer right now. Tell me what is . . . tell me what this told you. Explain to me what that told you.

3. *JJ:* That he traveled 35 miles per hour for 100 miles and then 45 miles per hour for 50 miles.

4. *PT:* Now, how is that like the turtle and the rabbit?

5. *JJ:* Um . . . well . . . um, he traveled a certain amount of miles per hour for a certain amount of miles and that traveled a certain amount of feet per second for a certain amount of feet.

6. *PT:* Okay. So it's, it's in that way, it's not very different. Is there anything that, um, that is different, that's about the situation described here and the turtle and the rabbit?

7. *JJ:* Um, well, the first time he went for 100 miles and the second he went for 50.

8. *PT:* Okay. Now is that, is that like the, the turtle and the rabbit or different from the turtle and the rabbit?

9. *JJ:* Well the turtle and the rabbit go, um, 100 feet both ways, or 200.

10. *PT:* Okay, and, this doesn't go 100 both ways. (Referring to "mileage" problem.)

11. *JJ:* Huh-uh. (No.)

12. *PT:* All right. What would you need to know, oh . . . I'm sorry, I'm getting ahead of myself. In this question, (pointing to "mileage" problem), it asks what was John's average speed on this trip. What does that mean?

13. *JJ:* Um . . . well . . . um . . . (pause). It means that . . . you want to figure out his average speed. (Giggles.)

14. *PT:* All right . . . I'm asking now what does average speed mean in this context?

15. *JJ:* How . . . for going one . . . how many miles per hour for . . . and just for 150 miles but just . . . and stay that same speed.

16. *PT:* Okay. And what's special about that speed?

17. *JJ:* It should come out the same as this. (Pointing to paper in front of her.)

18. *PT:* The same in what way?

19. *JJ:* Um, it should go the same amount of time or the same amount of miles.

20. *PT:* Just one of them or both of them?
21. *JJ:* Both.


JJ went on to calculate John's average speed, which was not surprising. However, a better question for assessing how well she had interiorized speed as a quantitative structure would have been, "John traveled at an *average speed* of 35 miles per hour for 100 miles and at an *average speed* of 45 miles per hour for 50 miles." Had JJ responded as straightforwardly to that question as she did to the one posed it would have been clear that she was *constituting* the situation by way of a scheme of mental operations, for only through mental operations could she have supplied quantities that were unmentioned in the text but were related structurally to those mentioned.

After the discussion in Excerpt 8 I asked JJ to think about a situation that was quite different from any she had seen. She replied in Excerpt 9.

Excerpt 9 (Session 5)
1. *JJ:* (JJ reads text.) Sue paid nine dollars and forty-six cents for Yummy candy bars at forty-three cents per bar and paid six dollars and eight cents for Zingy candy bars at thirty-eight cents per bar. What was her average cost of a candy bar?
2. *PT:* Now before even talking about how to answer the question, explain . . . can you tell me how this one, (points to "candy bar" text) is like the one you just did about John traveling?
3. *JJ:* Well, you want to figure out the average of the cost . . . of a candy bar and . . .
4. *PT:* Aside from the, okay, leave the question out, how is, how is this situation, (points to "candy bar" text), similar to the situation of John traveling at different speeds?
5. *JJ:* Because, um, it's, it's like, you could change that . . . it's . . . okay, it's, it's exactly just the same but with a different . . . it's with candy bars and it's, it's like forty-three cents per bar is like 35 miles per hour, and the nine dollars and forty-six cents for a Yummy candy bars and, uh, for 100 miles. . . So it's . . . and then that would be the same thing down, for this. (Points to "candy bar" text.)

6. *PT:* This . . . this meaning the Zingy candy bars? So how . . . how then would you, uh, find the average cost of a candy bar?

7. *JJ:* Um, you would divide forty-three cents into nine dollars and forty-six cents.

8. *PT:* And what would you find when you did that?

9. *JJ:* The . . . how many candy bars she got. And then you would do the same for these (Zingy).

10. *PT:* And how is that similar to what you did up here with John?

11. *JJ:* Because um . . .

12. *PT:* "That" meaning . . . being, um, finding the number of candy bars?

13. *JJ:* Because you want to find out how many hours it went (pointing to "mileage" text) and this, you want to figure out how many candy bars you could get. (Pointing to "candy bar" text.)

I take JJ's analysis of similarity between the two problems (average speed and average cost) as evidence that she assimilated both to a scheme of mental operations. One other possibility is that JJ's explanation of the similarity between the two could have been based on a textual correspondence of parts of one text and parts of the other (matching "per" phrases, for example). This turned out not to be the case, as a follow-up comparison between two problems had noncorresponding quantities mentioned explicitly, and yet JJ said they were "pretty much the same," and explained why.

*Concept of Rate.* To investigate further the nature of JJ's scheme, I presented her with additional correspondence tasks. The discussion is presented in Excerpt 10.[11]

Excerpt 10 (Session 8)

1. *JJ:* Carol has a swimming pool that holds 42,500 gallons of water. She fills it with two pumps, one pump put into the . . . one pump put water into the . . . water into the pool at 45 gallons per minute, the other pump put water into the pool at 70 gallons per minute. How long will it take Carol to fill the swimming pool while using both pumps at the same time if it is now empty?

2. *PT:* Okay, and how is that like the rabbit and the turtle? The rabbit or the turtle?

3. *JJ:* Well, that's like the turtle would go across at forty-five . . . um, feet per second, or whatever it is and then come back at 70 feet per second or minute, and it's like how long will it takes the turtle to . . . (Pause.)

4. *PT:* Okay, so the rabbit doesn't even enter into to it. . . . Could you write the arithmetic that you would do to answer that question. In another words, just write the division sentence, don't . . . don't actually do the division.

5. *JJ:* (Sighs. Long pause.)

6. *PT:* Did you run into a problem?

7. *JJ:* Um . . . Yes.

8. *PT:* Okay, what's that?

9. *JJ:* Well, I'm trying to figure out . . . (Pause.) . . . Well, . . . (Pause.)

10. *PT:* Do you still think this is like the turtle going over and back?

11. *JJ:* (Pause.) It's like the turtle going across at 45 feet per second or whatever it is, and the rabbit going at seventy feet per second, that's it. (Pause.)

12. *PT:* How was the 43,000 gallons like the distance?

13. *JJ:* It's like how . . . it's like how long . . . what the distance is between two points, but that's how much (long pause) the turtle and rabbit go together?

JJ evidently internalized the text as being similar to turtle situations because of the presence of two rates and one amount. When she attempted to interiorize her internalized situation (i.e., assimilate her initial comprehension of the text to her scheme of operations) to decide on the appropriate arithmetic, it did not fit. Most important, JJ apparently could identify the reason why it did not fit—it would be as if the turtle and rabbit were running together and both their distances were being accumulated as one quantity. I asked JJ to change the "Carol's Pool" problem so that it would be exactly like the turtle and/or rabbit, but also so that it would still be about the pool. We did this in Excerpt 11.

Excerpt 11 (Session 8)

1. *PT:* (Asks JJ to change "Carol's Pool" problem so that it is just like the turtle and the rabbit.)

2. *JJ:* That one pump would go on, for maybe a minute

and then the next pump, and then other pump would, and then they would switch on and off.

3. *PT:* And then they would switch on and off. When we had the turtle going, it went over at one speed and back at another, right? You're right, we could have them switch on and off, and it would be more like the turtle. But the turtle didn't go 70 feet per second for a little bit, and then 40 feet per second. It went over at one speed and back at the other.

4. *JJ:* Um. (Pause.) You could change it that, if it went for a minute for 45 feet or whatever, and then another minute for 70 . . . feet.

5. *PT:* You said that the 42,500 gallons is like the distance, is that right? . . . Is that like the distance one way, or the distance over and back?

6. *JJ:* Over and back.

7. *PT:* Okay, when the turtle went over and back at two different speeds, how far did it go at one speed, and how far did it go at the second speed.

8. *JJ:* 100 feet.

9. *PT:* Okay, what part of the turtle trip was that?

10. *JJ:* Half.

11. *PT:* Half, because sometimes you'd have it go 200 feet, didn't you? So you could have it go different distances, but it always went over at one speed and back at the others, and it always went half way at one speed, and half way at the other. Does that give you any ideas about now how to change this so it just like the turtle?

12. *JJ:* (Long pause.) You could . . . you could divide the, if you . . . (Pause.) If you divide that in half. (Points to text of problem.)

13. *PT:* That being the 42,500 gallons?

14. *JJ:* So it would be like the distance from one place to another and then back.

15. *PT:* All right.

16. *JJ:* But the half would be just one place to another and then the second half would be back. . . .And then the forty-five gallons would be . . . the speed . . . the speed going one way and that it would go 45 or whatever per minute and on the way back it would be the 70 gallons, so it'd be like 70 feet . . . feet back.

JJ saw that the turtle needed to go at one speed and then another, but evidently accommodated to that need by focusing on different pump rates for different times, "switching on and off." When I oriented her attention to the particulars of the turtle's situation, she then created explicit relationships among the capacity of the pool, the distance traveled by the turtle, the flows of water from the pumps, and the speeds at which the turtle ran.

The next problem differed from the turtle's situation on yet another characteristic: One knows the pool capacity, the pumps' rates, and the amount of time required to fill the pool. This was explored in Excerpt 12.

Excerpt 12 (Session 8)

1. *JJ:* Janna has a swimming pool that holds 93,000 gallons of water. She began using one pump to fill her pool, but it broke down, and she had to replace it. Her first pump put water into the pool at 73 gallons per minute. It took 1321 minutes of pumping time to fill the pool. How long did each pump work?

2. *PT:* Okay, how is that one like the turtle and the rabbit? Is this situation like the turtle and the rabbit? Or just the turtle or just the rabbit?

3. *JJ:* It's like one of them . . . the turtle not the rabbit . . . just one.

4. *PT:* Is it . . . is it like or different from the story you just wrote? (Referring to JJ's modification of "Carol's Pool".)

5. *JJ:* It's like because . . . she's used two different pumps and . . . they were both different, they both put waters into the pool at different gallons, different . . .

6. *PT:* The word is *rate*. At different rates.[12]

7. *JJ:* At different rates per minute, but . . . in this one you want to figure out how long each pump works?

8. *PT:* Okay, now, so how is that unlike the rabbit? [PT misspoke—he meant *turtle*.]

9. *JJ:* Well . . . (Long pause.)

10. *PT:* Can you tell, how it is unlike the story that you wrote? (Puts the two papers side by side.) Is there any real big difference?

11. *JJ:* No. (She studies the papers.) But, in this you know how long that one pump . . . You know, because in this one it was filled half full and then they switched. But in this one you don't know.

12. *PT:* You don't know how far up the pool was filled?

13. *JJ:* That's what you want to figure out.

14. *PT:* Oh, if you find out how much the pool was filled then you can answer this question?

15. *JJ:* Well . . . (long pause) . . . that would be the answer if you knew.

16. *PT:* So, if we knew that the pool was one-fourth filled when she switched, then that would be the answer?

17. *JJ:* That would be, umm, you could get the answer very easily from that because then you would just divide 73 into, well, how much . . . well you divide . . . then you could . . . find what's one-fourth of 93,000, and then divide that by 73.

JJ's remarks in Excerpt 12 again provides evidence that she possessed a scheme of mental operations into which she assimilated the situation. Not only did JJ recognize the similarities and differences between this situation and those she encountered with the turtle and rabbit, she knew what she needed to know to find out how long each pump worked. This suggests that she came to conceive the filled pool as being composed of two parts: the part filled by Pump1 running at Rate1 for duration Time1, and the part filled by Pump2 running at Rate2 for duration Time2 (see Figure 6.8).

Excerpt 10 showed JJ internalizing a text inappropriately, and experiencing disequilibrium when attempting to assimilate the product of that internalization into her scheme of mental operations. In the next excerpt, JJ again internalizes a text inappropriately. I could have promoted disequilibrium by asking her to decide upon appropriate arithmetic operations. Instead, I took a different approach, relating this problem to the "Carol's Pool" problem so that we would have an occasion to discuss the combination of two rates. This we did in Excerpt 13.

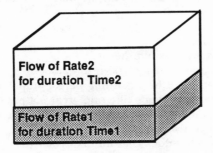

Fig. 6.8

Excerpt 13 (Session 8)

1. *JJ:* Jim has a swimming pool that holds 35,000 gallons of water. He has two pumps. One pump can fill his pool in 912 minutes. The other pump can fill his pool in 532 minutes. How long will it take Jim to fill the swimming pool using both pumps if it's now empty? (Long pause.) Well, you could do the same thing as you could do in the first one and you could divide that in half and . . . well, and then see how long it would take if it could fill his pool in 912 minutes and see how long it could fill half of his pool. And then you could switch and see how long you could fill half of his pool, and what, regularly, you could fill it in 532 minutes.

   (PT turns the page to "Carol's Pool" text: Carol has a swimming pool that holds 42,500 gallons of water. She fills it with two pumps. One pump put water into the pool at 45 gallons per minute. The other pump put water into the pool at 70 gallons per minute. How long will it take Carol to fill the swimming pool while using both pumps at the same time if it is now empty.)

2. *PT:* Let's go back to this one. She's got both pumps running. Is that correct? (JJ nods.) Suppose that this was just about one pump filling the swimming pool, 42,500 gallons, one pump is running at 45 gallons per minute, could you answer the question of how long it would pump to fill the swimming pool?

3. *JJ:* Yeah.

4. *PT:* And how would you do that?

5. *JJ:* You would divide 45 into 42,500.

6. *PT:* Okay. Suppose that I had a pump that pumped water at 100 gallons per minute. Could you answer the question how long would it take to . . .

7. *JJ:* Yeah.

8. *PT:* Same way?

9. *JJ:* Uh huh.

10. *PT:* And what would that be?

11. *JJ:* You would divide 100 into 42,500.

12. *PT:* Okay. So if he's running both pumps at the same time, then that's like one great big pump isn't it?

13. *JJ:* So you could keep adding, you could add 45 and 70 and you would get a number and . . .

14. *PT:* And what does that number refer to?

15. *JJ:* How many gallons for, well, per minute.

16. *PT:* Okay.
17. *JJ:* And then you could keep on adding those numbers until you get to 42,500.
18. *PT:* Okay. Or, could you do it an efficient way?
19. *JJ:* You could multiply. You could guess a number and multiply it.
20. *PT:* Now, let's go back to what I was asking you before. Suppose I told you that I had one pump that pumped water at 115 gallons per minute. How long would it take that pump to fill the pool? How would you answer that question?
21. *JJ:* You would divide that into 42,500. So you could add those and divide that into 42,500!

Each of the problems previous to "Jim's Pool" gave prominence to rates of flow, and JJ at first interiorized the text according to that image (Excerpt 13, paragraph 1). Instead of attempting to promote a state of disequilibrium in JJ so that she would reconceive the setting, I returned to the "Carol's Pool" problem (see Excerpt 10) to discuss the notion of a combination of flows (Excerpt 13, paragraphs 2–15).

JJ's remark, "And then you could keep on adding those numbers until you get to 42,500" (paragraph 17) suggests two conclusions about her emerging comprehension of the "Carol's Pool" situation: (1) She constructed a total rate of flow as a combination of the two rates of flow mentioned in the text, and (2) that she came to conceive of a filled pool as being made by a (measured) flow of water happening over a (measured) duration of time. In regard to the latter, it is as if JJ thought, "Add 45 and 70, then iterate that sum until you fill the pool (get 42,500 gallons)." Her later remark that she could use the strategy of guess and check with multiplication (paragraph 19) strengthens this interpretation.

JJ's approach here resembles her initial method of determining how many seconds it would take the rabbit to go a given distance at a given speed. She segmented distance in units of "speed-length," thereby constructing a corresponding duration (number of time units). In the present case, JJ evidently thought of segmenting the capacity of the pool by way of iterating the combined "flow amount," thereby constructing a duration over which the flow rate will fill the pool. It appears that with respect to combining two flow rates to produce a new flow rate, JJ recapitulated her construction of speed as a rate: segment the total amount of the extensive quantity being created in units of the "rate amount," thereby inducing a duration amount. In previous cases, the "rate amount" was a single quantity; in the present case, it was a *combination* of rates that she needed to construct as a rate. The fact that JJ evidently needed to recapitulate the construction of speed in the case of a combination of flow rates suggests that this was her general method of constructing rates of any kind. The fact that JJ so quickly constructed a combination of flow rates as a rate (relative to her construction of speed as a rate) suggests that recapitulation is a process by which reflective abstraction of the second kind happens—JJ became "skilled" at constructing rates, and the *process* of construction became suggestive of the kind of thing the process produces (Cellerier, 1972).

There remains the question of how structural was JJ's construction of combined flow rate. Was "combination of rates" a quantitative operation for JJ? Was the result of combining two flow rates of the same stature as an uncombined flow rate? That is, could she infer a combination in the context of complex relationships and then use it relationally as a single quantity? The latter portion of Excerpt 14 suggests that, yes, combining flow rates was a quantitative operation for JJ; the result of combining was indeed a single quantity.

Excerpt 14 (Session 8)

(PT turns page to "Jim's Pool" text: Jim has a swimming pool that holds 35,000 gallons of water. He has two pumps. One pump can fill his pool in 912 minutes. The other pump can fill his pool in 532 minutes. How long will it take Jim to fill the swimming pool using both pumps if it's now empty?)

1. *JJ:* Well . . . you could do the same and add those two together and divide that into 35,000.
2. *PT:* Okay. Now, how are these numbers different from the one's in the first problem (Carol's Pool)? (He puts the two papers side by side.) Look closer.
3. *JJ:* The 912 and the 532 is how long it would take to fill this pool, and this is how many gallons per minute it would take.
4. *PT:* Okay. Can you do the same thing here (Jim's Pool) that you did over there (Carol's Pool) and just add

these two numbers? Well, I mean of course we can add these two numbers. You can always add numbers. If we do add these numbers, 912 and 532, then we would get 1444. Okay, we would get 1444 minutes. If we were to say that's how long it takes something to happen, what is it that would happen?

5. *JJ:* How long . . . it would take to fill . . . two pools.

6. *PT:* Very good. That's right. That's how long it would take to fill two pools. This one would fill it in 912 minutes and this one would fill a second one in 532 minutes. Okay. Now, can you do something with these numbers so that you have exactly that situation (points to "Carol's Pool" text)?

7. *JJ:* (Long pause.) You could change that into how many gallons per minute and change that into how many gallons per minute.

8. *PT:* And how would you do that?

9. *JJ:* You would . . . you could divide . . . if one pump fills a pool in 912 minutes, then you could divide that into 35,000 because that's how much the pool holds. And then see how many gallons per minute it, the pump, can fill this pool. And then, that would be, you could, that would be that number and you would do the same for 532.

10. *PT:* All right, then you get the gallons per minute for each pump. Then what would you do?

11. *JJ:* (Long pause.) And then . . . you could . . . (Pause) . . . you could . . . add those two numbers so you get together . . . so then you'll get, so it will seem like one pump.

I dismiss JJ's inappropriate internationalization of the text (paragraph 1) as being largely unimportant.[13] The remainder of the excerpt contains three important episodes: The first is where JJ distinguished between the numbers in the two texts as being values of two different kinds of quantities (paragraph 3). The second is where JJ understood that adding the two numbers of minutes would evaluate the quantity "time to fill two pools" (paragraphs 4–5). The third is where JJ reconceives the "Jim's Pool" situation as involving two flow rates, each of which can be evaluated by division, and that those flow rates can be combined to produce a total flow rate (paragraphs 7–11).

The third episode (Excerpt 14, paragraphs 7–11) is important for understanding the extent to which JJ has interiorized rate as constant ratio. Her comment, "You could change that into how many gallons per minute . . ." (paragraph 7), her hesitation about how to make the change (paragraph 9), and her comment "see how many gallons per minute the pump can fill the pool" (paragraph 9) taken together suggests that she conceived of the situation in terms of rates. Her conception evidently involved an image of the pool being filled, as opposed to an image of breaking up a filled pool into 912 (or 532) portions, one for each minute. Had I been a better interviewer, I would have probed this further by asking JJ if she thought first of the pool being filled or first of the pool as already filled. Also, at the end of Excerpt 14, JJ again expressed a conception that the result of combining the two flow rates produces one flow rate: "so it will seem like one pump" (paragraph 16).

## DISCUSSION

My discussion of issues surrounding JJ's emerging conceptions of speed and rate is in two parts. My first remarks have to do specifically with JJ and with questions of conception. Second are my remarks relating events of this teaching experiment to issues that are more broadly associated with the field of multiplicative structures and mathematical development in general.

### Construction of Speed as a Rate

The only situations JJ could fully conceive early on had to do with determining the time required to go a given distance at a given speed. She could not conceive situations having to do with determining a speed at which to go a given distance in a given amount of time—unless the amount of time corresponded quite specifically with a whole number segmentation of the distance. This had nothing to do with a preferred strategy; it was not the case that JJ preferred to reason "within" one measure space (specifically distance) to draw conclusions about another (specifically time). Rather, JJ was constrained to reason primarily about distance because time was not a quantity of the same stature as distance; time of travel, for JJ, was implicit in how many "rate-lengths" (or

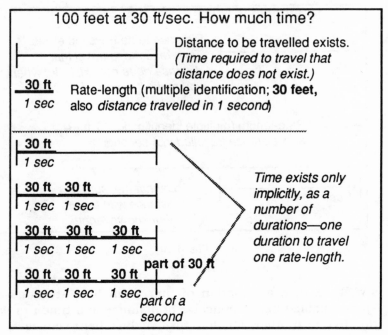

Fig. 6.9

parts thereof) were required to make a distance (Figure 6.9). With this interpretation of JJ's conception of speed we are able to explain her early inability to fully conceive situations in which she was required to determine a speed with which the rabbit would go a given distance in a given amount of time (Figure 6.10).

Only later (Session 3 on) did JJ begin to conceive of turtle and rabbit situations as involving covarying *accumulations* of the quantities distance and time. Evidently, in working on the homework assignment between Session 2 and Session 3, JJ abstracted time from her intuitions of speed so that it was a quantity in and of itself—one that could be measured in correspondence to distance, and one that varied in constant proportion to distance (Figure 6.11). This new conception of speed not only allowed JJ to answer questions that before were inaccessible to her, it also provided her with occasions to create situationally bounded mental structures that she would eventually interiorize as a constant ratio

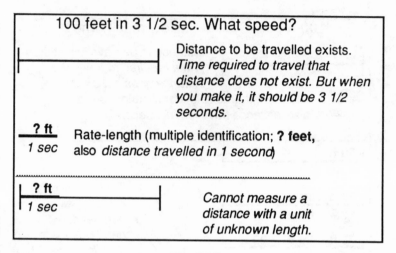

Fig. 6.10

structure. JJ's abstraction of time from her intuitions of speed enabled her to conceive of distance and time in relation to one another, which in turn enabled her to conceive of speed as the result of such a comparison.[14]

I would like to make an observation here lest it be missed. JJ's initial scheme for speed did not involve conceiving quantified motion as a quotient—it was not a ratio between distance and time. Rather, her initial conception of speed was that it was a *distance*, and her initial conception of time was that it was a *ratio*. JJ first had to construct speed as a ratio (i.e., as the result of a multiplicative comparison between distance travelled and duration of travel), and to do that she had to create time as an extensive quantity and abstract it and distance from her intuitions of motion. This, together with conceiving traveling as if it were completed, enabled JJ to construct speed as a ratio. JJ's ability to construct speed as a ratio had two ramifications: It enabled her to reason flexibly about determining an appropriate arithmetic operation to evaluate some quantity in a relationship, and it led to occasions where JJ reflected upon speed as constant ratio, thereby interiorizing speed as a rate.

An important lesson is to be learned from this teaching experiment. The standard method for introducing speed in schools is "distance divided by time." Although we cannot

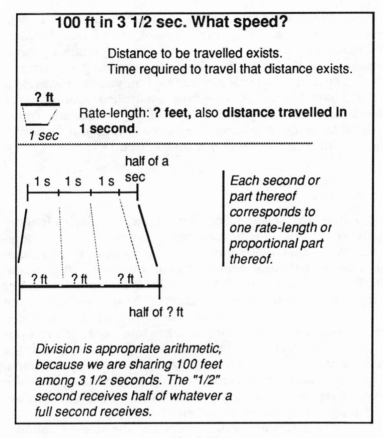

**100 ft in 3 1/2 sec. What speed?**

Distance to be travelled exists.
Time required to travel that distance exists.

**? ft**

Rate-length: **? feet,** also **distance travelled in 1 second.**

*1 sec*

half of a

1 s    1 s    1 s    sec

*Each second or part thereof corresponds to one rate-length or proportional part thereof.*

? ft    ? ft    ? ft

half of ? ft

*Division is appropriate arithmetic, because we are sharing 100 feet among 3 1/2 seconds. The "1/2" second receives half of whatever a full second receives.*

Fig. 6.11

know what the course of events would have been had I introduced speed as "distance divided by time," I say this with confidence: (1) "Distance divided by time" would have had little, if any, relevance to JJ's initial understanding of speed; (2) JJ would not have developed a concept of speed as a ratio, and therefore she would not have constructed speed as a rate; and (3) JJ would not have progressed as far as she did.

To tell students that speed is "distance divided by time" with the expectation that they will comprehend this locution as having something to do with motion, assumes two things: (1) They already have conceived of motion as involving two distinct quantities—distance and time, and (2) they will not take us at our word, but instead will understand our utterance as meaning that we move a given distance in a given amount of time and that any segment of the total distance will require a proportional segment of the total time. In short, to assume students will have any understanding of "distance divided by time," we must assume that they already possess a mature conception of speed as quantified motion. This places us in an odd position of teaching to students something that we must assume they already fully understand if they are to make sense of our instruction.[15]

## General Issues

*Arithmetic of Units*. Schwartz (1988) has made a lengthy analysis of what he termed *referent-transforming* operations. In his characterization, one can "transform" two quantities by an arithmetic operation to produce a resulting quantity. It is evident that JJ did not do this. The operands of her arithmetic operations were *values* of quantities, they were not quantities. The operations she performed on quantities were *quantitative* operations, and her inferences of appropriate arithmetic were based on relationships among quantities that, in turn, were created by her quantitative operations. When JJ "spoke arithmetic," it was evident that she was using it representationally, as she did not have a vocabulary to express her conceptions of situations.

Although it is not recommended by the science community, it is still common to see references to what is called *dimensional analysis,* or the arithmetic of units (e.g., miles ÷ miles/hour = hour) both to determine an arithmetical operation's resulting unit and to suggest appropriate arithmetic to do (e.g., Musser & Burger, 1991). JJ did not do arithmetic of units. Rather, she knew the quantities in the relationships, their units, and inferred a numerical evaluation for a quantity's value, knowing ahead of time the unit of a quantity's value. We should condemn dimensional analysis, at least when proposed as "arithmetic of units," and hope that it is banned from mathematics education. Its aim is to help students "get more answers," and it amounts to a formalistic substitute for comprehension.

*Problem Typologies*. The problems given JJ fell into the "between measure spaces" category (e.g., distance vs. time;

cost vs. number of candy bars; capacity vs. time), which made them "rate problems" according to several authors (Karplus, Pulos, and Stage, 1983; Nesher, 1988; Vergnaud, 1983, 1988, this volume). This view, however, emphasizes characteristics of texts. When we consider JJ's emerging conceptions of the situations depicted in texts we get a different perspective on the matter. JJ at first conceived problems concerning relationships between distance and time as what these authors would call *within measure space*—distance vs. distance. Only after JJ constructed time as an extensive quantity did she come to conceive of these problems as what these authors would call *between measure spaces*. It is my experience that any problem typology suffers this same deficiency; namely, that any given situation can be conceived in a multitude of ways that cut across the boundaries of the typology.

*Practice and the Development of Schemes.* In a quotation given earlier, Piaget expressed the strong position that assimilation is the source of schemes. One might interpret Piaget's position as a sophisticated form of Thorndike's Law of Effect or of the commonism "Practice makes perfect," but it would be misdirected to do so. Cooper (1991) has made the extremely important distinction between the notion of practice as it is normally used and what Piaget had in mind, which was the notion of repeated experience. *Practice* normally refers to an activity conceived by a designer of the practice; it is the repetition of observable behavior. We have known since the days of Brownell that what a child "practices" in a mathematics classroom often has little to do with developing habits of reasoning. "Repetitive experience," however, focuses our attention on the fact that what we want repeated is the constitution of situations in ways that are propitious for generalizing assimilations, accommodation, and reflection.

*Quantity vs. Number.* Although concepts of quantity and concepts of number most assuredly are highly related, both historically and ontogenetically, concepts of number quickly become confounded with matters of notation and language whereas concepts of quantity remain largely non-linguistic and unarticulated. I cannot say whether the inattention to matters of quantity is due to development or culture, but I suspect it is a result of unexamined cultural and institutional assumptions about mathematical learning and reasoning.

One aspect of the teaching experiment that is not evident from the excerpts in this chapter is JJ's unexamined assumption that she was obliged to use paper-and-pencil algorithms to perform numerical calculations. Moreover, JJ felt that she could not proceed in her reasoning *until* she had calculated any intermediate numerical result. If I contributed anything to JJ's progress, it was to relieve her of these constraints. For example, on one occasion JJ felt that it would be appropriate to divide by 100 by 3 but had no idea how to do it (see Excerpt 3). Had I not pointed out to her that she had a calculator, either I would have had to suggest we leave it at that or we would have had to digress from our discussion of relationships among speed, time, and distance. This was but one of many occasions where JJ would have been stymied had I demanded that all issues of calculation and numerical representation be settled before moving on in our discussions. This is not to say that matters of notation and numerical evaluation are unimportant. Rather, it is only to say that there are times and places for attention to issues of notation and calculation, but those times and places should not interfere with the development of foundational concepts (Thompson, 1992). In fact, the case of JJ would suggest that remaining oriented primarily to matters of conception has a salutary effect on the development of calculation skills, for it sets a context in which calculations normally considered "ugly" are seen as natural and necessary.

On this last matter, namely "ugly" calculations, I must agree with Confrey (this volume) that our interpretations of students' calculational difficulties and "misconceptions" concerning numerical operations are highly dependent upon customary foci of school instruction. Confrey observes that students' inclinations to associate certain operations with certain representations of number have more to do with their conceptual constitution of numbers and notations than it does with anything innate about numerical comprehension per se. Her observation is quite consistent with my observations of JJ. I saw JJ became "comfortable" with ugly numbers and figural distractions simply because her attention was oriented away from the activation of notationally based computational procedures. As Greer (this volume) and Harel,

Post, Lesh, and Behr (this volume) observe, we need to exercise great care to distinguish issues of notational convention and notationally based procedures from issues of meaning and application.

## NOTES

I wish to express my deep appreciation to Jim Kaput, Les Steffe, and Guershon Harel for the long conversations we have had on the issues addressed here. Research reported in this chapter was supported by National Science Foundation Grants No. MDR 89-50311 and 90-96275, and by a grant of equipment from Apple Computer, Inc., Office of External Research. Any conclusions or recommendations stated here are those of the author and do not necessarily reflect official positions of NSF or Apple Computer.

1. Von Glasersfeld (1989) provides a marvelous example of generalizing assimilation and subsequent accommodations. He discusses a child's assimilation of (what we would see as) a spoon to her "rattle scheme," subsequent distinctions (no rattle sound), and accommodations (produces a wonderful bang when hit on the table).

2. Judgments of similarity may range from figurative (situations "resemble" one another) to operative (the situations are, *in principle*, identical).

3. By *dimension* I mean a person's understanding of a quality as being measured potentially by any of a number of appropriate units. When a person's understanding of a quality includes the awareness of the possibility of using any of a multitude of mutually convertible units (i.e., an equivalence class of units), then this person possesses an "adult" conception of dimension.

4. Although you may not give any consideration to evaluating the difference between your height and your friend's height, you probably compare the two with full confidence that the difference could be evaluated by any of a number of methods.

5. Values of compared quantities need not be known to conceive them as being compared multiplicatively. The process by which one actually achieves a numerical value for the ratio is the quantification of the ratio. This is one more way in which quantitative reasoning differs from numerical reasoning.

6. This is one way to explain why a rate has the feel of being a single quantity.

7. Some might say the "50 mi/hr" can be instantiated also by a value for distance, thereby producing a value for time. I would argue that a person making this claim bases it on either of two reasoning processes: (1) an inference based on some form of the numerical formula $d = rt$, whereby one reasons that in, say, $125 = 50t$, $t$ must

be 2.5, or (2) that because the object goes 50 miles per hour, it must also go 1/50 hours per mile. The first inference, although valid, is not a quantitative inference. It is valid, formal, numerical reasoning. The second inference employs a sophisticated mental operation: inversion of a rate. Even in the second case, however, "50 mi/hr" is not instantiated with a value for a distance to produce a time. Instead, its inversion (measured in hours per mile) is instantiated with a value of a distance.

8. Unfortunately, this occasion is too often taught by teachers and recognized by students as the time to perform certain numerical rituals.

9. Kaput and West (this volume) raise an interesting extension of this way of thinking about rates. They speak about conceptions of "homogeneity" as characteristic of ratelike reasoning. Harel (1991) has used this concept of rate to explain students' difficulties with traditional "mixture" problems, where students must conceive of samples from a mixture as *always* being composed of constituent elements in constant ratio to one another, *regardless of the amount of mixture sampled*. In this conception, the notion of variability is not of accumulation, but instead is of random value. Accumulation still may be part of a person's conception of a given random value, but accumulation is not the predominant conception of samples of mixtures.

10. It turned out that JJ had already studied "average" as arithmetical mean, which caused much confusion at first, but we eventually straightened things out by agreeing that anytime I wanted her to think of "add up and divide" I would refer to it by the name of her teacher—"Mrs. T's average."

11. Sessions 6 and 7 were devoted to dealing with confusions JJ had about "average." Her sister decided to help JJ with her homework, reiterating to JJ that "average" means "add up and divide." It took two full sessions to deal with this confusion; those sessions could be the subject of another paper.

12. It turned out that the word for which JJ searches was *amount*, as in "different amounts per minute."

13. I suspect these inappropriate internalizations were more a function of schooling than of conception. Students (including JJ, according to her) come to expect new problems to be just like the ones they have recently solved.

14. This insight, which amazed me at the time JJ gained it, turns out to have been anticipated by Piaget. He stated, "The relation $v = d/t$ makes speed a relation and makes $d$ as well as $t$ two simple intuitions. The truth is that certain intuitions of speed, like those of outdistancing, precedes those of time. Psychologically, time itself appears as a relation (between space traveled and

speed . . . ), that is, a coordination of speeds, and it is only when this qualitative coordination is completed that time and speed can be transformed simultaneously into measurable quantities" (Piaget, 1977, p. 111).

15. Actually, we are telling them something new: Forget what you already understand; instead do this calculation anytime I ask you something about speed.

## REFERENCES

Bickhard, M. H. 1991. A pre-logical model of rationality. In *Epistemological foundations of mathematical experience*, ed. L. P. Steffe, 68–77. New York: Springer-Verlag.

Carpenter, T. P., and J. M. Moser. 1983. The acquisition of addition and subtraction concepts. In *Acquisition of mathematics concepts and processes*, ed. R. Lesh and M. Landau. New York: Academic Press.

Cellerier, G. 1972. Information processing tendencies in recent experiments in cognitive learning. In S. Farnham-Diggory (Ed.), *Information processing children*. New York: Academic Press.

Cobb, P., and L. Steffe. 1983. The constructivist researcher as teacher and model builder. Journal for research in mathematics education 14(2): 83–94.

Cobb, P., and E. von Glasersfeld. 1983. Piaget's scheme and constructivism: A review of misunderstandings. *Genetic Epistemology* 13(2): 9–15.

Cooper, R. G. 1991. The role of mathematical transformations and practice in mathematical development. In *Epistemological foundations of mathematical experience*, ed. L. P. Steffe, 102–123. New York: Springer-Verlag.

Greeno, J. 1987. Instructional representations based on research about understanding. In *Cognitive science and mathematics education*, ed. A. H. Schoenfeld, 61–88. Hillsdale, NJ: Lawrence Erlbaum.

Greer, B. 1988. Non-conservation of multiplication and division involving decimals. Journal of Mathematical Behavior 7: 281–98.

Harel, G. 1991. Presentation made to the University of Wisconsin Quantities Working Group, San Diego.

———, and J. Kaput. 1991. The role of conceptual entities in building advanced mathematical concepts and their symbols. In *Advanced mathematical thinking*, ed. D. Tall. Dordrecht: D. Reidel.

Hart, K. 1978. The understanding of ratios in secondary school. *Mathematics in school* 7: 4–6.

Hiebert, J., and M. Behr. 1988. Introduction: Capturing the major themes. In *Number concepts and operations in the middle grades*, ed. J. Hiebert and M. Behr, 1–18. Reston, VA: National Council of Teachers of Mathematics.

Hunting, R. P. 1983. Engineering methodologies for understanding internal processes governing children's mathematical behavior. *Australian Journal of Education* 27(1): 45–61.

Kaput, J., C. Luke, J. Poholsky, and A. Sayer. 1986. *The role representation in reasoning with intensive quantities: Preliminary analyses* (Technical Report 869). Cambridge, MA: Harvard University, Educational Technology Center.

Karplus, R., S. Pulos, and E. Stage. 1983. Early adolescents' proportional reasoning on 'rate' problems. *Educational Studies in Mathematics* 14(3): 219–234.

Kieren, T., and S. Pirie. 1991. Recursion and the mathematical experience. In *Epistemological foundations of mathematical experiences*, ed. L. P. Steffe, 78–101. New York: Springer-Verlag.

Lesh, R., T. Post, and M. Behr. 1988. Proportional reasoning. In *Number of concepts and operations in the middle grades*, ed. J. Hiebert and M. Behr, 93–118. Reston, VA: National Council of Teachers of Mathematics.

Lewin, P. 1991. Reflective abstraction in humanities education: Thematic images and personal schemas. In *Epistemological foundations of mathematical experience*, ed. L. P. Steffe, 203–237. New York: Springer-Verlag.

Musser, G., and W. Burger. 1991. *Mathematics for elementary teachers*, 2d ed. New York: Macmillan.

Nesher, P. 1988. Multiplicative school word problems: Theoretical approaches and empirical findings. In *Number concepts and operations in the middle grades*, ed. J. Hiebert and M. Behr, 19–40. Reston, VA: National Council of Teachers of Mathematics.

Ohlsson, S. 1988. Mathematical meaning and applicational meaning in the semantics of fractions and related concepts. In *Number concepts and operations in the middle grades*, ed. J. Hiebert and M. Behr, 53–92. Reston, VA: National Council of Teachers of Mathematics.

Piaget, J. 1965. *The child's concept of number*. New York: W. W. Norton.

———. 1967. *The child's concept of space*. New York: W. W. Norton.

———. 1970. *The child's conception of movement and speed*. New York: Basic Books.

———. 1971a. *Genetic epistemology*. New York: W. W. Norton.

————. 1971b. *Science of education and the psychology of the child*. New York: Viking Press.

————. 1977. *Psychology and epistemology: Towards a theory of knowledge*. New York: Penguin Books.

————. 1980. *Adaptation and intelligence*. Chicago: University of Chicago Press.

Powers, W. T. 1973. *Behavior: The control of perception*. Chicago: Aldine Books.

————. 1978. Quantitative analysis of purpose behavior: Some spadework at the foundation of scientific psychology. *Psychology Review* 85:417–435.

Schwartz, J. 1988. Intensive quantity and referent transforming arithmetic operations. In *Number concepts and operations in the middle grades*, ed. J. Hiebert and M. Behr, 41–52. Reston, VA: National Council of Teachers of Mathematics.

Shalin, V. L. 1987. Knowledge of problem structure in mathematical problem solving. Unpublished doctoral dissertation, University of Pittsburgh Learning Research and Development Center.

Steffe, L. P. 1991a. The learning paradox. In *Epistemological foundations of mathematical experience*, ed. L. P. Steffe, 26–44. New York: Springer-Verlag.

————. 1991b. Operations that generate quantity. *Learning and Individual Differences* 3(1): 61–82.

————. 1991c. The constructivist teaching experiment. In *Radical constructivism in mathematics education*, ed. E. von Glasersfeld. Boston: Kluwer Academic.

Steffe, L., E. von Glasersfeld, J. Richards, and P. Cobb. 1983. *Children's counting types: Philosophy, theory, and application*. New York: Praeger Scientific.

Thompson, P. W. 1982. Were lions to speak, we wouldn't understand. *Journal of Mathematical Behavior* 3(2): 147–165.

————. 1985a. Computers in research on mathematical problem solving. In *Learning and teaching mathematical problem solving: Multiple research perspectives*, ed. E. A. Silver, 417–436. Hillsdale, NJ: Lawrence Erlbaum.

————. 1985b. Experience, problem solving, and learning mathematics: Considerations in developing mathematics curricula. In *Learning and teaching mathematical problem solving: Multiple research perspectives*, ed. E. A. Silver, 189–236. Hillsdale, NJ: Lawrence Erlbaum.

————. 1989. A cognitive model of quantity-based reasoning in algebra. Paper presented at the annual meeting of the American Educational Research Association, San Francisco.

————. 1991. To experience is to conceptualize: A discussion of epistemology and mathematical experience. In *Epistemological foundations of mathematical experience*, ed. L. P. Steffe, 260–281. New York: Springer-Verlag.

————. 1992. Notations, conventions, and constraints: Contributions to effective uses of concrete materials in elementary mathematics. *Journal for Research in Mathematics Education* 23(2): 123–147.

————. in press. Quantitative reasoning, complexity and additive structures. *Educational Studies in Mathematics*.

———— and T. Dreyfus. 1988. Integers as transformations. *Journal for Research in Mathematics Education* 19: 115–133.

Vergnaud, G. 1983. Multiplicative structures. In *Acquisition of mathematics concepts and processes*, ed. R. Lesh and M. Landau, 127–173. New York: Academic Press.

Vergnaud, G. 1988. Multiplicative structures. In *Number concepts and operation in the middle grades*, ed. J. Hiebert and M. Behr, 141–161. Reston, VA: National Council of Teachers of Mathematics.

von Glasersfeld, E. 1980. The concept of equilibrium in a constructivist theory of knowledge. In *Autopoesis, communication, and society*, ed. S. Benseler, P. N. Hejl, and W. K. Koeck, 75–85. Frankfurt: Campus.

————. 1989. Cognition, construction of knowledge, and teaching. *Synthese* 80: 121–140.

————. 1991. Abstraction, re-presentation, and reflection: An interpretation of experience and Piaget's approach. In *Epistemological foundations of mathematical experience*, ed. L. P. Steffe, 45–65. New York: Springer-Verlag.

# 7 Missing-Value Proportional Reasoning Problems: Factors Affecting Informal Reasoning Patterns

*James J. Kaput*
*Mary Maxwell West*

## INTRODUCTION

### A Context for Our Research

The standard way to approach understanding the complex processes of learning and doing some type of mathematics has been to treat the key activities as tasks, which is how they are typically designed and experienced in school and in psychological experiments. The tasks are treated as collections of variables, and the students likewise are defined by sets of relevant variables. Finally, the observable performance is understood as cognitively mediated, the result of thinking. This then locates explanation of observed performance in identifiable relations between hypothesized cognitions and observed behaviors. Such an approach, which has been the norm in the domain of proportional reasoning, ignores much of significance to understanding or accounting for authentic, situated activity—affect, belief, motivation, and especially, social context and the forms of embeddedness of activity. Ultimately, explanations will necessarily need to encompass wider spheres beyond the "purely cognitive" as Schoenfeld (1987) has put it and as Lave (1988) and Saxe (1990) have so convincingly argued. Indeed, the characterization of what is to be learned as a "mathematical domain" is already a commitment to the standard paradigm.

Nonetheless, for the types of activities to be discussed in this chapter, and the narrow and artificial contexts in which we have studied the mathematical activity, a purely cognitive approach may be sufficient as a start to more ambi-

235

tious studies in richer contexts. In fact, we will offer some analyses of task variables and cognitive task analyses of a fairly traditional sort—but with an important difference: *our analyses are done relative to forms of reasoning that we believe are naturally occurring*. Most analyses of mature performance have previously been organized around patterns of reasoning that depend on a formal algebraic representation of a proportion. Further, we hope that our attention to semantic factors embodied in the tasks and "naturally" occurring informal reasoning patterns will help inform work in the wider sphere, where the important work needs to be done. And, in this way, as discussed near the end of the chapter, help provide the basis for a curriculum that does a better job of respecting and building on naturally occurring forms of reasoning in the conceptual field of multiplicative structures. This would parallel the results of recent work in the field of additive structures.

## Our Starting Points

Standard *missing-value proportional reasoning word problems* are problems where one instance of a ratio is given, either implicitly or explicitly, and one of the two values of a second equivalent ratio is given—where the equivalence of ratios may also vary in the degree of explicitness. The task is to determine the corresponding value of the second ratio. This is a particular case of more general proportional reasoning tasks that involve "reasoning in a system of two variables between which there exists a linear functional relationship" (Karplus, Pulos, and Stage, 1983, p. 219). These are also referred to by Vergnaud (1983) as isomorphism-of-measures problems.

Harel and Behr (1989), and Harel, Behr, Post, and Lesh (1991, 1992) have offered a very careful and detailed task-variable taxonomy of missing value proportional reasoning word problems, the most comprehensive to date. Their analysis yields 256 types of problems, partially ordered to reflect relative levels of difficulty. Their work shares, with that of many others in this area of research, an approach to the mathematical content organized by the formal algebraic structure of missing-value problem solutions—the writing and then the manipulation of an equation representing equivalent ratios as equivalent fractions. This is reflected at all levels of their analysis, and especially in the language used to describe the task variables.

The approach taken in this chapter is complementary to that analysis in two ways: first, because we base our task-variable analysis on more primitive, less algebraically organized solutions; and second, because we attend to aspects of problem context and semantics that, although mentioned by others, are not central to their analysis. Our analysis, much less ambitious in scope and level of detail, relates more closely to the analysis of primitive multiplication and division offered by Kouba (1989), Steffe (this volume), Thompson (this volume), Carraher and Schliemann (1985), Carraher, and Carraher (1988) and analyses of primitive proportional reasoning strategies documented by Hart (1984) and explored by Lamon (this volume). Indeed, Harel et al., (1991; in press) and Behr, Harel, Post, and Lesh (this volume) have supplied some of the underlying analysis of unit formation that also informs our work.

## A Rationale for Our Approach

A strong message from recent psychological research in learning and cognition, not only in multiplicative structures but in most areas that have been studied in detail by education researchers, is that patterns of reasoning developed prior to formal instruction have effects on student approaches to problems that outlast or intrude in powerful ways on the formally taught approaches Resnick (1983). Hence our strategy here is to concentrate on the basic patterns of multiplicative and proportional reasoning rooted in counting and unit formation that predate instruction, but that can be incorporated into potentially effective instructional strategies.

## Chapter Plan

This chapter is organized into three parts. Part I lays out our theoretical framework regarding (a) the nature of intensive quantities as mental constructions and the different forms of these constructions; and (b) the different informal patterns of reasoning observed among students prior to and after instruction, contrasting these with formal, equations-based approaches typically taught in schools. Part II describes our empirical work in relation to the theoretical framework by (a) describing our teaching intervention, including the representations used, and the results that led us to a further study of the problem characteristics associated with problem difficulty; (b) analyzing problem characteristics that seem to underlay observed variations in problem difficulty; and (c) describing the reasoning we observed among students in interviews that attempted to control certain of these problem characteristics while varying others. Part III briefly draws curricular design implications of our work as well as lays out suggestions for further research.

## I. THEORETICAL FRAMEWORK

### Levels of Understanding of Intensive Quantities: Particular Ratios vs. Rate-Ratios

*Extensive and Intensive Quantities.* Our view of the *formal* properties of quantities and operations on them is given by Schwartz (1988). That is, we use the convention that the "extensive quantities" mentioned in a problem statement consist of numbers and referents, where the referent identifies the measure, in some unit, of some aspect of an entity, situation, or event (length, area, temporal duration, monetary value, weight). Thompson (1990) defines a *quantity* as a conception of a particular "quality" of an object that, in its fully formed schema, "is composed of the object, a quality of the object, an appropriate unit or dimension (i.e., equivalence class of units), and a process by which to assign a numerical value to the quality" (p. 3). This is another, fuller way to capture what we mean by a *quantity* including a number and a "referent," where the referent includes a unit that measures an aspect of the entity, situation, or event being considered.

A special case of measure is a count of some discrete set (seven pieces of silverware), where the unit is based on the separate identity of the objects in the set (no claims are made here about the conceptual nature of the unit, which may vary across individuals according to their experience and development).

Two extensive quantities can be used to construct an *intensive* quantity such as 3 pounds per cubic foot, 5 miles per hour, 3 pounds per 5 dollars, 3 parts of oil for 5 parts vinegar. We therefore use the phrase *intensive quantity* as a blanket term to cover all the types of quantities typically described in our culture as rates (speed, density, price), all manner of ratios (e.g., seven pieces of silverware for every four pieces of china), unit conversion factors (e.g., 3 ft/yd), scale conversion factors (1 ft/in), and so on. Most of these can be described using the "*X* per *Y*" locution.

*Particular vs. Rate Conceptualizations of Intensive Quantities in Terms of Their Referents.* Intensive quantities can be conceptualized and used in two fundamentally different ways, which we will refer to as *particular intensive quantity* (or *particular ratio* and *rate intensive quantity* (or) *rate-ratio*. A rate-ratio is a conceptualization of an intensive quantity as a general description of an entity, situation, or event, as when we use *seven silverware for every four china* as a general description of *all* the placemats that a restaurant sets out. Similarly, if *3 pounds per 5 dollars* refers to the price of *any* sized quantity of peanuts, then it is acting as a rate-ratio.

On the other hand, if we are describing a *particular purchase* of 6 pounds for 10 dollars or 3 pounds for 5 dollars, then we are using the phrase as a particular ratio description, a particular instance of the rate-ratio. Similarly, we are using a particular ratio to describe the (multiplicative) relation between the number of pieces of silverware and china on a particular table, say twenty-eight pieces of silverware to sixteen pieces of china. This particular vs. rate distinction similarly applies to other forms of intensive quantity, for example, scale or unit changes.

Unfortunately, these two very different conceptualizations of intensive quantity are designated by the same notation. Even in ordinary language, we typically refer to the general price of peanuts and the particular price of a particular purchase using the same terms, relying on context to distinguish between them. Indeed, depending on whether we are thinking of a sale situation in terms of "getting" or "giving," we might describe the price either as pounds per dollars or as dollars per pound, respectively.

A rate intensive quantity applies to a situation only when the attribute of the situation being described by that intensive quantity is possessed by that situation homogeneously. Hence a rate understanding also involves understanding of homogeneity, as will be further discussed.

*The Particular vs. Rate Distinction in Mathematical Terms.* We have distinguished the particular-vs.-rate conceptualizations as two ways of conceptualizing the *referents* of the quantities—the entities, events, or situations. We can also distinguish them in mathematical terms. As already noted, Karplus et al. (1983) described proportions as standing for linear functional relationships. They viewed an intensive quantity (he referred to such as *rates* or *ratios*) as the multiplier $m$ in the linear function $y = mx$. Thus, they assumed a rate-ratio conception underlying equations of the form

(1) $y$(dollars) = [5 (dollars)/3 (pounds)] * $x$ (pounds), or
(2) $y$ (dollars)/$x$ (pounds) = 5 (dollars)/3 (pounds)

In these terms, a particular purchase corresponds to a particular input-output pair of this function, which can be realized via particular substitutions for $x$ and $y$. In the second equation this yields a particular ratio.

Thus we can identify a rate-ratio with a uniquely defined linear function between the two measures involved, for example, as in (1), from numbers of pounds to numbers of dollars. And then one can regard a rate-ratio as having particular values, which we have termed *particular ratios*. Note that here we are speaking of functions on quantities rather than pure numbers, an important distinction. Just as quantities can be thought of as having two "layers," namely, numbers and referents, functions on quantities can be described as two layered, with one layer being functions on numbers and the second layer being a systematic correspondence between referents. See Schwartz (1988) for a discussion of how a formal mathematics of quantity can be put together.

*Building Rate Intensive Quantities from Particular Intensive Quantities: Experience with Equivalence and Homogeneity.* We believe that many rate intensive quantities are built cognitively from experience with particular intensive quantities. Regarding a particular intensive quantity as a mental operation relating a specific pair of quantities, a series of such mental operations must precede the generalization to a rate intensive quantity. The student should be able to create and act upon many particular ratios in a given context before being expected to conceptualize a rate-ratio describing that situation in more abstract terms. After all, one must experience more than one particular instance before those instances can possibly be experienced as sharing anything.

Our everyday cultural experience provides extensive practice in particular instances of certain rates, such as price. In the case of speed, the biology of perception may enable us to encode this as a rate without needing to build on particular instances. In both price and speed, the language also helps by providing a ready word that encapsulates the ratio as an entity. Much more can be said about this process, its relation to biological, cultural and linguistic patterns, semantic factors, and so forth. See Thompson (1991; and especially, this volume) for further analysis.

Rate conceptualizations of many intensive quantities are thus built on repeated experience of the particular and are in this sense more "advanced." An important feature of the rate understanding is its complex, interlocking deployment of three major ideas:

1. *Numerical* equivalence across particular intensive quantities,
2. *Semantic* equivalence across situation descriptions, and
3. *Homogeneity* of the intensive quantity's referent in the situation being described.

The notion of equivalence of intensive quantities therefore has two dimensions, one strictly numerical, and the other semantic, having to do with equivalent descriptions of the same situation. Most discussions of equivalence deal with the numerical dimension and usually boil down to discussions of equivalent fractions or equivalence of pure numerical ratios. However, to know that one might provide different, but equivalent descriptions of, say, the same price, is a somewhat different matter. This point is dramatized by saying that "for every four students absent, twelve members of the class are present" when the class has thirty-two students. This is both numerically and semantically equivalent to saying that the ratio of absent to present members of the class is one to three. It would be even more jarring if the class had twenty-eight students. (Of course, both of these are equivalent to the part-whole description saying that one-fourth of the class is absent.) But the fact that we would not normally use these descriptions points out that another, not strictly numeric, factor is involved in our idea of equivalent descriptions, having to do with the relation between the numbers and their referents. Our description-schema seem to have two modes: Either describe the relationship between absent and present students explicitly for the whole collection, or describe it for a minimal representative sample—but do not describe it for an arbitrary-sized subset. Attention to an a particular *sized* subset seems to be reserved for a *particular* subset. Perhaps this should be regarded as a linguistic pattern issue rather than a schema issue.

The third, closely related ingredient of a genuine rate conception is understanding of homogeneity—the idea that *all* purchases are governed by the same price, no matter how that price may be stated, or all samples of homogeneous mixture have the same ratio of ingredients. Equivalence, numerical and semantic, applies only when the property of the situation being modeled is homogeneous in that situation. This may be problematic, and homogeneity may not obtain for a certain situation. But that is a property of the situation, not the description. Nonetheless, understanding homogeneity, and hence when it applies, is an important ingredient of rate-ratio understanding. This can sometimes involve understanding the scope of the description and other subtle matters, such as sampling. Scope occurs as an issue when, for example, a price may change for different sized purchases. Sampling is an issue in density situations and is even more complex in situations involving sampling and averages. The preceding description of the ratio of students absent to those present might be a general description of the entire school system on a particular day or an average across days. In either case, questions of sampling are intertwined with the complex idea of average.

It has been noted by Bell (1989) that, among isomorphism of measure problems, rate problems are generally more difficult than other types. We suspect that this fact may be based on the same underlying conceptual operations involved in understanding rate-ratio as previously described. If one is given a piece of rate information—someone averages 35 mi/hr, say—and then one is given a *particular* quantity of time or distance, one must relate these two types of information using the idea of equivalence and an understanding of the homogeneity—the rate applies across all particular intervals. Thus rates as ordinarily used involve the same conceptual operations that we have discussed. Indeed, this commonality is behind our choice of "rate" terminology.

We are tempted to add yet a fourth factor bearing on a full conceptualization of rate-ratios—the idea of variation. After all, equivalence across different samples, or across different sized purchases, or the idea of a speed of 30 mi/hr applying to different sized intervals of time or distance, all involve conceptualizing multiple instances from among a larger set of possible values. This requires the underlying idea of variable.

### Broad Categories of Task Variables and Reasoning Patterns

Although we will not be reporting on controlled experiments involving systematic variation of task variables (we are reporting a post-hoc examination of data gathered in classroom-based teaching experiments), it will be useful to lay out the categories of variables we will be considering before moving on to a discussion of reasoning patterns. These, in part, are affected by the task variables.[1]

There are four broad categories of task variables to take into account:

1. The semantic structure of the situation depicted in the problem statement;
2. The numerical structure of the problem;
3. The tools and representations available to the problem solver;
4. The form of the text (and perhaps other representations or media) in which the problem statement is presented (Goldin and McClintock, 1984).

Of course, each "task" variable simultaneously serves to define a *student* variable in the sense that, for example, *se-*

*mantic structure* in particular problem-solving episodes, eventually must be interpreted as "semantic structure as understood by the student." Similarly, the *tools and representations available to the problem solver* must be interpreted as "cognitively available," whether or not they are physically present. This said, within each of these categories, our given problems offer a particular value of the task variable.

It is beyond the scope of this chapter to analyze each class of variables at the level of detail provided by Harel and Behr (1989) and Harel et al. (1992), so instead we will examine each as it relates to the three types of reasoning patterns as identified later. The physical tools available to students for the pretests described here were ordinary paper and pencil and the form of text is reasonably well controlled in the problems that we will deal with. We will concentrate first on the semantic variables and then on the numerical structure variables as well as interactions between semantic and numerical variables.

*Reasoning Patterns Constituting Competent, but Informal Proportional Reasoning.* By *competent, but informal proportional reasoning,* we mean patterns of reasoning that support the solution of missing-value problems without reliance on the syntactic manipulation of formal algebraic equations (e.g., cross multiplication or formal division to help isolate a variable). Competent informal reasoning might, in its most sophisticated form, involve writing algebraically correct equations, but not performing syntactically organized operations on them.

We see three broad types of competent, but informal proportional reasoning, the first of which we take as the most fundamental:

1. Coordinated build-up/build-down processes.
2. Abbreviated build-up/build-down processes using multiplication and division.
3. Unit factor approaches.

In the case of discrete quantities and certain whole number contexts, the second of these reasoning patterns is a direct cognitive descendant of the first based on the repeated addition version of multiplication, which in turn is rooted in counting by units other than 1. The primitive roots of the build-up strategies are reflected in their spontaneous appearance independent of instruction that we and others have observed, for example, Hart (1984) and Lamon (this volume). The unit factor strategy seems to be evoked in slightly different tasks, although it can be related to build-up strategies as well.

Each of these forms of reasoning can take place in one or more representational systems. The most frequently used are tables of data and pictorial representations, each of which supports build-up processes most directly. They can appear at varying levels of formal organization. Direct numerical calculations, usually in the form of division and multiplication, frequently appear in association with the second two approaches. Last, they can occur in either discrete or continuous contexts.

A well-documented *incorrect* reasoning pattern described by (Karplus et al., 1983) and much earlier by Piaget and Inhelder (1974), involves an additive rather than a multiplicative approach. It is a kind of truncated build-up process, where the difference between the unit sizes, as will be described, is applied to construct the unknown quantity. After describing these informal reasoning patterns, for purposes of contrast we will describe the traditional formal equation-building approach to solving missing-value problems. We will see strong differences between the ways conceptualization and computation relate to each other in the informal vs. the formal approaches.

## Informal vs. Formal Reasoning Patterns

We will now examine the several forms of competent, but informal, reasoning that have been commonly observed in missing-value problem situations, and then contrast these with the more formal equation-building approach typically taught in schools. To help focus discussion at a finer level of detail, let us consider several specific problems.

- *Placemat Problem* (1): A restaurant sets tables by putting seven pieces of silverware and four pieces of china on each placemat. If it used thirty-five pieces of silverware in its table settings last night, how many pieces of china did it use?

- *Placemat Problem* (2): A large restaurant sets tables by putting seven pieces of silverware and four pieces of china on each placemat. If it used 392 pieces of silverware in its table settings last night, how many pieces of china did it use?
- *Italian Dressing Problem:* To make Italian dressing you need four parts vinegar for nine parts oil. How much oil do you need for 828 ounces of vinegar?
- *Park Problem:* The Boston Park Committee is building parks. They found that fifteen maple trees can shade twenty-one picnic tables when they built the Raymond Street Park. On Charles Street, they will make a bigger park and can afford to buy fifty maple trees. How many picnic tables can be shaded at the new park?

*Build-up Processes.* The simplest build-up processes, as identified by Hart and others, including Lamon, involve the following type of coordinated incrementing of quantities, illustrated by a modal solution to Placemat (1):

For seven silver, there is four china;
for fourteen silver, there is eight china;
for twenty-one silver, there is twelve china;
for twenty-eight silver, there is sixteen china;
for thirty-five silver, there is twenty china.

This amounts to a form of coordinated double skip counting, by sevens and by fours. It might be facilitated by writing a table of some sort, such as those that follow:

| Silver | 7 | 14 | 21 | 28 | 35 |
|--------|---|----|----|----|----|
| China  | 4 | 8  | 12 | 16 | 20 |

| Silver | China |
|--------|-------|
| 7      | 4     |
| 14     | 8     |
| 21     | 12    |
| 28     | 16    |

Very primitive operations underlie this pattern of reasoning, some of which are required before any incrementing can begin. In particular, the solver must distinguish between the referents involved to be able to construct the two quantities involved at a gross level—in this case, counts of silver pieces and counts of china pieces. The solver must then be able to form groups or segments that are the referents for the incrementing quantity steps and finally must be able to coordinate the two types of groups or segments to coordinate the dual incrementing acts.

In the following we list a set of processes that appear to be at the heart of the build-up pattern of reasoning. These need not occur in the exact order given, and some of them may be replaced by the abbreviated versions described in the next subsection. We refer to forming *units* to cover both the discrete and continuous contexts simultaneously. In the discrete case one forms groups, and in the continuous case one forms segments. We break the processes into two categories, initial conceptualization and computation.

### Initial Conceptualization

1. Distinguish between the two referents *A* and *B* to be quantified in the problem solution.

2. Construct a semantic correspondence relation between the classes of referents *A* and *B* at a gross level.

3. Form units within each of the referents, *A* units, *B* units.

4. Construct either a correspondence relation between respective units at the group level (matching an *A* unit with a *B* unit), or construct a higher order group containing the *A* unit and the *B* unit as its two elements.

5. Distinguish between the third given quantity and the fourth, unknown, quantity by linking each to its respective referent type, *A* or *B*. (This is part of setting the goal for steps 6 and 7).

### Computation

6. Increment (decrement) or skip count both quantities until the third given quantity is reached, coordinating on the basis of either the match between replicated *A* units and *B* units or on the basis of replications of the higher order unit consisting of the *A* unit and *B* unit joined together.

7. Identify its corresponding element of the other quantity as the problem's solution.

The two versions of steps 4 and 6 are closely related to the distinction made by Kouba (1989) between matching and grouping approaches to multiplication and division involving discrete quantities. Coordination of the incrementing process across the two quantities has been shown by Conner, Harel, and Behr (1988) to be challenging, even for college students.

*Abbreviated Build-up Processes.* The abbreviation or consolidation of repeated addition into multiplication leads to a corresponding abbreviation or consolidation of the dual build-up/build-down processes described previously. Associated with the more efficient handling of the incrementing-decrementing process using multiplication and division, there seems to be an abstraction of the unit-forming process away from the semantically organized referents and toward the pure numerical values of the quantities involved.

The first five steps in the process forming the initial conceptualization are essentially the same as those in the incrementing-decrementing process. The difference is in the computations used to derive a solution. The push toward a computationally more efficient process may come from a growth in sophistication on the part of the student, or it may come from an increase in the numerical complexity of the problem.

The last part of the abbreviated version solution to Placemat 2 amounts to the following, where we assume that the first five steps of the buildup process have been completed: We are given 392 pieces of silverware, so 392 silverware divided by 7 silverware per placemat gives 56 placemats. There are 4 pieces of china per placemat, so there were 4 china per placemat times 56 placemats, which gives 224 pieces of china. More symbolically, we can write this two step process as follows:

$$\frac{392 \text{ silverware}}{7 \text{ silverware/placemat}} = 56 \text{ placemats}$$

$$56 \text{ placemats} \cdot 4 \text{ china/placemat} = 224 \text{ china}$$

We will describe these two steps at 6A and 7A, amounting to the abbreviated versions of 6 and 7 of the pure buildup process.

6A. Divide the total given quantity by the quantity per unit to obtain the number of units.

7A. Multiply the number of units by the corresponding quantity per unit to determine the total unknown quantity.

Note that this process consists of a quotitive division that determines the number of known-quantity increments, followed by a multiplication of that number of increments by the size of the unknown-quantity increment. It is clearly a conceptual elaboration of a build-up strategy in the sense that it is based in the same initial conceptualizations. Its computational execution is based on a prior conceptualization of multiplication and division as abbreviations of the coordinated incrementing process. Despite its conceptual sophistication, the abbreviated build-up process is additive at heart. It models replicative growth rather than true multiplicative growth.

*Divisibility Complications of the Build-up and Abbreviated Build-up Processes.* The major complication of interest here has to do with divisibility of the known quantity's unit size into the total known quantity. This is illustrated in the Park Problem, where the unit size of maple trees is fifteen, and the total number of maple trees given is fifty. In a pure incrementing process, the unreduced unit size may yield incrementing steps that are too large to "hit" the given quantity. In the abbreviated build-up process, the quotient might not be a whole number, hence not correspond to an integral number of units. The two common strategies for dealing with a divisibility failure are (1) to make a unit-size adjustment early in the process and then carry through with the process with the adjusted unit, or (2) to make an adjustment late in the process. We will illustrate each, for both the pure and the abbreviated build-up processes.

For both processes one may reduce the unit sizes of the corresponding quantities by dividing each by an integer so that the resulting known-quantity unit size does divide the known quantity. This amounts to a step 3.5, between steps 3 and 4. One then carries through with the adjusted unit sizes. The application of this strategy in the Park Problem would involve dividing by 3, to yield a unit size of 5 for groups of maples and $21/3 = 7$ for the corresponding table unit size. Then either the pure or abbreviated incrementing process would be executed as described already.

A late-adjustment alternative for the incrementing process is to carry through the increment by the given unit size until that step which would overshoot the given quantity. For the Park Problem, one would increment trees from 15 to 30 to 45, and corresponding picnic tables from 21 to 42 to 63. One then reduces the unit size of the given quantity so that additional increments would reach the target. For the Park Problem, the target of 50 would be achieved by reducing the tree unit to 5 (and hence the tables unit to 7). Thus one would add on one more increment to reach 50 trees and 70 tables. This amounts to an adjustment of step 6 to include an additional action 6.5.

A late-adjustment alternative for the abbreviated incrementing process involves analyzing the remainder after dividing. For the Park Problem, one would divide 50 trees by 15 to get 3 groups of 21 tables, with a remainder of 5 trees. Now 5 trees is 1/3 of 15 trees and 7 tables is 1/3 of 21 tables. By a unit-reduction argument as used earlier, one adds 7 tables to the $3 \cdot 21 = 63$ tables to complete the solution, 70 tables. This amounts to adding a step between 6A and 7A, and following 7A with a final adjustment.

*The Unit Factor Approach.* The unit factor approach was also observed to occur spontaneously among the students tested before instruction. It can be developed as a natural elaboration of the divisibility adjustment strategy. In fact, this was the basis for introducing the unit factor idea, along with unit pricing, in the teaching experiment. It is typically used in continuous contexts, where whole number quotients are not required. Again, the four initial conceptualization actions underlying the unit factor approach are essentially the same as for the incrementing strategy. But once the two initial units are matched, a decision needs to be made regarding which quantity's unit size will act as the divisor and which will act as the dividend in forming the quotient that will be the unit factor. This requires distinguishing between the known and unknown quantities, so that the unit size of the known quantity can act as the divisor. But this is merely the fifth step with a particular prefigured unit size. Hence the first five steps of the unit factor approach are essentially those of the build-up processes. The actual computations that complete the solution amount to dividing to determine the unit factor and then multiplying the quotient by the given quantity. We will label these as steps *6U* and *7U*, respectively. Notice that the division here is partitive rather than quotitive, which was the case in the abbreviated build-up processes.

6U. Divide the unit size of the unknown quantity by the unit size of the known quantity to determine the unit factor.

7U. Multiply the unit factor and the given total quantity to determine the total amount of the unknown quantity.

In the case of the Italian Dressing Problem, these steps amount to dividing the 4 parts vinegar by the 9 parts oil, to get .44 parts vinegar per one part oil. Then multiply the .44 parts vinegar per one part oil by 828 ounces of oil to get the solution, 368 ounces of vinegar. More formally, we might write this as follows:

$$\frac{4 \text{ vinegar}}{9 \text{ oil}} = .\overline{44} \text{ vinegar/oil}$$

$$.\overline{44} \text{ vinegar/oil} \cdot 828 \text{ oil} \cong 368 \text{ oz. vinegar}$$

*The (Incorrect) Additive Approach and True Multiplicative Reasoning.* A student may distinguish the quantities, construct units, and correctly identify the unknown quantity, but then go on to assume that the difference between the unknown quantity and the total known quantity is the same as the difference between the respective unit sizes. For example, in Placemat 1, we were given that there are seven pieces of silverware for every four pieces of china, and there are thirty-five pieces of silverware. A student using the additive approach would assume that since there are three fewer pieces of china (per placemat), the total number of pieces of china should be three fewer than thirty-five, the total number of pieces of silverware. That is, there should be thirty-two pieces of china.

The additive approach seems to have several sources, depending on the state of the student and certain task variables. Sometimes it appears as a default procedure when the student is confused—a way to do *something* in the face of confusion, as occurred in certain of our interviews. Other times, especially when the difference between the unit sizes is relatively small compared to the unit sizes themselves, it is a deliberate strategy based on an additive conceptualization. While the Placemat problem did not elicit additive errors, the Chairs problem did: "Joan used exactly fifteen cans of paint to paint eighteen chairs. How many chairs can she paint with twenty-five cans?"

Finally, in cases involving geometric similarity and linear measurement, it is a strongly adopted strategy that dominates all others in frequency of occurrence. In this case its prevalence seems to be tied to a lack of understanding of the quantitative implications of similarity and the pull to compare linear measurements additively. This pull seems to be based on extensive experience using additive differences in linear measurement comparisons—we almost always answer questions about which is longer, who is taller, which is farther, and so on, with a subtraction of lengths or distances.

One other factor, which is primary to the relation between additive and multiplicative conceptions, involves two very different conceptions of growth: replicative, as underlies the informal approaches discussed in this chapter, and multiplicative, where the growth is distributed continuously through the quantities, as in scale change. We suspect, and agree with Confrey (this volume), that this latter conception has fundamentally different roots in experience and cannot be reduced to the former. Part of our suspicion is based on the very different results we obtained in the geometric similarity problems, which require the multiplicative conception, vs. the other problems, for which a replicative notion of quantity change suffices. This distinction deserves more attention than it receives in this chapter.

*The Formal Equation-Based Approach.* For contrast, we will examine briefly the standard school approach to solving missing-values problems. To set up a proportion equation, one needs first to identify and distinguish the quantities in-

volved. This amounts to the first two steps in the conceptualization stage described previously. However, at this point, depending on the method taught, one can begin to write an
equation directly—either a within-measure or an across-
measure comparison (Tourniaire and Pulos, 1985). In either
case, one is constructing two multiplicative comparisons,
one of which is entirely known and the other of which must
have equal value. Interestingly, a within-measure comparison involves a pair of part-whole relationships, whereas an
across-measure comparison involves a pair of part-part relationships.

For example, in the Placemat Problem 2, we need to compare number of pieces of silverware and china. If one is to
write a within-measure comparison equation, one need only
keep amounts of silverware and china data on separate sides
of the equation. But, to complete the equation-writing process, one needs to know how to match the two pairs constituting the two ratios that need to be declared equal. In this case
two pieces of silverware data are given (7 silverware per placemat and 392 silverware altogether) and only one of china is
given (4 china per placemat). However one writes the silverware side of the equation, the china side should match in
terms of part-whole relationship. Suppose we write a part/
whole statement, (7 silverware)/(392 silverware). Then the
part-whole relationship for china must be written in the same
direction, (4 china)/(? china). Usually, the unknown is represented by a literal, say X. This yields the following equation,
where we have included referent units:

$$\frac{7 \text{ silverware}}{392 \text{ silverware}} = \frac{4 \text{ china}}{X \text{ china}}$$

What minimal conceptual understanding of the situation is required to preserve the part-whole or part-part order
in the writing of such an equation? It appears that part of the
answer to this question is something akin to the conceptualization described in step 5: to distinguish known from unknown quantities, and to relate these to the two types of
quantities involved (in our example, silverware and china).
But an additional ingredient may also be required; namely, an
understanding of the invariance of that relationship across
the two multiplicative comparisons.

It further seems to be the case that one can create the equation *without* this understanding—by using a gross distinction based on the relative size of the numbers involved. In our example, one could simply note that the small-to-large comparison of the silverware needs to be maintained by a small-to-large comparison of china. In this case, one can substitute this heuristic for a genuine quantitative understanding of the situation and successfully create an equation.

Whatever means are used to create the equation, at this point the syntactic maneuvers that constitute the computation of the solution can begin. These will vary according to the position of the unknown in the equation. The fact that we wish to emphasize is that *those maneuvers are not guided by one's conceptualization of the quantitative content of the situation. They are quantitatively vacuous.*

Note further that the intermediate quantities formally generated in the course of these maneuvers do not normally have sensible referents. For example, if one were to multiply both sides of the equation by "$X$ china," or if one were to "cross multiply" (which is actually the composite of several quantitative operations), the referents would not have a sensible interpretation in the situation being modeled. This is very much in contrast with the informally based computations, especially the build-up approach, where each intermediate state derives its existence from and is tied to the initial conceptualization.

*Relations Between Initial Conceptualization and Computation of the Solution and the Role of Heuristics.* We will now contrast the relation between initial conceptualization and solution computation in the informal reasoning patterns with the corresponding relation in the formal equations approach. In the informal approaches, the relation is simple and natural—the computation is a natural extension of the conceptualization. *One is operating directly on the structures produced by the initial conceptualization.*

This is strong contrast to a formal, equations-based approach, where the switch from initial conceptualization to computation is abrupt, and where *one acts upon the formal system of symbols, using its syntactical rules rather than one's conceptualization of the situation.*

Although we have not dwelled on heuristics, a useful "looking back" heuristic for the informal approaches is to compare the relative size of the unknown and known unit sizes with the relative sizes of the computed and known total amounts to see if they are in the same direction. This amounts to an eighth step in the sequence.

This, as well as the previously mentioned size-preservation heuristic, amounts to a switch in mode of thinking from computation back to conceptualization. However, as we have described them, each of the informal reasoning patterns involves a move from conceptualization to computation, so the heuristic amounts to a return to the conceptual mode. Another general heuristic for the informal approach is to track referent units to determine that the composite units produced during the solution process make referential sense, especially in the unit factor approach (this is sometimes called *units analysis*). Note that this does not quite apply in the formal equations approach after the equation has been set up, but it does apply in the setting-up process.

We have witnessed experienced teachers advocate using the larger-smaller pairing heuristic with students, including the use of units or descriptors in the equation, with considerable success—as measured by student ability to solve routine problems. The boundaries of their understanding and competence are easily revealed, however, by their poor performance on variant problems involving totals or part-whole comparisons.

Note the contrast between the number-size direction heuristic in the context of building an equation and the number-size direction heuristic as a check on the outcome of the informal approaches. The former is intended to be used as a *substitute* for a quantitative understanding, whereas the latter is intended as a *supplement*.

## Revisiting the Rate Concept in the Context of Problem Solving

*On the Relations Between Unit-Formation and Rate-Ratio Conceptualization.* Let us briefly examine how the underlying mental operations of unit formation and the development of the rate conception may relate to each other. We take as a starting point that two basic processes are involved in the construction of a rate intensive quantity conception. One is the two-stage process of unit formation, and the second is the development of the rate conception from experience with particular intensive quantities. Each of these takes place in a context where there is some understanding of homogeneity in the situation being modeled.

It appears that the unit formation process must precede the rate abstraction simply because the paired units are the cognitive objects that are then applied to abstract the rate conception.

First, let us quickly review the process of unit formation associated with solving the placemat problems. One must form units of seven silverware pieces and units of four china pieces. Then one forms higher order units of these units; namely, units of units of silverware and units of china pieces. The collection of these higher order units constitutes the rate-ratio correspondence and hence serves as the basis for computing the solution—in either build-up strategy. In this particular case, a semantic factor helps form the higher order units; namely, the placemat that produces cognitively a semantic "holder" of the two lower level units. There also happens to be a strong separation of identities of the two lower level units, a separation maintained through the situation depicted. Beyond the placemat organizing our conception of the lower level units into the higher level ones, there is also the functional semantic relation between silverware and china—they "go together" to help one eat a meal. Hence they are associated, but they maintain this separation. We regard all this as the functional role of an appropriate restaurant schema (or "frame" or "script") to organize one's thinking.

The separation of quantities in the placemat situation contrasts with the Italian dressing situation, where both quantities are continuous, *and* they are mixed together. Hence they lose their separate identities. Nonetheless, they are strongly associated as ingredients of the same foodstuff. However, there is no obvious higher order unit holding a share of oil with a share of vinegar, as would be the case if the given information were stated in terms of producing, say, a bottle of salad dressing. Notice the wording "four parts oil to five parts vinegar." Although the other quantitative information is given (and presumably requested) in particular units, namely, ounces, the size of the segments is given *abstractly*—a *part* can consist of any number of ounces. Hence there is a pull toward a rate-ratio conceptualization of the situation in a way not provided in the placemat problem.

*The Roles of Rate vs. Particular Conceptions of Intensive Quantity.* What level of understanding of intensive quantity is required or developed by the four different patterns of reasoning described earlier? We suggest that the different approaches differ in the conceptions of intensive quantity they require and that these conceptions are differentially generated in response to different situations, that is, different task variables.

The most primitive informal strategy, the build-up strategy, appears to have no real requirement for a true rate conception because, using it, one can generate a solution based on a series of particular ratios. This is especially the case if one employs a table or organized list of some kind as described earlier. Further, the idea of homogeneity is not likely to play a large role except to govern the replication process. On the other hand, in using a build-up process, one is likely *constructing* a rate conception, because the result of the process is exactly the kind of sequence of experiences with particular ratios that is required to build the rate conception as described earlier.

The cognitive construction of a correspondence between the two entities in a situation can be regarded as the key step in the construction of a rate-ratio between the measures of these entities. This conceptualization seems to take place in two ways. One of these is a "bottom-up" or term-wise construction based on coordinating pairs of segments or groups. Such may be regarded as relatively unsophisticated and supports the pure build-up strategy. Its scope is limited to situations involving small numbers and relatively simple numerical relationships.

A more sophisticated wholistic conceptualization is one that regards the given relationship as applying to all possible segment or group pairs at once. Here homogeneity seems to play a stronger role. This rate-ratio conceptualization of a situation enables one to use the "all-in-one" division and multiplication process that does *not* require attending to all the individual pairs, as in the build-up strategy. It is both conceptually and computationally more efficient.

Nonetheless, we suspect that the abbreviated build-up strategy places a relatively low demand on a rate conception, because one can can divide by the size of a unit without necessarily understanding the universality of that unit as participating (matched with the other unit) in a rate description of the situation. On the other hand, if there is a divisibility failure, requiring one of the "repairs" to either version of the build-up process, then a rate conception likely plays a considerably larger role. One must be able to recognize that a given pair of units constituting an intensive quantity, as a descriptor of the situation, is equivalent to another, "reduced," pair. Appreciating this equivalence and knowing that it applies across all instances of the situation being modeled are at the heart of the rate conception, so without a rate-conception, students are unlikely to perform the necessary reduction adjustments except perhaps in special cases where halving is involved.

The unit factor strategy, for the reason just cited regarding divisibility adjustments, seems to require a rate conception. After all, the unit factor *is* a rate.

The formal, equation-based strategy does not seem to require a rate conception. In reviewing the steps involved, one can build the equation with only (1) a gross distinction between the quantities, (2) a relation between the total amount and the unit amount, where this latter distinction can usually be based on a simple size discrimination (the unit is the smaller quantity), and (3) the preservation of this relation across the two quotients. In fact, as noted, it is possible to construct an equation without a full quantitative understanding of the situation. Of course, the formal solution process is even more independent of quantitative understanding.

## II. EMPIRICAL WORK

### Overall Organization and Methods of the Study

We report here the results of a teaching experiment and a follow-up study designed to clarify characteristics underlying problem difficulty as observed in selected pretest problems.

We executed a teaching experiment involving the entire sixth grade ($n$ = 138) of a suburban middle school in an upper-middle income area, in which about 20 percent of students were bussed from the nearby inner city. With the help of two experienced teachers we designed curriculum materials for four units that were integrated into the existing curriculum over a period of five months. Unit A covered partitive and quotitive division using concrete representations; Unit B covered ratio and proportion and used object-based and tabular representations (see examples in Figures 7.1 and 7.2); Unit C focused on comparison of ratios and introduction of coordinate graphs; and Unit D focused on integrated applications of multiplicative thinking (scale changes in astronomy contexts, bicycle gear ratios).

The new materials were used in four experimental classes at the three existing course levels, and three additional classes at the upper and middle levels served as comparisons. These three classes were taught by the same two teachers as the experimental classes and followed the same topics as covered in the Houghton-Mifflin grade 6 text, except for the honors classes, which used a grade 7 version of the text.

Written pre- and post-tests were administered at the beginning and end of each unit and beginning and end of the entire year. In addition, a subsample of twenty-seven experimental and control students were closely tracked in nine interviews over the year and with classroom observations.

The written tests were designed to facilitate examination of student's reasoning on problems. Therefore tests included a variety of problem types, which required shifting reasoning strategy from one problem to the next, and credit was given for showing work. Follow-up interviews with the closely tracked students after each test, in which they were asked to redo and explain their solutions, confirmed that students' major approach to a problem could be identified from their written tests. Therefore it is reasonable to consider the written tests as valid information on students' reasoning on these problems. A description of the coding procedures is available from the authors. A report of the particular strategies used on each problem is in preparation.

Unit B focused on ratio reasoning and provides the most extensive data on students' reasoning on missing-value problems. The Unit B pre- and post-tests included problems requiring simple multiplication and division, missing-value proportion problems using part-part, part-whole and totals of ratios, and "trick" problems that were additive situations presented in wording closely resembling that of proportion problems.

The data presented comparing experimental and comparison groups controls for course level in that it includes the four middle level classes only. The additional data on problem difficulty include the six classes at the upper and middle levels and the pretests for Unit B and for the overall year. There was only one class at the lower level, and this class has been excluded from both analyses because many social and behavioral factors affected the performance of these students and a good understanding of their reasoning requires a more intensive one-on-one interviewing approach than we could pursue.

### Representations Used in Unit B
### for the Experimental Classes

The two experimental classes were taught by Teacher A and were very similar in overall instruction. During this unit students first used the object representation alone as in Figure 7.1, and then used it while linked to a tabular representation. About half of class time was on-computer activities, each child at a separate computer. The activities occupied about 11 days of class time.

A typical problem might be the following (which was not actually used, however): Suppose that Noah decides to give three umbrellas to each pair of animals coming onto his ark (assuming that each pair will eat one over the next 40 days and 40 nights). If he has twenty-one umbrellas left, how many more animals should he allow onto the ark?

*Object-Based Representation.* In doing such a problem, students first chose icons from a menu of choices (screen not shown), in this case triangles for umbrellas and dogs for animals. They then entered the given quantities, twenty-one umbrellas and the ratio of umbrellas to animals, into the problem setup window in the upper right part of the screen (see Figure 7.1). They then went on to build each cell (correctly or incorrectly) according to their understanding of the

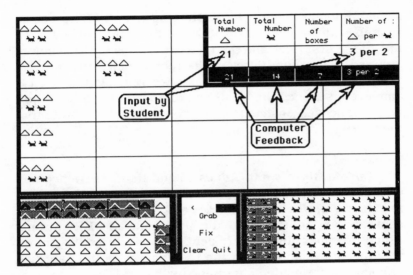

Fig. 7.1
*Annotated Icon-Manipulation Software Screen*

problem by dragging icons into the cells. As they worked they could observe the changing total quantities reported on the screen, given in inverse video as shown in Figure 7.1. Students could grab and drag icons from the reservoirs into the cells in any order they wished and array them any way they wished. They could also adjust their choices at any time by moving icons from one cell to another or back to the reservoir at the bottom of the screen. We have shown a modal screen for a correct solution. A detailed analysis of the processes of setting up and executing such a solution is provided in Kaput (in press).

Note that in the previous intervention, Unit A, the students worked with simple multiplication and division of discrete quantities using a similar object-based reasoning environment, but with only a single set of icons. In that environment problems took the following form: There are three umbrellas per animal; if there are nine animals, how many umbrellas are needed? (rate multiplication); or, there are three umbrellas per animal, and twenty-one umbrellas, how many animals? (quotitive division); or, there are twenty-one umbrellas and seven animals, how many umbrellas per animal? (partitive division).

The representation in Figure 7.1 was deliberately designed to foster the dual level grouping or unitizing process, whereby two animals form a unit and three umbrellas form a unit at the first level, and then the two groups are themselves grouped at a second level in the rectangular cells, respectively. Various scaffolding options were available that, for example, constrained the "grab" action to whole groups, eliminated unneeded objects from the reservoir at the bottom of the screen.

*Linked Objects and Tables.* In addition, experience was offered in manipulating linked representations such as given in Figure 7.2, where both a table and an object-based representation are controlled by the MORE and FEWER buttons, which provide a direct incrementing or decrementing of either or both quantities. In the case shown, both quantities are being incremented simultaneously.

This system models the build-up process directly. As one clicks on the More button, both the table and the object representation increment simultaneously. However, an option also allowed for one or the other representation to become invisible, which put a student in the position of predict-

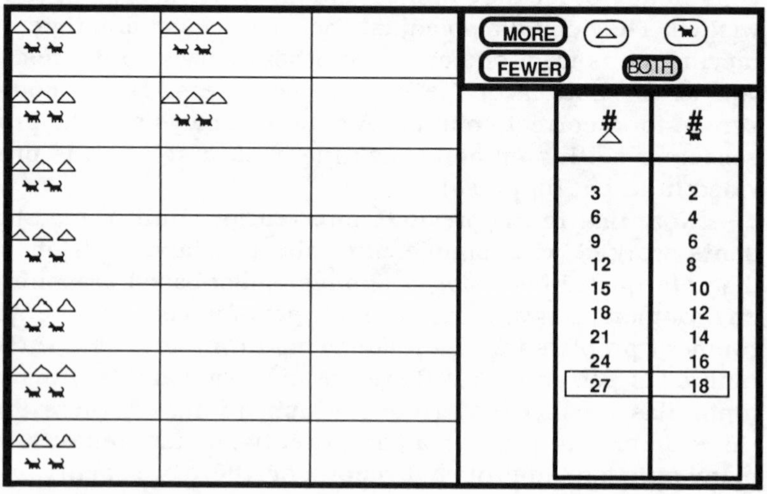

Fig. 7.2
*Typical Tables-Based Software Screen*

ing appropriate values on the basis of what was seen in the visible representation.

Yet another, nonincrementing, version of the table-object linked system requires the students to enter the corresponding value to a given value of one of the variables. For example, the student might be given that there are twenty-one umbrellas and requested to type in the corresponding number of animals or the number of "boxes" (rectangular cells). A correct response would be entered into a table, while an incorrect response would merely yield an object-representation with the corresponding number of animals. This provides immediate visual feedback that the number of animals is not correct, because animals or umbrellas would be left unpaired. This type of activity was intended to promote the abbreviated build-up strategy, which the class referred to as the *boxes approach.*

Note that except in this last case, where the computer generated problems using constrained random numbers, all problems were provided off-line in the form of worksheets and challenges from the teacher. Students were required to write out all solutions, whether they were done on the computer or with pencil and paper.

Although the Unit B activities involved only discrete quantities, Unit C activities later moved beyond to continuous quantities, including one day's work with similar figures, with concentration on different sized, but proportionally designed clothes patterns. This work was preceded by problems involving the need to reduce ratios before an incrementing process would suffice, referred to earlier as the *early-adjustment strategy,* as well as two days involving unit rates. A small amount of C-unit time (less than a class hour) was devoted to coordinate graph descriptions of intensive quantities.

## Comparative Results from Unit B

The written pre- and post-test data for Unit B show that students using the experimental materials performed better on the test problems in several ways, described later. It is important to note, however, that the curriculum of the experimental and comparison classes differed in several dimensions, including the number of class periods of instruction on the topic. In the experimental classes instruction occupied about eleven class periods. The two comparison classes followed the text chapter on ratio and proportion, one (C1, Teacher B) providing five periods from the text and then about eight class periods of computer-based activities, *Function Machines*, designed by W. Feurzeig and J. Richards, on rate and profit, and the other (C2, Teacher A) providing the minimal possible coverage of the topic from the text in three class periods. Therefore overall differences in test performance must be interpreted in relation to the entire treatment and not to an effect of the concrete representations only. The clinical interviews show how the concrete representations are used by students in reasoning; a detailed example is reported in West et al. (1989). The significance of the teaching experiment is to indicate the extent of use of various strategies before and after a given implementation of the materials in the classroom setting.

The Unit B pre- and post-tests were designed to be very difficult for students in grades 6–8, as predicted by previous research (Karplus et al., 1983; Hart, 1984; Tourniaire and Pulos, 1985). That is, the problems used nonintegral ratios and had no obvious unit measure. They covered a wide variety of situations, including geometric figures (known to be difficult) and excluding price (known to be easy).

The experimental approach appears to be natural and learnable, and it yielded more correct and near correct solutions than the traditional approach used in the comparison classes, which relied on proportion equations solved via cross multiplication—the formal approach, including the size—comparison heuristics, described previously. The build-up strategy occurs in 12–15 percent of pretest problem attempts and increases to 42 percent on the posttest in experimental students. Of problems attempted by that strategy, the rate of correct solutions is 90 percent. The formal equation-based strategy does not occur on pretests, increases only to 26 percent on posttests, and has a correctness rate of 70 percent. Thus the overall effectiveness of the traditional approach is lower, confirmed by students' posttest scores shown in Table 7.1.

More important, the experimental approach allowed students to overcome the widely documented tendency to use incorrect additive strategies on hard multiplicative problems

TABLE 7.1.   PERCENT OF CORRECT SOLUTIONS ON PRE-
AND POSTTESTS (*number of students*)

|  | Pretest | | | Posttest | | |
|---|---|---|---|---|---|---|
|  | *Exp* | *C1* | *C2* | *Exp* | *C1* | *C2* |
| 75–100% | 0 | 0 | 0 | 7 | 2 | 0 |
| 50–74% | 0 | 0 | 0 | 9 | 4 | 2 |
| 25–49% | 1 | 2 | 0 | 7 | 1 | 0 |
| 0–24% | 31 | 16 | 13 | 9 | 11 | 11 |

Exp = experimental classes (Teacher A)
C1  = comparison class 1 (Teacher B)
C2  = comparison class 2 (Teacher A)

(Tables 7.2 and 7.3). Sixty-three percent of experimental students, as compared with 39 percent of comparison students,
increased use of correct multiplicative approaches on the post-
test (whether by unit factor, build-up, or formal strategy)—
see Table 7.3.

Additional evidence of stronger understanding of multiplicative structures exist in students' explanations in interviews, and in the experimental students' ability to solve different problem types such as part-whole and totals variants
of missing-values problems. We shall not present this data
here in order to concentrate on missing-value problem features underlying problem difficulty.

TABLE 7.2.   STUDENTS' PREDOMINANT STRATEGY ON
MULTIPLICATIVE WORD PROBLEMS ON PRE- AND POSTTESTS

|  | E1 + E2 | | | C1 | | | C2 | | |
|---|---|---|---|---|---|---|---|---|---|
| Pretest | *A* | *X* | *M* | *A* | *X* | *M* | *A* | *X* | *M* |
| A | 5 | 3 | 8 | 3 | 0 | 1 | 4 | 1 | 1 |
| X | 1 | 5 | 9 | 0 | 2 | 6 | 2 | 3 | 0 |
| M | 0 | 0 | 0 | 0 | 0 | 1 | 0 | 0 | 0 |
| ? | 0 | 0 | 1 | 3 | 0 | 2 | 1 | 0 | 1 |

A = additive
M = multiplicative, including scalar, functional, and cross multiplication
X = mixed additive and multiplicative
E1 & E2 = experimental classes
C1 & C2 = comparison classes
[1] number of students with the specified strategies on the pretest and the
postest.

TABLE 7.3.    STUDENT GAIN IN MULTIPLICATIVE STRATEGY

| Class | Who did not increase use | Who increased use |
|-------|--------------------------|-------------------|
| E1 | 4 | 9 |
| E2 | 7 | 12 |
| C1 | 9 | 9 |
| C2 | 10 | 3 |

$\chi^2 = 4.02$, $df = 1$, $p < .05$, $n = 63$

## Problem Difficulty Analysis: Empirically Derived Rank-Order Hierarchy

Observing that some problems were clearly harder than others on Unit B pretests and post-tests and that students differed in the extent to which the iconic and tabular representations in Unit B helped them grasp the underlying multiplicative structure of a problem (Table 7.2), we pursued a further study to understand what features of problems "make problems easier," which we take to mean help students to construct and apply a rate-ratio conception. We likewise examined features that seemed to make problems harder.

We first undertook a post-hoc analysis of the characteristics underlying problem difficulty for all pure missing-value problems on the two pretests given to the six middle and upper level classes *before* any instruction in ratio and proportion.

The percentage of students in the six classes with correct or near correct solutions (regardless of computational success) on each of fifteen problems was computed, and the resulting ranking of problem difficulty was then applied in a within-student analysis for confirmation. (The strategies that would lead to correct or near correct solutions observed on these pretest problems were the build-up and unit factor strategies with a number of variations, as described in the theoretical section and Unit B results.) The rank order of difficulty across students is shown in Table 7.4.

To examine whether this hierarchy would also apply within individual students, the problems were arranged in this order and the data were examined for deviations from this specific order for each student. Basically, a student was expected to solve correctly all problems below the most diffi-

TABLE 7.4.   PRETEST MISSING VALUES PROBLEMS RANK
ORDER OF DIFFICULTY ACROSS SIXTH GRADE STUDENTS
(before instruction in ratio or proportion; $n = 115$)

| Rank | Sum of % correct in six classes | Problem |
|------|------|---------|
| 1 | 451 | 1) Simon worked 3 hours and earned $12. How long does it take him to earn $36? |
| 2 | 419 | 2) A car of the future will be able to travel 8 miles in 2 minutes. How far will it travel in 5 minutes? |
| 3.5 | 394 | 3) In a certain school there are 3 boys to every 7 girls in every class. How many girls are there in a classroom with 9 boys? |
| 3.5 | 393 | 4) Judy earns $63 in 6 weeks. If she earns the same amount of money each week, how much does she earn in 4 weeks? |
| 5 | 190 | 5) A large restaurant sets tables by putting 7 pieces of silverware and 4 pieces of china on each placemat. If it used 392 pieces of silverware in its table settings last night, how many pieces of china did it use? |
| 6 | 153 | 6) The Boston Park Committee is building parks. They found out that 15 maple trees can shade 21 picnic tables when they built the Raymond Street Park. On Charles Street, they will make a bigger park and can afford to buy 50 maple trees. How many picnic tables can be shaded at the new park? |
| 7.5 | 125 | 7) A printing press takes exactly 12 minutes to print 14 dictionaries. How many dictionaries can it print in 30 minutes? |
| 7.5 | 122 | 8) A men's clothing store sells Hanes™ dress socks in gift boxes. Every gift box |

(*continued*)

TABLE 7.4.    (*Continued*)

| Rank | Sum of % correct in six classes | Problem |
|------|------|------|
| | | has 2 pairs of solid socks and 3 pairs of striped socks. If the store sold 270 pairs of solid socks last week, how many pairs of striped socks were sold? |
| 9 | 116 | 9) To bake donuts, Jerome needs exactly 8 cups of flour to make 14 donuts. How many donuts can he make with 12 cups of flour? |
| 10 | 101 | 10) Joan used exactly 15 cans of paint 18 chairs. How many chairs can she paint with 25 cans? |
| 12 | 78 | 11) The West Middle School has a bake sale. The baggies of cookies contain chocolate chip cookies mixed with oatmeal cookies. All the baggies are exactly alike. Altogether, the baggies on the end table have 16 chocolate chip cookies and 20 oatmeal cookies. On the middle table, the baggies have a total of 24 chocolate chip cookies. How many oatmeal cookies altogether are in the baggies on the middle table? |
| 12 | 78 | 12) Tyrone's Toy Store sells toy cars in boxes that contain sports cars mixed with trucks. Every box is exactly alike. Together, the boxes in the first aisle have 9 sports cars and 15 trucks. The boxes in the second aisle have a total of 12 sports cars. How many total trucks are in the boxes in the second aisle? |
| 12 | 78 | 13) To make Italian dressing, you need 4 parts vinegar for 9 parts oil. How much oil do you need for 828 ounces of vinegar? |
| 14 | 45 | 14) The two sides of Figure A are 9 cm high and 15 cm long. Figure B is the |

(*continued*)

TABLE 7.4.  (*Continued*)

| Rank | Sum of % correct in six classes | Problem |
|------|------|------|
|  |  | same shape but bigger. If one side of Figure B is 24 cm high, how long is the other side? |

9 cm.    A    15 cm.    24 cm.    B    ? cm.

| 15 | 31 | 15) The two sides of Figure are 35 cm. high and 30 cm long. Figure B is the same shape but smaller. If one side of Figure B is 21 cm high, how long is the other side? |

35 cm.    A    30 cm.    21 cm.    B    ? cm.

cult one solved correctly, and expected to fail all those above. The extent to which the data fit this model is shown in Table 7.5, by class.

A sample of students from one class is shown in Figure 7.3. The definition of deviation is illustrated in this figure; we marked the point at which the student missed more than two problems. If the student failed an "easy" problem before that point, it was termed a deviation. And if a student solved a "harder" one after that point, it was likewise termed a deviation.

Table 7.5 shows that the percentage of fit to this model is 86–97 percent, indicating that this hierarchy applies within students as well as across students for this set of problems. The hierarchy is quite robust.

TABLE 7.5.  PRETEST PROBLEMS RANK ORDER
OF PROBLEM DIFFICULTY: FIT OF ACROSS-STUDENT MODEL
TO INDIVIDUAL STUDENTS ($n$ = 106)

| Class | N | Total possible fits | No. of deviations | Percent of fit to model* |
|-------|-----|-----|-----|-----|
| RA | 22 | 330 | 33 | 90 |
| RE | 23 | 345 | 45 | 87 |
| RC | 20 | 300 | 21 | 93 |
| BD | 12 | 180 | 10 | 94 |
| BE | 18 | 270 | 20 | 93 |
| BF | 11 | 165 | 4 | 97 |

*This percentage is the same as Guttman's (1941) statistic to describe sca-
lability of items. Although these data fit a hierarchical model well using
Torgerson's (1958) criterion of .90, we are not using this analysis to define a
scale, but only to differentiate hard from easy problem types, which should
be further investigated by clinical methods.

## Problem Difficulty Analysis: Analytically Derived Hierarchy

*Task Variables Facilitating Problem Solution.* We developed an interpretation of problem characteristics underlying
the observed rank order. Although the problems differ in
many features, most have in common certain features that
have been shown to underlie problem difficulty in other studies (see Karplus et al., 1983). In most problems, quantities
were introduced in ascending order (as opposed to descend-
ing order, which is known to tax processing capacity), and the
unknown quantity was introduced lastly. Therefore these
problem-presentation features were not the main ones differentiating this particular group of problems and will be ignored.

We hypothesized that the following features would make
problems easier by making them more likely to be conceptualized as a multiplicative comparisons and more amenable
to some form of build-up strategy (either a pure build-up, or
an abbreviated version, as discussed earlier). We note that
each of these factors relates either to the initial concep-

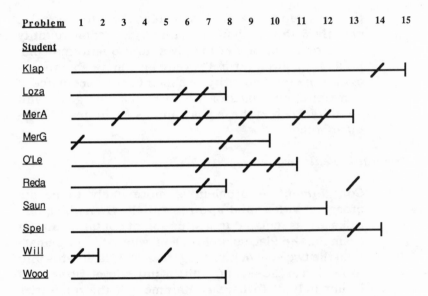

Horizontal line means expected (consecutive) correct sol'n
Slash means deviation from expected correctness value

Fig. 7.3

*Sample of data, showing individual deviations from group problem hierarchy.*

tualization stage or to the computation stage of solving the problem.

## Numerical Features

- *Reduced form of ratio:* Beyond making problems more likely to be conceptualized as a multiplicative comparisons and more amenable to some form of a buildup strategy, success is also more likely because the repair processes associated with divisibility problems described earlier would not likely be required (we provided divisibility challenges only in the context of unreduced ratios).

- *Familiar multiple:* A problem is more likely to be approached with a multiplicative strategy if the quantities in either scalar or functional relationship are a multiple familiar to students; that is, in the multiplication table up to 9 × 9. This enables them to recognize directly, without calculation, that a given total quantity is a product of one of the given unit quantities comprising the given ratio. They are also more able to recognize the total quantity as a member of a sequence of increasing multiples of the corresponding given unit quantity. Each of these possibilities facilitates a build-up approach.

## Semantic Features

- *Containment:* A problem is more likely to be approached with a build-up strategy if the given unit quantities are associated in a situation of containment, for example, the Placemats Problem, where the placemats hold the two sets of items together. This facilitates unit formation at the second, units of units, level. However, it is not helpful to have containment if the quantities contained are not specified (as in the Socks, Cookies and Toy Store Problems; more on this shortly).
- *"For every/each" statement:* A situation is more likely to be conceptualized as a rate-ratio if underlying quantities are associated by an explicit "for every" or "for each" statement, because it makes the matter of homogeneity explicit.
- *Familiar rates:* A situation will be understood as a rate-ratio situation if the underlying quantities are associated in the rates most children are familiar with, such as speed and price. Then it is also more likely that a unit-factor approach would be utilized, which adds to the set of available solution methods.

*Task Variables Debilitating Problem Solution.* Certain problem features either the initial situation conceptualization or the solution computation more difficult.

## Numerical Features

- *Divisibility failure:* If the two given quantities did not evenly divide one another, as in the Park Problem (15 trees does not divide evenly into 50 trees), then a "repair" or elaboration process was required, involving considerable additional computational processing.

- *Small differences between quantities making up the given ratio:* In our discussion of the additive error pattern, we noted that there was an "additive pull" in cases where the quantities making up the given ratio were numerically relatively close to one another. This seemed to invite an additive rather than a multiplicative comparison between the quantities, as in the Chairs Problem (15 vs. 18).

## Semantic Feature

- *Ambiguous groups:* This variable can manifest itself in several ways. One, as in the Italian Dressing Problem, involves ambiguity in the respective identities of the extensive quantities in the situation. When the ingredients are mixed, their separate identities are lost, which in turn, we hypothesize, creates an additional challenge to keeping their measures conceptually separate in the initial conceptualization stages of essentially all the strategies discussed earlier. Also, the lack of easily imaged containers or other markers by which the two units might be identified or constructed (a consequence of the fact that continuous units are involved) may contribute to the quantity-identity problem. A second way that grouping actions might be ambiguated is through the introduction of additional, nonfunctional grouping elements in the situation depicted. Such occurs in the Socks Problem, where two levels of pairs of socks are introduced—the "natural" pairing arising from humans being bipeds, and the "artificial" pairing created for the sale. A third level exists in this case due to the use of the boxes in which paired pairs are sold, which might be regarded as a functional grouping element. A fourth level of organization arises from the unit of time for which the given sale data applies, which is one week.

The relation of these features to the empirically derived problem hierarchy is shown in Table 7.6, where we combined in a simple linear fashion the positive and negative features

TABLE 7.6. CHARACTERISTICS OF MISSING VALUES PROPORTION PROBLEMS

| | | Numerical Features | | Situat. Features | | | | Numer. Features | | Situat. | | |
| | | Reduced Ratio? | Easy Multiple? | Quantified Containment? | "for every" Stated? | Price/ speed Rate? | Total Facilitating | Uneven Multiple? | Close Instances | Ambiguous Groups? | Total Detracting | Sum |
| Problem | Quantities | | | | | | | | | | | |
|---|---|---|---|---|---|---|---|---|---|---|---|---|
| Simon-pay 8A | 3:12/x:$36 | 1 | 1 | | | 1 | 3 | | | | 0 | 3 |
| Car of future 11A | 8:2/x:5 | 1 | 1 | | | 1 | 3 | | | | 0 | 3 |
| Class 4A | 3:7/9x | 1 | 1 | | 1 | | 3 | | | | 0 | 3 |
| Judy pay 15aA | $63:6/x:4 | 1 | 1 | | [1] | 1 | 2 | | | | 0 | 2 |
| Placemat 4C | 7:4/392x | 1 | | 1 | [1] | | 2 | | | | 0 | 2 |
| Park 1C | 15:21/50x | | | | | | 0 | -1 | | | -1 | -1 |
| Print 10C | 12:14/30x | | | | | 1 | 1 | -1 | | | -1 | 0 |
| Sox 11C | 2:3/270x | 1 | | 1 | | | 2 | | | -1 | -1 | 1 |
| Donut 14C | 8:14/12x | | | | | 1 | 1 | -1 | -1 | | -2 | -1 |
| Chairs 5C | 15:18/25x | | | | | | 0 | -1 | -1 | | -2 | -2 |
| Cookies 7C | 16:20/24x | | | | | | 0 | -1 | -1 | -1 | -3 | -3 |
| Tyrone 12C | 9:15/12x | | | | | | 0 | -1 | -1 | -1 | -3 | -3 |
| Italian dressing 15 | 4:9/$28x | 1 | | | | | 1 | -1 | | -1 | -2 | -1 |
| Figure 3C | 9:15/24x | | | | | | 0 | -1 | -1 | -1 | -3 | -3 |
| Figure 13C | 35:30/21x | | | | | | 0 | -1 | -1 | -1 | -3 | -3 |

by weighting the positive and negative features as ±1, respectively, and then summed their values for each problem.

This weighting system is obviously not meant to be quantitatively precise as it does not take into account whether certain features may be much more significant than others in their effect on reasoning, nor does it attempt to account for the many possible interactions of features. Rather, this scheme is meant only as a heuristic to support a rough analysis of problem difficulty for this particular set of problems.

The scheme shows a general concordance between empirical and a priori rankings, with two notable exceptions that may be due to problem order on the test. The Park Problem, which was of intermediate difficulty by feature analysis, ranked sixth easiest empirically. However, it was the first item on the pretest. The Italian Dressing Problem, likewise of intermediate difficulty on the basis of feature analysis, was third hardest, but was the last on the pretest. Because the problems were not counterbalanced for order, the amount of time and attention a problem received across and within students was affected by its position in the test.

We observed that certain additional problem features may make a problem more vulnerable to additive errors. The geometric figure problems were the most vulnerable to additive error, before and after the intervention. The concrete representations appeared to have no effects on students' reasoning on these problems; students appeared to construe the comparison of figures additively rather than proportionally.

### Interviews Isolating Numerical and Semantic Feature Variation

To further our understanding of the effect of these problem characteristics on reasoning, we undertook a small interview study with six sixth grade students who had not yet been instructed in ratio and proportion. (A parallel study of six students who participated in the previous teaching experiment was also conducted but is not reported here.)

Two interviews were conducted with each student: The first was a diagnostic interview designed to detect the approximate level of student's competence by beginning with an easy problem and then presenting harder and harder problems similar to those in the problem hierarchy described previously. The second interview began with the first problem failed in the first interview and presented a problem whose context was hypothesized to be "easier," or the same problem with "easier" numbers, where *easier* in both cases is according to the problem features described earlier.

Over the six interviews of the second set, there were twenty-one instances of such shifts to easier problems. Of these, eleven resulted in the student taking a new correct multiplicative approach, and the remaining ten did not. Although eleven out of twenty-one instances is not a high proportion of instances, when shifts in reasoning do occur, they are clinically impressive and do suggest to observers that a particular change in problem feature did elicit a change in reasoning. Obviously the interview technique is critical to ensure that the shift in thinking is not due only to having more time to consider one type of problem or to researcher's clues.

In general, the interviews indicated the following:

1. The analysis of problem features was generally supported: problems with containment and "for every" statements more easily elicited multiplicative approaches than those without, and time and money rates were also readily perceived as multiplicative situations. Problems without these clues were more confusing to students. The hardest problems were those where the given quantities involved uneven numerical multiples and close instances of the underlying ratio or where there were confusions in perceiving underlying groups. Apparently, in the discrete context the critical semantic features that students need to apprehend in order to conceptualize a situation as multiplicative are homogeneity and units of units, expressed here as sets of repeated identical sets of objects.

2. Students differed in the evidence needed to understand a situation multiplicatively. For some, explicit containment or a "for every" statement is necessary and ambiguous rates are not sufficient—by *ambiguous rate* we mean an intensity quantity statement without mention of *rate* or use of *for every*. For others, ambiguous rates served as the basis for rate-ratio conceptualizations and multiplicative reasoning, and only situations with ambiguous *groupings* were failed.

3. Students also differed in their vulnerability to frustra-

tion and confusion, which affected their ability to reason through problems productively. The second, "hard-to-easy" interview purposely started with a problem the student was expected to fail; two students were especially overwhelmed by their confusion and were unable to focus on the problems, even failing problems similar to ones they had successfully reasoned through on the first, "easy-to-hard," interview.

We present here discussion of two cases illustrating these points.

*Case 1. Anne.* Anne was a sixth grader interviewed before instruction in ratio or proportion. Interview 1, the "easy-to-hard" interview, consisted of three simple multiplication or division problems followed by eight proportion problems with two "trick" addition problems interspersed among these. (Interspersing addition problems was necessary to detect whether children were really reasoning quantitatively about a situation or just depending on a set algorithm). The proportion problems began with three problems that had containment or "for every" statements and easy multiples, then the Park Problem, a time-rate problem similar to the Printing Dictionaries Problem, then the Placemats Problem, and finally an addition problem. Anne solved correctly the multiplication and division problems and all proportion problems up to the Parks problem. She used a build-up strategy. The first proportion problem shows her approach. Problem: In a certain school there are four boys to every six girls in every class. How many girls are there in a classroom with twelve boys?

> *Anne:* So, "Twelve boys." So I'd draw groups . . . I'd sort of divide the twelve by four so that would be three, so I'd draw groups of four so one, two, three . . . three groups of four. (Pause). So there'd be six girls and then . . . There're four boys here, so there'd have to be six girls here. (Long pause). Eighteen?
> *Interviewer:* Okay, so then tell me what you did when you drew those pictures—what were you thinking?
> *Anne:* Well I was thinking—these were boys, and there're in like three groups, and here's one group of boys, here's another and another. And then . . . sort of like . . . if four boys—like you could do four boys would equal six girls

and so I just did that and then I drew six little circles for girls. And since there's six and there's three groups I did six times three is eighteen.

On reaching the Parks and Rate problems ("Jean's math problems") she attempted the abbreviated build-up strategy, but became confused when her initial division left a remainder in the Parks Problem. On the time rate problem, she first attempted a unit factor approach, then switched to the scalar approach that had been successful on prior problems. Again she became confused about how to deal with the remainder resulting in either of these divisions. She went on to easily solve the Placemats Problem with the abbreviated build-up strategy.

On Interview 2, one week later, she was given first the Cookies Problem (see Table 7.4, Problem 11). She again made drawings of the situation, but this time offered an incorrect additive strategy. The statement within the problem, "all the baggies are alike," although providing containment and homogeneity, did not help her to construe this as a multiplicative situation.

> *Anne:* And then altogether the baggies on the end table have sixteen chocolate chip cookies and twenty oatmeal cookies, so I drew the end table and then I put sixteen chocolate chip and twenty oatmeal cookies, and then I just wrote that on the table and then I drew a middle table cause it doesn't say anything else about (inaudible) and then it has a total of twenty-four chocolate chip cookies and sixteen chocolate chip cookies [on the end table] so I put twenty-four over here for the middle table and sixteen over here for the end table and then there's twenty oatmeal cookies and then the difference is four. So if there's twenty-four here— the difference would be four so I put twenty-eight because it's four more oatmeal cookies on each table than chocolate chip cookies.
>
> *Interviewer:* Okay, Anne, how did you know to subtract or add, why didn't you multiply or divide? How did you know to add or subtract?
>
> *Anne:* Because . . . it said, well first of all it said you don't really—in this sentence you don't really—it said altogether the baggies have sixteen and twenty, so I knew here not to add because you wouldn't do sixteen plus twenty, because they already tell you altogether sixteen and twenty.

*Interviewer:* Okay, but why didn't you divide these numbers?

*Anne:* Because . . . because . . . (pause)

*Interviewer:* Can you tell me?

*Anne:* Cause it . . . cause this wouldn't be right because if you divided—because sixteen goes into twenty like once with a remainder of four and that would be one remainder four, but there's twenty-four here, so if you divided twenty-four into something then it wouldn't be right because only having one baggie [on] each table.

Anne floundered when she lacked a clear understanding of the distribution of objects in this problem, then justified her additive strategy as due to numbers that did not divide evenly. She persisted in this incorrect additive strategy on three similar problems given subsequently. Note that in all of these the difference between the two numbers in the given ratio is relatively small, a problem feature frequently leading to an additive strategy.

Chairs: 15:21::50:X
Chairs: 16:20::24:X
Cookies: 4:6::12:X

Apparently she saw no evidence in these problems that objects were distributed in repeated sets (the discrete version of homogeneity). The grouping of paint with chairs in the Chairs Problems is semantically problematic in the sense that the actual association of paint with chairs is not likely thought of in terms of discrete cans and discrete chairs, but in a more embedded way, with paint becoming part of the chairs, thereby yielding a quantity differentiation difficulty. As discussed earlier, this affects the initial conceptualization stage of problem solution. Even changing the quantities to present easier multiples in the second version of the Cookies Problem did not move her to a multiplicative strategy.

Only on the next problem, which contained a "for every" statement, did she switch to a multiplicative strategy. Problem: In a parking lot there are fifteen trucks for every twenty-one cars. If there are fifty cars, how many trucks are there?

This problem provides no easily divisible quantities. However, Anne perceived this as a multiplicative situation and proceeded with a unit factor approach to the problem.

Having difficulty with that calculation (involving the fraction 5/7), she then proceeded to a scalar division (50/21). In spite of its computational awkwardness, she arrived at an approximate answer. Therefore she persisted in her multiplicative approach in spite of difficult numbers in this problem, in contrast with her previously justifying an additive strategy because numbers were not evenly divisible. We take this as grounds for believing that she behaved additively as a means for accomplishing *something*, rather than giving up entirely. In the next problem, Parks 7:6::42:X, she used the same build-up strategy as she did in Interview 1, speaking again of "groups" of seven trees and six benches.

In these problems, Anne searched for cues that objects were distributed in repeated groups and, without evidence of this, resorted to additive strategy. The Chairs Problem, which does not offer a "for every" statement, is not a time or money rate, and embodied a quantity-differentiation difficulty, was hard for her and apparently did not include sufficient evidence for her of multiplicative structure. On the other hand, when she was able to form stable units of units, even difficult numbers did not confound her, because her semantically based conceptualization of the situation scaffolded the numerical aspects of her reasoning.

*Case 2. Kathleen.* Kathleen also was a sixth grader who was interviewed before instruction in ratio or proportion. In Interview 1 Kathleen completed all the problems with containment or "for every" statements correctly using first the buildup strategy and then the abbreviated buildup strategy. These problems also contained easy multiples. However, on reaching a problem without such cues, she floundered. We recall the Park Problem: The Boston Park Committee is building parks. They found out that fifteen maple trees can shade twenty-one picnic tables when they built the Raymond Street Park. On Charles St. they will make a bigger park and can afford to buy fifty maple trees. How many tables can be shaded at the new park?

> *Kathleen:* I don't know how to do this. I don't know how it's laid out.
> *Interviewer:* What if there were five trees could shade seven picnic tables? Does that help?

> *Kathleen:* No, five tables and seven trees (reversing quantities) . . . at a table there would be one and one-quarter trees . . . that can't be right. For fifteen trees and twenty tables . . . you have to know how many tables and 15 doesn't go into 50, so I don't know how to do it.

Although the interviewer offered a version with easy multiples by offering the quantities 5 and 7, Kathleen could not envision this situation as one of repeated groups. The lack of even divisibility is too much for her in the absence of a stable unit conceptualization. However, she went on to use build-up approaches to solve easily the Placemat Problem and Rock Concert Problem, which have containment and a "for every" statement, respectively. She also approached a time rate problem with unit factor strategy, even though this problem does not specify "for every." On reaching the Cookies Problem, she said she "need[s] to know how many per bag." But in the next problem, Trucks and Cars, which includes a "for every" statement, she easily reasoned using the build-up approach.

On Interview 2 she used the additive strategy on Cookies 16:20::24:X, but said she thought it was not right. When the same numbers were used in the Chairs Problem, however, a rate problem with no statement of "for every," she tried, unsuccessfully, to use a unit factor approach. Finally, when the Cookies Problem was offered with numbers 4:6::12:X, she used the build-up strategy and successfully solved the problem.

Kathleen needs less strong groupings than Anne in perceiving a situation as multiplicative and has a stronger concept of rates. Her statement in the Parks Problem "I don't know how it's laid out" suggests that she searched for evidence of the distribution of objects. She perceived the Chairs and Parks Problems as multiplicative, stumbling only because of calculation difficulties. Yet she succumbed to the additive error when confused.

## Discussion of Interviews

Why were students so often led to a strategy by divisibility, by whether "numbers go"? Students have reason to examine whether numbers are easy multiples in considering the appropriate operation in word problems because textbook presentations often group problems of one type together and offer easy numbers that will facilitate computation for that type. However, this results in students' reasoning taking the form of type detection in order then to apply an algorithm learned for that problem type, rather than attending to problem situations themselves and reasoning through them. In this respect, textbooks have led students *away* from situational reasoning rather than toward it.

Additive errors in multiplicative situations presented another problem. As noted, students differed in the kind of evidence they needed to diagnose a situation as embodying a rate-ratio and hence a multiplicative comparison. In some situations the additive error resulted simply from trying to do *something*, whereas in others it resulted from an incorrect analysis of the problem situation. Because the additive error is common, the curriculum needs to present both additive and multiplicative situations and give students exercise in discriminating between them. Most textbooks have a separate chapter on ratio and proportion and give students no exercise in this discrimination.

The concrete representation we used matches students' natural build-up strategy and is applied by them to situations with containment and homogeneity. Time and money rates are approached by students with unit factor approaches and appear to engage a distinctly different schema.

## III. IMPLICATIONS FOR CURRICULUM DESIGN AND FURTHER RESEARCH

One can discuss curriculum design matters at varying levels of revision depth. At a shallower level, one can comment on the curriculum structure and content as given, concentrating on existing texts, how they are organized, the kinds of problems that they contain, and so on. In particular, we have noted that students have reason to examine whether numbers are easy multiples in choosing the appropriate operation in word problems, because textbook problems are so often designed to facilitate numerical computation rather than quantitative reasoning. This situation is aggravated by the fact that textbook presentations group problems according to types, thereby factoring out the critical initial conceptualization stage in quantitative reasoning. We revised our own curriculum to provide students exercise in discriminating additive from multiplicative situations after noting students' confusion.

However, we believe it is necessary to pursue a much deeper level of curriculum revision, one that deals with the organization of content, representations, and activities across the entire elementary and middle school curriculum from a cognitive and situational-semantic perspective. From this perspective, we see a strong need for the kinds of concrete representations that support and extend students' natural build-up reasoning patterns rooted in counting, skip counting, and grouping. We have reason to believe that these make sense to children in the upper elementary grades, grades 3–5. Hence we suggest that a long-term approach to multiplicative reasoning should begin in grade 3 and build upon the naturally occurring strategies discussed in this chapter and that of Lamon (this volume). We have seen that they can be learned by average students, and built upon to more sophisticated strategies. The amount of time used in our interventions was not adequate to establish the level of understanding and competence that we would want, but it was sufficient to prove feasibility of the enterprise considered as a multiyear effort across several grades culminating at grade 6.

Not discussed in this chapter are the graphical representations utilized in Units B and C of the teaching experiment and the associated notion of linear function that we feel should play an important organizing role in students' thinking about multiplicative structures. Basically, we feel that the formal algebraic approach to proportions should be taught *after* quantitative reasoning competence has been established in the informally based styles described in this chapter.

Transition to algebraic representation can be based on the use of algebraic equations to represent equivalent ratios across instances, where the instances can be based on the build-up approach. In this way one introduces a new notation to represent what one already understands, well before manipulating that notation to produce new representations.

Further, we feel that the equation-based approach should be taught in the context of determining the values of linear functions rather than as a technique for generating a "magic equation" that can be syntactically solved to yield an answer whether or not the student understands the quantitative content of the situation being modeled. Also, the actual contexts for learning multiplicative reasoning should obviously be richer and larger than those discussed here. Unit C in our teaching experiment included project-oriented activities extending across more than a single class. Given a longer term approach extending across grade levels, such should play the dominant role in setting contexts for multiplicative reasoning. And likewise, the assessment of multiplicative reasoning should match. In this sense, our "teaching experiment" did not fully represent our preferred approach, which assumes a much longer time frame.

### A Major Unaddressed Issue

A major question not addressed in this chapter is how to deal with multiplicative change situations that are not well modeled by build-up patterns, change situations that are *not* inherently replicative. These include the geometric similarity problems handled poorly by our students. The larger, re-scaled figure is not the join of several smaller ones. Rather, each of the infinitely subdivisible parts of the smaller figure "grows" by the same amount to produce the larger as discussed by Confrey (this volume). This form of multiplicative growth likely has different primitive conceptual roots and is likely to require a different curriculum strand and different types of concrete representations. We expect that one of these may involve stretching and shrinking activities directly controllable by the student and the numerical output of measures to tables, graphs, and equations. Currently, such rich concrete experiences are rare for students, but they need not be. We have suggested a longitudinal "scaling curriculum" to help serve this end (Kaput, 1993).

Last, the two forms of growth or change—the replicative and the more truly multiplicative—can be modeled by the same formal representations. And this is the likely place where the two strands should intertwine. But this should be quite late in the curriculum, probably after algebra has been well established as a modeling tool, in the upper middle grades.

The rush to put formal computational tools in students'

hands before they understand their quantitative foundations is one major factor in the widespread incompetence and alienation from mathematics among students across the land. Although the tradition of teaching ratio reasoning in the formal style is very long-lived, we should not assume that it should be venerated, or continued. The preponderance of data indicates that it does not work and may in fact do actual harm.

## NOTES

We wish to thank Clifton Luke, Joel Poholsky, Laurie Pattison-Gordon, and Shielah Turner for their diligence and insight, without which this study would not have been possible. All software utilized was cheerfully and brilliantly developed by Laurie Pattison-Gordon.

Work in this paper was supported by the Apple Computer Corporation External Research Program, the National Science Foundation grant #8855617 and the Department of Education OERI grant #R117G10002. Views expressed are those of the authors and not necessarily those of the the supporting agencies.

1. It may be worth noting that the methodologically driven needs of systematic task variation and the ecological validity needs of teaching experiments are almost diametrically opposed. To test authentic competence at the end of a teaching experiment, one needs to vary tasks in far more ways than can be systematically manipulated. It is of little value, for example, to control problem wording tightly to isolate other effects if the point of the intervention was to teach ability to recognize situations requiring multiplicative reasoning across a wide variety of situations and descriptions of situations.

## REFERENCES

Bell, A. 1989. Multiplicative word problems: Recent developments. In *Proceedings of the eleventh annual meeting of the PME-NA*, ed. G. Goldin, C. Maher, and T. Purdy. 198–204. New Brunswick, NJ: Rutgers University Press.

Carraher, T. N., and A. D. Schliemann. 1985. Computation routines prescribed by schools: Help or hindrance? *Journal for Research in Mathematics Education* 16: 37–44.

———, and D. W. Carraher. 1988. Mathematical concepts in everyday life. In *Children's mathematics: New directions for child development*, ed. G. Saxe and M. Gearhart. 71–88. San Francisco: Jossey Bass.

Conner, G., G. Harel, and M. Behr. 1988. The effect of structural variables on the level of difficulty of missing value proportion problems. In *Proceedings of the tenth annual meeting of the PME-NA*, ed. M. Behr, C. Lacampagne, and M. M. Wheeler. 65–71. DeKalb: Northern Illinois University Press.

Feurzeig, W., and J. Richards. 1991. *Function machines.* Cambridge, MA: BBN Labs.

Goldin, G., and C. McClintock. 1984. *Task variables in mathematical problem solving.* Philadelphia: Franklin Institute Press.

Guttman, L. 1941. The quantification of class attributes: A theory and method of scale construction. In *The prediction of personal adjustment*, ed. P. Horst. New York: Social Science Research Council.

Harel, G., and M. Behr. 1989. Structure and hierarchy of missing value proportion problems and their representation. *Journal of Mathematical Behavior* 8: 77–119.

———, T. Post, and R. Lesh. 1991. Variables affecting proportionality. *The Proceedings of the 15th Annual Meeting of the PME*, Vol. 2, F. Furinghetti. 125–132. Assisi, Italy.

———, T. Post, and R. Lesh. 1992. The blocks task: Comparative analyses with other proportion tasks, and qualitative reasoning skills among seventh grade children in solving the task. *Cognition and Instruction.* 9: 45–96.

Hart, K. 1984. *Ratio: Children's strategies and errors.* Windsor, UK: The NFER-Nelson Publishing Company.

Kaput, J. 1993. The urgent need for proleptic research in the graphical representation of quantitative relationships. In *Research in the graphical representation of functions*, ed. T. Carpenter, E. Fennema, and T. Romberg. Hillsdale, NJ: Lawrence Erlbaum.

Kaput, J. (in press) Creating cybernetic and psychological ramps from the concrete to the abstract: Examples from multiplicative structures. In Making sense: Teaching for understanding with technology. ed. D. Perkins, J. Schwartz, and S. Wiske. London: Oxford University Press.

Karplus, R., S. Pulos, and E. Stage. 1983. Proportional reasoning of early adolescents. In *Acquisition of mathematical concepts and processes*, ed. R. Lesh and M. Landau. 45–90. New York: Academic Press.

Kouba, V. 1989. Children's solution strategies for equivalent set multiplication and division word problems. *Journal for Research in Mathematics Education* 20(2): 147–158.

Lave, J. 1988. *Cognition in practice.* New York: Cambridge University Press.

Piaget, J., and B. Inhelder. 1974. *The child's conception of quantities*. Boston: Routledge and Kegan Paul.

Resnick, L. B. 1983. Mathematics and science learning: A new conception. *Science* 220: 477–478.

Saxe, G. 1990. *Culture and cognitive development: Studies in mathematical understanding*. Hillsdale, NJ: Lawrence Erlbaum.

Schoenfeld, A. 1987. Cognitive science and mathematics education: An overview. In *Cognitive science and mathematics education*, ed. A. H. Schoenfeld, 1–31. Hillsdale, NJ: Lawrence Erlbaum.

Schwartz, J. 1988. Referent preserving and referent transforming quantities. In *Number concepts and operations in the middle school*, ed. J. Hiebert and M. Behr. 127–174. Reston, VA: National Council of Teachers of Mathematics; and Hillsdale, NJ: Lawrence Erlbaum.

Thompson, P. 1990. A theoretical model of quantitative reasoning in arithmetic and algebra. Paper presented to the annual meeting of the AERA, San Francisco. Available from the author, San Diego State University, Department of Mathematical Sciences.

———. 1991. Quantitative reasoning, complexity, and additive structures. Paper presented to the annual meeting of the AERA, Chicago. Available from the author, San Diego State University, Department of Mathematical Sciences.

Torgerson, W. 1958. *Theory and methods of scaling*. New York: John Wiley.

Tourniaire, F., and S. Pulos. 1985. Proportional reasoning: A review of the literature. *Educational Studies in Mathematics* 16: 181–204.

Vergnaud, G. 1983. Multiplicative structures. In *Acquisition of mathematics concepts and processes*, ed. R. Lesh and M. Landau. 127–174. New York: Academic Press.

West, M., Luke, C., Poholsky, J., Pattison-Gordon, L., Turner, S., & Kaput, J. 1989. Proportional reasoning strategies: Results of a teaching experiment. In Proceedings of the eleventh annual meeting of the PME-NA, ed. C. Maher, G. Goldin, and R. Davis, 80–86. New Brunswick, NJ: Rutgers University Press.

# IV
# MULTIPLICATIVE WORLDS

# 8 Splitting, Similarity, and Rate of Change: A New Approach to Multiplication and Exponential Functions

## Jere Confrey

An abstract, intellectual understanding of deep time comes easily enough—I know how many zeroes to place after the 10 when I mean billions. *Getting it into the gut is quite another matter.* Deep time is so alien that we can really only comprehend it as metaphor. And so we do in all our pedagogy. We tout the geological mile (with human history occupying the last few inches); or the cosmic calendar (with *Homo sapiens* appearing but a few moments before *Auld Lang Syne*). A Swedish correspondent told me that she set her pet snail Bjorn (meaning bear) at the South Pole during the Cambrian period and permits him to advance slowly toward Malmo, thereby visualizing time as geography. John McPhee has provided the most striking metaphor of all (in *Basin and Range*): Consider the earth's history as the old measure of the English yard, the distance from the king's nose to the tip of his outstretched hand. One stroke of a nail file on this middle finger erases human history.

—Stephen Jay Gould, *Time's Arrow—Time's Cycle*

## PROBLEM STATEMENT AND SIGNIFICANCE

In our research to date, we have documented the roots of an understanding of exponential functions in the development of multiplication and division. By reanalyzing the multiplication and division literature in light of our teaching experiment results, we have argued that the majority of the current approaches to multiplication (Fischbein, Deri, Nello, and Marino, 1985; Greer and Mangan, 1984; Bell, 1983; Nesher, 1988; Steffe, 1988) rely too exclusively on a model of multiplication as repeated addition with an underlying basis in counting. *We are proposing an alternative experiential basis for the con-*

291

*struction of number in a primitive cognitive scheme we label* splitting (Confrey, 1988, 1989a).

The basis for this conjecture is the following claim: Models of multiplication based on counting or repeated addition do not adequately explain many of the actions of young children that can be seen as multiplicative and do not explain the contextual situations typically modeled with exponential and logarithmic functions. We have developed an alternative model of multiplication, which is independently constructed from, but complementary to, repeated addition, and which we believe is more appropriate as an explanatory model for these exponential situations. In particular, we believe it is a precursor to a more adequate concept of ratio and proportion and subsequently of multiplicative change and exponential functions.

Among the actions we observed in children that can be interpreted multiplicatively are some related to addition (affixing, joining, annexing, and removing) and others that seem relatively independent of addition (sharing, folding, dividing symmetrically, and magnifying). In these latter actions we find the basis for splitting. In its most primitive form, *splitting* can be defined as an action of creating simultaneously multiple versions of an original, an action often represented by a tree diagram. As opposed to additive situations, where the change is determined through identifying a unit and then counting consecutively instances of that unit, the focus in splitting is on the one-to-many action (Dienes, 1967; Kieren, 1976). Closely related to this primitive concept are actions of sharing and dividing in half, both of which surface early in children's activity. Counting need not be relied on to verify the correct outcome. Equal shares of a discrete set can be justified by appealing to the use of a one-to-one correspondence and in the continuous case, appeals to congruence of parts or symmetries can be made.

Splitting can also be differentiated from counting (and repeated addition) by its geometric connections to similarity. Similarity forms the basis for our depth perception as we maintain the identify of objects as they move toward (magnification) or away (shrinkage) from us. Splitting's geometric character can be captured by its image in the tree diagram, the embedded figure, or the growth spiral. Pothier and Sawada (1983) worked on partitioning in geometric contexts and provided evidence that can be interpreted as support for the splitting conjecture. They recognized the ties between partitioning, sharing, and rational number; the robustness of the halving construct; and its usefulness in even and oddness; and they predicted that composition of functions could play a significant role. Whereas they explored the relationship between partitioning and congruence (through translation and rotation), they did not explore their relationship to similarity (which requires dilation, also). Furthermore, although they laid the groundwork for the splitting conjecture, they did not explicitly differentiate and contrast counting and splitting. Perhaps this is because (1) they did not place their work in the context of repeated splits, such as is suggested when one works with exponentiation or recursion, and hence did not see splitting as creating a multiplicative structure; and (2) they were not considering the parallels between the construction of linear and exponential functions as we were. Their work, which we read after making our initial conjecture (Confrey, 1988), has proven very useful in the further development of this conjecture.

After examining student learning carefully, informed by constructivist teaching experiments and historical reanalysis (Smith and Confrey, 1989; Confrey and Smith, 1989), we offer a splitting conjecture that can radically alter the K–12 curriculum, as is widely demanded (Kaput, 1988). We suspect the construct of splitting receives inadequate nurturing throughout the curriculum and that its introduction is inappropriately delayed until addition and subtraction are well underway. Splitting needs to be developed as a complement to the counting worlds. Building a curriculum that allows and encourages young children to develop and distinguish actions related to counting from actions related to splitting should help them simultaneously build two concepts of multiplication: one based on counting out repeated actions of addition, and one based on splitting as a primitive action. Such a curricular change may ameliorate the problem accurately identified by Hiebert and Behr (1988, p. 9) that "Competence with middle school number concepts requires a break with simpler concepts of the past, and a reconceptualization of number itself." Numbers need not be developed uniformly; even in the early grades, their pluralistic character needs to be carefully integrated into the curriculum. As written succinctly by Spengler (1925), "There is not, and cannot be, number as such. There are several number worlds, as there are several cultures . . . " (from Confrey, 1980, p. 124).

## WHY EXPONENTIAL FUNCTIONS?

Exponential functions are used to model numerous scientific and social phenomena. Disputes about international issues such as global warming, population control, disease prediction and control, radioactive waste, inflation rates and national debt, environmental tolerance to acid rain, and so forth often revolve around understanding the mathematical behavior of exponential functions. Their importance for an educated citizenry is indisputable.

Furthermore, conceptually and empirically, exponential functions offer a unique opportunity to explore the relationship between mathematics and nature and, in doing so, can make mathematics relevant and accessible, qualities demanded in most cries for curricular reform, such as the National Council of Teachers of Mathematics's *Curriculum and Evaluation Standards* (1989). In the early 1990s, D'Arcy Thompson's *On Growth and Form* (1917, reprinted, 1961) postulated elegant ties between geometric growth and biological form as he sought ways to "search for relations between things apparently disconnected, and for 'similitude in things to common view unlike'" (p. 6). As he donned the epistemic lenses by which he viewed natural phenomena, he professed: "Their [natural phenomena's] problems of form are in the first instance mathematical problems, their problems of growth are essentially physical problems, and the morphologist is, ipso facto, a student of physical science" (p. 8). In doing so, he offered a compelling example of how to view an organism (and the trace of its development) in multidisciplinary ways. His description of the nautilus shell was, "The shell, like the creature within it, grows in size *but does not change its shape;* and the existence of this constant relativity of growth, or constant similarity of form, is of the essence, and may be made the basis of a definition of the equiangular spiral" (p. 179). We believe that this key recognition provides the basis of an alternative view of multiplication as well.

Theodore Andrea Cook wrote in *Curves of Life* (1914, reprinted 1979): "Nature abhors mathematics." His admonition argued not for the dismissal of mathematics but for the further recognition "that simple mathematics, as we have hitherto developed them, can never express the whole complex truth of natural phenomena." He advocated that, to understand natural phenomena, one needed to examine their deviance from simplicity, not reduce them to their well-understood patterns. He later contrasted his book with one written earlier by Colman (previously unknown to him) that, he contended, was concerned with "proportional form." He wrote, "[Colman] may be said to be dealing with morphology apart from physiology, with form separate from function, whereas, in my judgment, considerations of function and growth are essential to the right understanding of form and its proportion" (p. 440). We interpret this as support for our claim that proportional reasoning can be viewed more productively within the framework of growth.

Both Thompson and Cook understand the value of the integration of biology and mathematics through the study of growth and its relationship to form. These ideas, we suggest, are woefully absent from our mathematics curricula and this absence contributes to students' perception that mathematics is irrelevant to forms of human activity. The concern and call for a public understanding of the exponential curve was taken up by the Club of Rome in the book *The Limits to Growth*, wherein they argued that a failure to disentangle exponential and linear growth would have disastrous social and environmental impact (Meadows, Meadows, Randers, and Behrens, 1974). And yet, here in the last decade of the century, we still see evidence of general inability to comprehend the implications of these functions.

This failure is due largely to a difficulty in comprehending the impact of a multiplicative rate of growth. In a recent example, a computer science graduate student at Cornell University crippled a nationwide computer network by introducing a virus (or more technically a worm). He claims to have been only mischievous and to have erroneously programmed a higher rate of multiplicative growth than he had intended. Our pH scales, which transform the multiplicative relationships among the changes in hydrogen ion concentrations into our more familiar counting numbers, lull unsuspecting citizens into complicity in large scale environmental damage. As stated by Bill McKibben in his devastating exposition in *The End of Nature:* "the logarithmic scale that we use to determine the overall composition of our soils or

waters—pH—distorts reality like a fun-house mirror, for any-
one who doesn't use it on a daily basis. For instance, 'normal'
rainwater has a pH of 5.6, but the acidified rain that falls on
the Adirondacks has a pH between 4.6 and 4.2—that is, it is
ten to forty times as acid" (1989, p. 7).

The concern for an understanding of exponential growth
may derive from its ties to natural and social phenomena but
it extends into other disciplinary skills and insights. An understanding of exponential relationships is essential for understanding scale and for gaining some perspective on the
vastness of geologic time (as indicated in the opening quotation) or astronomical space (Gould, 1987; Morrison, Morrison, Eames, and Eames, 1982). Dawkins, in the *Blind
Watchmaker* (1987), makes the argument that small incremental changes that multiply and reproduce can account for
large transformations of evolutionary viability. Exponential
functions, scientific notation, and logarithmic scales are all
tools for understanding key cultural ideas. They allow us to
transcend the brevity of our personal existence and view ourselves within a larger framework of space and time. As David
Hawkins described: "To know the universal in its own right is
first and foremost to enjoy it as an achievement. . . . Kant
called it the mathematically sublime. It is the experience in
which our grasp of things suddenly outruns our imagination,
like the vastness of space and time. More properly: we catch
ourselves in the act of going—of knowing—beyond imagination, and thus we first know ourselves as knowers" (1974,
p. 103). He further provided an example of such an achievement of imagination: "There was a classroom of 10-year-olds
a few years ago in New England which had almost spun itself
into a cocoon of paper tape, with a foot to the hundred thousand years of geological history. This was a second start, the
first having given a foot to the century; and I think they were
about to invent logarithms. The starry heavens may dwarf the
imagination, as may a sandbox, but the truth is we have resources to stretch the imagination" (1974, pp. 104–105).

## EXPONENTIALS IN THE CURRICULUM:
## THE NEED FOR A NEW APPROACH

If the idea of exponential growth is indeed so important, then
why is it treated in such a cursory way in the curriculum?

Does it indeed represent a critical barrier to learning (Hawkins, 1974) or an epistemological obstacle (Brousseau, 1986)? That is, is it an obstacle that demands a departure from or rejection of our established forms of conceptualization? From our past work on the topic (NSF Grant No. MDR-8652 160), we believe we can support the claim that, whereas exponential growth is poorly understood in the general population, much of the difficulty results from its treatment in the curriculum. We claim that our curricular treatment of multiplication as *exclusively* or *predominantly* an abbreviated version of repeated addition accounts for underdeveloped understanding of multiplication as scaling, magnification, or growth. This leads to an inadequate differentiation among the different interpretations of multiplication in the K–12 curriculum, which has a deleterious impact on students as they encounter the study of ratio and proportion and, subsequently, exponential functions. Reducing multiplication to repeated addition alone may result from an unwarranted desire to produce and present a singular construction of the real numbers, rather than to celebrate and promote the possibility of multiple numeric systems.

The grounds for the assertion that there is a curricular explanation for students' difficulty in learning exponentials comes from our experience in working with students in teaching experiments. We have documented that students, when confronted with challenging problems (those that can be connected into human activity and lead to multiple ways of approaching and solving a problem), develop effective, viable (albeit alternative) understandings of mathematical phenomena (Confrey, 1989a, 1989b). We have witnessed students' methods that, as we attempt to frame them in familiar forms, at first appear tentative and chaotic to us, but that, on reexamination, prove inventive, provocative, and thoroughly capable. These inventions provide us with genuine opportunities for reconsidering our mathematical ideas. The examination of mathematical ideas within pedagogically rich settings provides the impetus for the reconceptualization of fundamental mathematical ideas in ways that can transform the curriculum and the processes of teaching and learning. *Therefore, we claim that pedagogical inquiry can lead to a serious epistemological argument that can alter the character of the mathematics taught and learned in schools.*

## THE SPLITTING CONJECTURE

A major theoretical result of the two years of work has been the proposition that there is a nearly exclusive emphasis on *counting* in the elementary and secondary curriculum and an associated neglect of an alternative construction of the positive reals, which we label *splitting*. We make the basic conjecture that, by developing the construct of splitting in the curriculum, one can establish a more adequate and robust approach to such traditionally thorny topics as ratio and proportion, multiplicative rate of change, exponential functions, and so on. Therefore we contend that multiple constructions of numeric quantity are possible and that uniformity in the treatment of the concept of number is detrimental to quantitative maturity of an educated citizen.

Although we recognize, through exemplary work on the basic schemes involved in counting (Steffe, von Glasersfeld, Richards, and Cobb, 1983; Steffe and Cobb, 1983; Steffe, 1987, 1989; Carpenter and Moser, 1983), that students will gain competence and insight into multiplication as repeated addition, we also warn that this approach fails to take advantage of children's intuitive insight into splitting, reproduction, and similarity relations. Our preliminary studies of young children and our review of the literature suggest that this arena of cognitive potential is sorely underdeveloped in our current curricular design. The curriculum contains isolated explorations into measurement topics and partitioning of circles and rectangular regions in the introduction of fractional parts, but to our knowledge, no one has connected the disparate pieces into an integrated view of the construction of the positive real numbers and advocated for a careful development of its fundamental building blocks. We would like to undertake such a theoretical and empirical investigation in pursuit of establishing a more adequate basis for the understanding of ratio and proportion and exponential functions.

### Preliminary Evidence of the Development of the Concept of Splitting in Young Children

Preliminary, informal interviews with young children indicate that repeated halving is developed early and capably. One child, aged 8, was careful to distinguish between obtaining 8 and not 6 as the result of three repeated two splits. This shows that young children may be able to imagine repeated splits prior to exposure to repeated addition. This same child was once offered \$100 to eat 5 lbs. of salt potatoes. He spontaneously asked if he could get \$50 for eating only half of them. He continued to divide the reward into 2 until he reached, with help, \$6.25. On another occasion, he spontaneously discovered that he could make a triangle of triangles and proceeded to do so on MacDraw, creating the figure shown in Figure 8.1. Also in an informal setting, a 5-year-old child showed her fascination with scale as she held her thumb and her index finger a set distance apart and close to her face so she could look through them and measure the relative size of adults' heads with her fingers. She also held her hand in front of her eyes with her fingers splayed to frame an object. She then walked toward or away from objects, adjusting the amount of separation either until her fingers closed or until they were forced open beyond the extension of her fingers. When she first looked out of a twenty-second story window of a large urban hotel, she exclaimed, "Yikes, this is a good place for measuring" and she held up her thumb and index finger. These anecdotes indicate that early concepts of splitting may be a provocative site for further investigation.

We label our basic conceptual primitive *splitting* and argue that it is conceptually linked with the geometric transformation, similarity. The split, with its ties to work on parti-

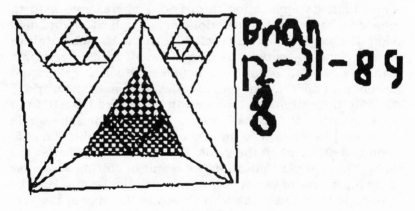

Fig. 8.1

tioning, is an alternative basis for the construction of a number system and possesses strong explanatory potential for interpreting children's methods. It is a precursor to a more adequate concept of ratio and proportion and subsequently to a multiplicative rate of change and exponential and logarithmic functions. A split is an action of creating equal parts or copies of an original. Splitting is a primitive operation that requires only recognition of the type of split and the requirement that the parts are equal. (Note that we use the term *operation* as an internalized action (Piaget, 1970) and hence do not intend to reference the mathematical operations of multiplication and division at this primitive stage.) A split is successfully accomplished at the point of establishing equal or identical parts or copies. Naming the result is a separate act, and the result of splitting is, in fact, named to be consistent with the counting system rather than the splitting system, a practical decision, but one that has contributed significantly to the confusion of the two systems of number construction. The successor action of the split could have been used as the basic unit to create an entirely separate number world.

The independence of splitting from counting can be demonstrated by showing that the requirement of equal-sized partitions can be obtained by arguments of symmetry and congruence by folding continuous planar objects or, in the case of discrete objects, by testing for one-to-one correspondence. Furthermore, we purposefully choose to avoid the use of the term *partitioning*, due to researchers' tendency to limit its application to division, whereas we wish to provide a basis for the operations of multiplication *and* division. The word *split* is also preferable due to its rich, intuitive connections to everyday language. Wood, roads, decisions, bananas, hairs, infinitives, couples, stocks, cells, and seconds all split, in our everyday description of the world.

Multiplication and division can be defined in relation to this primitive. Multiplication is established when the whole is defined in relation to the objects after the split, and division is defined when the whole is not reinitialized after the split. Thus, in multiplication the result of cutting a circular cake (by means of two successive perpendicular cuts) can be named *four*, because the whole used is the size of the sections created, or, in division, as *one-fourth*, because the whole remains the whole cake. Multiplication and division are mutually constructed as inverses of each other.

A splitting structure is constructed of repeated splits. Thus, when faced with the question of the numerosity of a pile of objects, one might count the objects sequentially or divide the piles repeatedly into two equal piles until the starting amount is recognizable, quickly calculating the total by repeated doubling. This action is labeled a *sequence* of two splits. There is cross-cultural evidence that this method is developed both in determining numerosity (Carraher, 1986; Gerdes, 1989) and in composing music in Java (personal communication, Becker, 1989). Millroy (1990) provides examples of an expert furniture carpenter using splitting to locate and size dovetail joints in a drawer and for locating holes to riempie a chair. Another example that demonstrates the two-split is the choice one faces if one were to build a picket fence. One could specify the spacing and then proceed with a slat-space-slat-space to the other end, or one could place a slat in the middle, and then in the middle of the two sides, and so on until obtaining the required density. These disparate strategies produce very different views of numerosity or quantity. Both produce a finite, predictable numeric algorithm describing the desired outcome.

To construct the set of positive rational numbers, one could create a splitting structure for each prime, thus defining a number by its prime factorization. Nonprimes would be constructed from compositions of splits. (A six-split can be achieved as a composition of a two-split followed by a three-split or vice versa.) Practically, we rely on only a few types of primitive splits. In our culture these include the two, three, five, and ten. (Exceptions include the splitting of pies in restaurants into seven pieces. This has created a market for a specialized tool: a pie splitter.)

Naming the result of the splitting action is done to be consistent with the number names that have evolved from counting structures. For example, when naming the result of three two-splits, people name the result in a manner consistent with the naming of the result of an act of counting, thus counting the result in four groups of two as 1, 2, . . . , 3, 4, . . . , 5, 6, . . . 7, 8 (Figure 8.2).

This is an efficient decision because it creates a uniformly named numbering system. It is also, in some ways, an

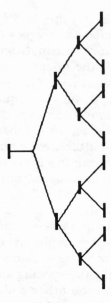

Fig. 8.2

unfortunate decision, for it dramatically increases the probability that all multiplication will be treated as repeated addition. That is, to name the result of a split, one engages in
iterative counts, as so eloquently described by Steffe et al.
(1983). This masks the integrity and independence of the
prior action: the creation of equal-sized groups by an invariant process, a process that is the basis of a splitting view of
multiplication.

Much of our insight into splitting resulted from the use
of the context of exponential functions for examining multiplication. This led us toward an examination of repeated
multiplication and the properties of associativity (Piaget,
Berthoud-Papandropoulou, and Kilcher, 1987). Vergnaud, in
discussing his product of measures, also examined such an
approach (1983, 1988).

Steffe, by contrast, is led to focus almost exclusively on
the distributive property in that his focus on the construction of iterative and composite units derives from acts of
counting. Thus his interview tasks frequently rely on having
the students add an additional number of groups of equal
quantity to a given set of such groups, for which the student

has already constructed a total, and then asking them to predict the number of iterated units, a set of sets. Steffe carefully distinguishes the student's construction of the set from the construction of the set of sets, an action that requires a coordination of counts. This leads to a view of multiplication in which the roles of the multiplier and multiplicand are carefully distinguished, a concern raised by Dienes and Golding (1966, p. 34). The distinctions raised by these two researchers are key in the development of multiplication from repeated addition.

Other researchers have examined the operations of multiplication and division (Fischbein et al., 1985; Greer and Mangan, 1984; Bell, 1983; Nesher, 1988; Steffe, 1988; Resnick, Nesher, Leonard, Magone, Omanson, and Peled, 1989). In the work of Fischbein et al. operations are tied to implicit, unconscious, and primitive models. They argued that repeated addition was the primitive for multiplication and that partitive and quotitive approaches applied to division. Their investigations relied on analyzing the successes and failures of students in predicting (not actually carrying out) the results of problems in which the multiplier and multiplicand varied. They documented the difficulty that children have when the multiplier is a decimal, especially one that is less than 1.

These results have been replicated and refined in studies that systematically varied the type of multiplier, problem structure, and response mode (De Corte, Verschaffel, and Van Coillie, 1988), and the context and multiplier-multiplicand order (Luke, 1988). The results of these replications and refinements indicate that when the multiplier is a decimal less than 1 and the context is asymmetrical (the multiplier and multiplicand are not interchangeable such as in rate, but not in area), the student performance is dramatically curtailed.

Greer (1988) built on this work by exploring the ways in which students will revise their choice of operations when presented with a multiplier that is a decimal less than 1. Greer labeled students' failure to maintain an operation under such a transformation of number type a "nonconservation of operation." His assumption is that a student with a mature understanding of a formal operation would, when solving problems with the same wording but different numbers, consistently predict that the same operation would apply, disregarding any change in the numbers (from integer, to decimal greater than 1, to decimal less than 1). He suggests that the students' failure to conserve the operation is the result of a misconception. That misconception is described as "multiplication makes bigger; division makes smaller" (MMBDMS).

Greer also takes the position that distinctive primitive models guide multiplication and division. "It seems clear that multiplication's natural origins lie in repeated addition of natural numbers, and division's in equal sharing of a collection of discrete objects," (1988, p. 289). If one puts these three assumptions together, one sees how Greer's argument evolves: (1) Multiplication and division have primitive models that are cognitively distinct; (2) hence, a switch in prediction from multiplication to division indicates a significant conceptual shift; (3) both those primitive models will lead one to believe that multiplication makes bigger and division makes smaller. So, when we witness a shift from multiplication to division, it indicates a shift in the student's primitive models, and probably implies the presence of MMBDMS.

We believe alternative analyses can be more suitably argued. (1) Numbers and operations involve complementary cognitive processes. Numbers are constructed as the result of operations. Operations are tied to situations and actions. (2) Therefore, there can seem to be a misfit between a particular contextual problem and a number if that number has been constructed from a conflicting operation. (3) When placed into such a situation, the student must *either* modify the operation or the operational character of the number. (4) In the case of multiplication by decimal numbers greater than 1, a process similar to Fischbein et al.'s absorption effect occurs. This might be perhaps better described as a bounding process, where the values are seen as between two integers, and slightly more than the smaller amount or slightly less than the larger one. MMBDMS is affirmed, and the operation character of numbers for multiplying is slightly modified. (5) In the case of multiplication when the multiplier is less than 1, either modifying the operation or modifying the operational character of the number can produce a successful result. We presume that students identify decimals less than 1 with division. One choice is that they can "conserve the operation" by admitting decimals less than 1 as operationally multiplicative. This might be accomplished by viewing decimals less than 1 as simply another name for a smaller unidimensional quantity that can vary in size from zero to infinity. MMBDMS is denied by such a modification of the number concept.

Alternatively, they may recognize that the decimal value less than 1 is created by an act of division and want to keep their belief in MMBDMS as a viable construct. This can be done legitimately by recreating the decimal operator as a ratio $(a/b)$ and decomposing it into multiplication by $a$ and division by $b$, in either order. Then, multiplication will make bigger (since it will always be integral) and division will make smaller (for the same reason).

Unfortunately, in the interview excerpt provided in Greer (1988, p. 286–287), the interviewer's insistence on confronting the student with his lack of conservation of operations undermines any viable alternative model that might have emerged. If the interviewer had taken the student's proposal more seriously and allowed the student to attempt to resolve the conflict that resulted by dividing by .85 and getting a larger product, then one could see how these competing beliefs were resolved. However, to label the student's response prematurely as a misconception makes such an investigation of the development of alternative strategies unlikely. Such an attempt to articulate the close operational ties between multiplication and division might be illustrated by the quote in Sowder (1988), when a child said "So, its kinda confusing about whether I should divide or multiply. Because sometimes when you multiply, like let's say you're multiplying by point seven five. The number comes out smaller than the original number. So it's kind of like dividing, when you multiply by decimals" (p. 232).

*What Greer's argument ignores is the ambiguous operational character of the decimal numbers themselves.* For Greer, it seems that decimals ought to be completely and nonproblematically assimilated into the category of numbers, and hence choice of operations should be impervious to changes in the numeric character of the quantities. The question that is not addressed is what is meant by *number* such that membership in the class of numbers can be represented equally as decimals or as integers. How one understands the extension of the natural numbers to the rational decimal numbers (ignoring for now their further extension into the real numbers) is a key question. At least three models of decimals compete in a sophisticated understanding of decimals, none of which is entirely satisfying in its own right. One can appeal to a correspondence between the number line and the set of decimal numbers with the following claims: Every terminating or repeating decimal can be expressed as a magnitude (often relying on ratio form), hence decimals are simply magnitudes or quantities. Or, one can treat the rational decimals in their ratio form as operators in which the intimate ties between multiplication and division become evident, but ordering becomes problematic. Or, one can treat the decimals computationally, as strings of integers, needing only the careful placement of the decimal point (which coordinates their ten-split quality with their count quality), because their formal mathematical rules permit it. The implications of each of these models for the meaning of multiplication and division, MMBDMS, and conservation of operations need to become a more extensive part of the research discussion.

Greer assumes that the wording of problems and kinds of quantities involved, independent of the numbers, determine the operation that must be used. This assumption can be challenged simply on linguistic grounds. We say "three times a number" or even "3.2 times a number," but often say ".85 *of a number*." Students avoid the conflict by memorizing the command, "*of* means multiply," but this is more of a mnemonic device. We suggest that the researchers in this tradition, (Fischbein et al., 1985; Greer, 1988; Bell, 1983; Luke, 1988; Graeber and Tirosh, 1988; Nesher, 1988; Peled and Nesher, 1988; Resnick et al., 1989) have all focused too exclusively on the use of decimal numbers as strings of integers (or as calculator productions), which has contributed to their likelihood of identifying students' actions simply as inadequate, neglecting the viability of the students' alternative frameworks. In our work, we seek to examine the operational construction of the decimal, ratio, or fractional number simultaneously with the construction of operations, and we believe that this will lead to a useful understanding of those interactions.

Nesher and Peled (Nesher, 1988; Peled and Nesher, 1988) proposed three types of multiplication (a mapping rule, multiplicative compare problems, and Cartesian multiplication) and examined the linguistic basis of the distinctions. They found that most Israeli children can differentiate among types of multiplication in creating and representing problems and that the multiplicative compare problems were constructed most readily, followed by mapping problems and then, only occasionally, by Cartesian multiplication. They proposed that the Hebrew language has an unusual linguistic expression for scalar multiplication that perhaps would account for the high percentage of "compare" responses. In exploring the students' understanding of the constraints on the use of multiplication, Peled and Nesher (1988) identify what we consider to be the most essential characterization of the scheme for splitting, whether multiplication or division: *there must be equal-sized groups.* In their study, students demonstrated their understanding of this for mappings, container problems, and rectangular arrays. Scalar multiplication, combinatorics, and area problems with continuous quantities were more difficult for students to discuss in relation to this requirement.

Mechmandarov (1987) applied Nesher and Peled's results to an instructional setting to examine to what extent instructional treatment would influence the "primitive models" of Fischbein. She found that when she introduced the mapping approach with Vergnaud's four-place "isomorphism of measures" (Vergnaud, 1983; 1988), students were less likely to experience the errors in operations identified by Greer and Bell. Thus, of the distinctions identified by Schwartz (1988), she found a successful intervention with the extensive times intensive quantities (three books per shelf × four shelves) using a 2 × 2 matrix:

| Books | Shelves |
|-------|---------|
| 3 | 1 |
| ? | 4 |

Using this approach, she found that students were more likely to conserve the operation across number types. These and other studies suggest that multiplicative structures are not simply reducible to forms of repeated addition. They further suggest the productivity of visualizing the composite actions of a ratio quantity as distinct operations.

According to Schwartz (1988), a dimensional analysis of

multiplication shows it to be referent transforming (in relation to quantity), whereas addition is referent preserving.
This work can be interpreted as support for an independent
approach to multiplication and as a suggestion that students'
understanding of multiplication be examined in contextual
and linguistical settings that promote a greater diversity of
forms of multiplication. At the current time, the social, educational dominance of the repeated addition model makes it
difficult to assess the viability of alternative approaches. As
Peled and Nesher put it, "It is not clear whether the observed
difference in problem 'popularity' is due to some primitive
structures (as claimed by Fischbein) or to traditional instruction. The result implies that for instructional purposes we
need to put more effort into the construction of models,
which will support the less familiar multiplicative structures" (1988, p. 261).

In our work we have concentrated on the act of *repeated
multiplication*, as did Piaget et al. (1987). Only in repeated
multiplication does the role of the whole unit become visible.
This whole unit plays a key role in our work. When asked to
multiply three factors ($a \times b \times c$), students have demonstrated two distinct strategies. Some will treat each element
individually as singletons and $c$-split each of the $a \times b$ elements. Others will "reinitialize" the product ($a \times b$), that is,
treat it as a whole, and then $c$-split it. These two strategies
may form the basis for understanding similarity relations
(i.e., the same operations are carried out on different objects).
We describe the first one as using singletons, and the second
one as reinitializing (see Figure 8.3).

## The Construction of a Splitting World as Contrasted to a Counting World

Our construct of the natural numbers is inexorably linked to
our understanding of counting. Using counting, the repeated
addition of 1, as a basic structure, we build the operations of
addition, subtraction, and multiplication as repeated addition and division as its inverse. As we construct the rational
numbers, we introduce the idea of a partition and then we
count these partitions to construct a view of fractions as
"counted partitions," parts of wholes. In this world, multiplication is strongly shackled to repeated addition. Division

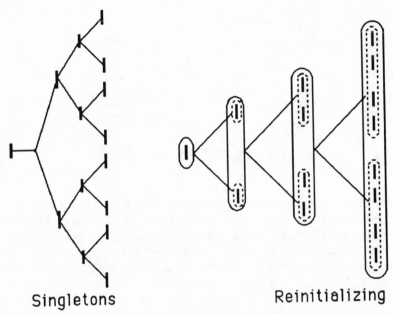

Singletons                              Reinitializing

Fig. 8.3

enters in creating equal divisions of integral values and hence more precise and smaller units with which to count and repeatedly subtract.

We conjecture that this construction of the rational numbers is insufficient for the development of a strong concept of ratio and proportion and for the development of multiplicative rates of change (Confrey and Smith, in press a). We propose to describe in this section how one can compose an alternative experiential basis for the construction of numbers in which ratio functions as the primary descriptor of change and the development of the rational numbers using multiplication and division is accomplished without recourse to addition and subtraction.

We reserve two roles for the counting numbers in the construction of the splitting numbers, neither of which has an impact on the repeated operation itself: (1) The counting numbers are used for convenience as an *index*, counting the number of splitting actions that have taken place. Thus, we use the counting number, 4, to identify that four two-splits are represented in the Figure 8.4; this use of the counting

Level index   1   2   3   4

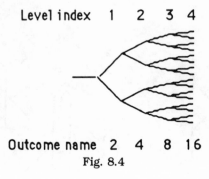

Outcome name   2   4   8   16
Fig. 8.4

numbers as an index is largely an issue of convention and convenience. Using Peano's axioms with an initial term and a successor action, one can build many number systems, and any of these would suffice as an index. Using the counting numbers to index the splitting numbers is equivalent to mapping the positive whole numbers onto positive geometric sequences and, in this form, forms the genesis of the exponential function. (2) The counting numbers are used to *name* the result or outcome of a split. Thus, in Figure 8.4, the name given the output after three two-splits is 8. Recall that originally in creating the counting numbers, the repeated action was adding 1—and the names, 2, 3, and 4 were invented to describe the results of each action. This could have been done for splits—but then we would have multiple, noncoordinated number systems. Therefore, for convenience, the result of a split is named as a count. In this action of naming the result of a split as a count, we witness the segmented count that then leads progressively to multiplication as a sophisticated form of repeated addition. We would like to emphasize that the cognitive act of recognizing a situation as multiplicative and displaying it appropriately occurs prior to this counting action. In our theoretical exposition of these ideas, Table 8.1 emphasizes the distinctions and parallels in the ideas in the splitting and counting worlds.

By making the distinctions in this table, we are suggesting that human beings confront situations in which their desired quantitative actions are best described by splitting rather than by counting. These situations are ones in which the starting point is conceptualized as a whole. For example, in compound interest problems, the initial principle is con-

TABLE 8.1.    A COMPARISON BETWEEN COUNTING AND
SPLITTING STRUCTURES

| Counting | Splitting |
| --- | --- |
| Zero is the origin | 1 is the origin |
| Adding 1 is the successor action | Splitting by $n$ is the successor action |
| The unit is 1 | The unit (of growth) is $n$ |
| Addition and subtraction are basic | Multiplication and division are basic |
| Difference is used to describe the interval between two numbers | Ratio is used to describe the interval between two numbers |
| Rate is the difference per unit time | Rate is the ratio per unit time |

ceived of as a single quantity that is retained and increased by the amount of interest earned. The origin of such a process is a 1, rather than 0. In such a setting, the successor action is not an incremental count as it is in linear functions: going up stairs, traveling at a constant speed, and so forth. In such a setting, the whole is transformed multiplicatively into more (or less): it is stretched, magnified, replicated, reproduced or expanded (compacted, shrunk or grouped). These actions are all tied to the primitive version of such actions, the split.

To create a way to describe the concept of unit in the splitting world, *we have defined* unit *in any world as the invariant relationship between a successor and its predecessor; it is the repeated action.* In a counting world, the unit is 1; for the constant action that is attached to a count is adding 1. From 0, we both count by 1 and create the number 1, *simultaneously* achieving its mutual character as ordinally and cardinally 1. This definition ties the concept of unit to the operational constructive process of the system. Thus, if we vary the successor action, we vary the unit.

From this concept of an additive unit of 1, we can see how other notions of constant additive units are built: counting by 2s, 3s, and so on. (See Steffe et al., 1983, for the most complete description of these acts and the development of multiplication as repeated addition by these actions.)

We argue that in an $n$-splitting world, where the successor action is multiplication by n, *the unit is* n. In splitting, n can be viewed as the invariant multiplicative action between successor and predecessor. From 1, with our first split, we create the unit of n and the first number in the sequence as n (which may be indexed by the counting number 1). In our work, to assist our readers, we call the unit in the splitting world a *unit of growth.*

To describe the ordered progression 1, 2, 4, 8 as having a constant unit of 2 appears awkward to the first-time reader. It feels as if the unit size is changing. However, we emphasize that when numbers are conceived of as the results of constructive actions and units describe the invariance in the action between successor and predecessor, such awkwardness subsides. If one imagined a tree diagram of two splits, it is readily apparent that an invariance in the structure needs to be preserved by the similarity of each split and the overall similarity. We are claiming that this primitive and intuitive structure can be useful in creating a numeric system, and that it has a unit concept embedded in it.

Multiplication and division are constructed as repeated splits and are basic operations, constructed as inverses of each other. This is parallel to the construction of subtraction (take away) as the inverse or undoing of adding. In order to describe change or difference, we need to establish how to compare quantity. To compare numeric quantities, we create the idea of "interval" as sets of continuous units (composite units in Steffe's, 1988, terms).

In counting, we create addition of intervals greater than 1 by combining repeated successor actions together into composite units that can be added as a single successor action of the same kind rather than counted individually. We do this as an act of curtailment, first by creating the starting amount as a composite group. This allows for the development of counting on. Then as we learn the addition facts, we curtail this into single production acts of $a + b$, where $a$ and $b$ are integers. In a traditional number line representation, this allows us to affix intervals as sums and collections.

Difference, the basic descriptor for change in a counting world, is constructed from either counting backward (subtracting 1) or comparing the size of counting intervals, and counting the nonoverlap. Thus, negative numbers can be constructed by continuing to decrement by 1 beyond the origin of 0 or by subtracting two intervals $a - b$, where $b > a$ and $a$ and $b$ are integers.

In splitting, the 2 is constructed by the action of a two-split. This is the first construction of a splitting number. The two-split then creates the oneness of the origin, the initial amount. That the initial amount is conceptualized as 1 is a strong indicator of a context for splitting. The repeated action of multiplying by 2 creates the concept of a unit (of growth). An example of how one student constructed a timeline with a multiplicative unit of growth for the powers of 10 is provided in the section "Origins in Student Work."

What we are proposing is that the operations of multiplication and division come from the construction of the $n$-split in the splitting world. This implies that the actions of sharing, dealing, magnifying, and creating multiple sets of equal groups are the precursors and the roots of those operations. Once one has the simple unit stabilized by its invariant role as a unit in a sequence, one can begin to explore the creation of composite units (Steffe, this volume) in the splitting world (Carlstrom, 1992). That is, using a two-split, one goes 1, 2, 4, 8, 16, . . . and from this one can explain eight as a unit of three two-splits from 1.

The construction of the splitting world is not just an intellectual exercise; we encounter contexts in daily activities in which a multiplicative unit is the preferred means of comparison. In the counting world, first-order change implies difference in the sense of subtraction. In the splitting world, these comparisons are made from ratios. For instance, if we were to describe the distance from the earth to the moon as compared to the distance from the earth to the sun, we would not subtract the distances, but would tell how many times further it is to the sun than it is to the moon. If we wish to compare acid or base concentrations, as we do in our common pH scale, we compare them multiplicatively (although the pH scale is written to look like a count).

If we effectively have established the importance of a multiplicative unit (see Confrey and Smith, in press a,b for a more developed account of this), then we can move on to discuss the construction of an interval, a simple composite action that can be seen as "the same kind" as the basic $n$-split itself (Carlstrom, 1992). This will allow for multiplication or division by powers of $n$. We suggest using the terminology of a $c$-sized interval, where $c$ is used to remind us that we will describe the size of the splitting interval using counting numbers. Thus, in the case of the two-split, a three-sized interval is 3 units of growth, or times 8. To understand any ratio other than the basic $n$-split itself requires one to build this composite unit. A ratio is the invariant relationship between pairs of splitting numbers for any equal-sized interval. The simplest ratio is that of the $n$-split itself. Ratios within a single splitting world will be of the form $n^p$, where $p$ is an integer. This means that to claim that $16/2 = 64/8$ in a two-split structure requires only the proof that both pairs $(2, 16)$ and $(8, 64)$ are the same number of successor actions apart or the same number of units of growth (in this case, 3 units of a two-split). That is, equal-sized intervals in the splitting world correspond to equal proportions in a counting world. Therefore, we claim that the ratio construct is the underlying construct in equal proportions and that proportional reasoning starts with this sense of invariance and identity and then is expanded to include greater than and less than relations among ratios.

We can construct the unit fractions ($1/n$, where $n$ is an integer) from division and the ratio concept. Division, the inverse of multiplication, can be carried out past the origin, 1, to create these numbers. Alternatively, unit fractions can be created by comparing two splitting intervals, $a$ and $b$, where $b > a$ and $a$ and $b$ are $n$ to a power. (This can be thought of as going out $a$ places on a splitting structure and coming back $b$ places, where $b > a$; hence $n^a/n^b$.)

Finally, in the table we describe how rate in the counting world is viewed as the difference per unit time. However, to explain what is meant by saying that a population is increasing at a constant rate of 3 percent per year, we need to also create a concept of rate as a ratio per unit of time that is held constant across values. This is the concept of rate in the splitting world, and it allows one to claim a constant rate of change in multiplicative situations.

Constructed in this manner, splitting structures will produce geometric sequences, whereas counting structures will produce arithmetic sequences. To make these geometric structures more dense, we need a method to place new values in between existing values to whatever density desired. This method requires that we be able to insert values while maintaining consistency with the primitive actions that built the structure. Whether splitting or counting, we can change the size of the "unit," but not the kind of action that creates the unit. Thus, we increase the density of a counting structure, such as $\{1, 2, 3, 4, \ldots\}$ by inserting new elements between existing elements such that, when completed, successive elements still have an invariant unit based on the action of addition, that is, $\{1, 1.5, 2, 2.5, 3, 3.5, 4, \ldots\}$. Likewise, the density of a splitting structure, such as $\{2, 4, 8, 16, \ldots\}$ is increased by inserting elements such that the consistency of the type of action, multiplication, is maintained; that is, $\{2, 2\sqrt{2}, 4, 4\sqrt{2}, 8, 8\sqrt{2}, 16, \ldots\}$. Such actions, where any number of elements may be inserted between pairs of successive elements, are sufficient to construct structures with the density of the rational numbers. As we know from number theory, constructing the real numbers in counting structures, and likewise, the positive real numbers in a splitting structures requires more advanced techniques, beyond the requirements of this discussion.[1]

### Future Work on the Conjecture

The implications of the splitting conjecture could be far reaching if the conjecture is confirmed further in work with young children. It suggests that, with appropriate activities, the concept of splitting could be introduced together with counting. Multiplication and division would become as elementary as addition and subtraction, with care taken to see that multiplication derived from splitting was carefully coordinated with multiplication as repeated addition. Finally, it is possible that a concept of ratio could be introduced in parallel with the concept of difference in subtraction. This is in marked contrast to the current situation, where, as stated by three textbook writers for elementary teachers, "Although multiplication problems provide opportunities to discuss ratio informally, the concept of ratio is typically not studied until fifth grade" (Reys, Suydam & Lindquist, 1989, p. 201).

The rational number project (Behr, Lesh, Post & Silver, 1983; and Kieren, 1976) has documented a rich variety of concepts of rational numbers. The splitting conjecture can gain or lose support on the basis of its predictions for the treatment of these concepts. For example, it predicts that fractions and ratios need to be distinguished. A fraction (as a part of a whole) is a number whose numerator is constructed through an act of counting, while its denominator results from a split. Determining the common denominator will be an issue of coordinating splitting structures. Addition and subtraction of fractions will become straightforward once they have common denominators, creating a common counting unit. In a ratio, which is interpreted as a split, both the numerator and denominator are the result of splits. Multiplication and division of ratios should be relatively straightforward within a splitting world context. These ideas will need clarification in future work.

Certain topics will require the coordination of counting and splitting worlds, and promoting a conceptual understanding of these should pose the greatest challenge. One would expect difficulty with place value and decimal notation. These topics would be particularly difficult in a curriculum that neglects a sufficient treatment of splitting.

We suspect that the relationship that will evolve between counting and splitting is not best described as one of independence but as interdependence or complementarity. Otte (1990) argues for the importance of complementarity in understanding mathematical development. In his discussion of the development of function, he states: "The development of function theory and analysis since Cauchy is usually even described as an escape from both the deceptive evidence of spatial intuition and the fetters of the arithmetical or analytic expression. It seems appropriate to qualify such views by stating that the 'escape' amounted rather to a deepening of the related perceptions by employing each to develop the other in a complementarist fashion" (p. 58).

We believe that splitting and counting are complements or themata (Holton, 1973) and that they have their roots in the complementarity of arithmetic and geometry. We further conjecture that the early introduction of splitting into the curriculum could lead to an analogous deepening of each of the number concepts.

## TIES TO GEOMETRY

The development of the splitting structures as parallel to counting structures raises an interesting question. If splitting structures are formally isomorphic to counting structures (i.e., if we map the group of the positive real numbers over multiplication onto the group of the real numbers over addition), then how are these two worlds distinguished? One answer lies in the recognition that our understanding of numbers is coordinated with other fields of human experience. We have claimed in the last section that splitting is differentiated from counting because the actions related to splitting (dividing, enlarging, partitioning, magnifying) are different from the actions related to counting (adding, affixing, joining, annexing, subtracting, removing). Geometry provides a fertile ground for exploring, refining, and illustrating these distinctions.

Counting, we suggest, can be described in terms of translation, and it can be very useful in modeling actions that can be conceived of as movement across our field of vision. Splitting, we suggest, has its connections in the geometric concept of similarity and hence is useful in describing motion toward and away from us. In particular, we have been engaged in researching how students perceive the construction of such geometric figures as tree diagrams, embedded figures and spiral growth (see Figure 8.5). All of the figures tend to lead towards the concept of recursion. "An object is called recursive if it contains itself as a part, or is defined by means of itself" (Otte, 1990, p. 41). Alternatively, "Recursive designs [are those designs] because these figures contain subparts which are in some sense equivalent to the entire figure" (Abelson and diSessa, 1982, p. 87). Recursion is appropriate for modeling repeated multiplication because it al-

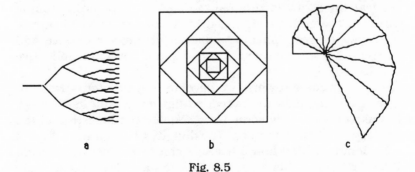

a b c

Fig. 8.5

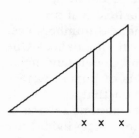
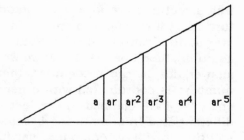

Fig. 8.6

lows for something that "grows in size *but does not change its shape;* and the existence of this constant relativity of growth, or constant similarity of form, is of the essence . . . " (Thompson, 1917, p. 179).

The two figures shown in Figure 8.6 demonstrate the strength of the claim for a fundamental connection between recursion and multiplicative relations. Both figures seem to be recursive, they both seem to meet the requirement that the object contains itself as a part, and yet one is essentially additive (that is, despite the similarity among all the triangles, the figure grows with an additive unit) and the other is essentially multiplicative with a multiplicative unit. The second, however, is recursive in a way that the first is not. Consider in each a figure of a triangle larger than the smallest triangle so that at least two vertical sides are included in the chosen figure. Now if one magnifies that triangle with two or more vertical divisions so that one of those two divisions lands exactly on a longer vertical line of the unmagnified figure, then two different outcomes occur. In the second figure, all the other lines of the magnified figure will lie on vertical lines of the original, consecutive to the matching lines. In the additive figure, this is not the case. We would describe the multiplicative case as strongly recursive and suggest that strong recursion is a signal for a multiplicative context.

In our current work, we have developed a tentative path from tree diagrams to embedded figures to spirals through the use of geometric sequences. We use the insertion of the geometric mean to increase the density of the figures. Figure 8.5c demonstrated how a triangle can be devised to grow on itself, where, in this case, the hypotenuse of a right triangle

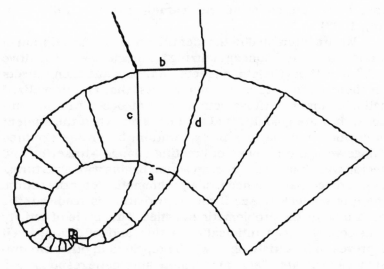

Fig. 8.7

becomes the longer leg of its successor and the process continues at a constant multiplicative rate of growth (i.e., the ratio per unit time remains constant). If one considers the comparison between two rates of growth, one can conceive of the growth of the quadrilateral wherein two opposite sides represent the growth from a to b (a rate of b/a) and the other two sides represent the growth from c to d (a rate of d/c) (see Figure 8.7).

This process generates a plane of geometric growth whose properties complement and contrast with those of the Cartesian plane. The spiraling quadrilaterals converge to a point. The growth factor (unit of growth) outward is $b/a$ in one direction and $d/c$ in the other. We are investigating the potential of this representation for exploring exponential growth and ratio and proportion.

## ORIGINS IN STUDENTS' WORK

Lest we give the impression that this theoretical analysis sprang full-blown from an analysis of the literature or through mathematical contemplation, we need to emphasize that its origin lies in the interpretation of teaching interviews. Our conduct and interpretation of this literature is cast within the constructivist paradigm (Steffe, 1989; Confrey, 1988).

We are interested in the identification and description of student schemes, conceptualizing a scheme to be a cognitive habit of action. The student acts on things and then reflects on those actions to create operations, that is, internalized actions. However, those actions do not occur without purpose; they are goal-directed actions. Even when the student is curious and engages in exploration, he or she explores things with the purpose of creating change—tinkering and seeing what happens. Thus, an action begins with a perturbation, a discrepancy, a need, or a problematic. After the action, there is a check to see if the perturbation is resolved, the need met, or the problematic satisfied. This cycle of activity thus consists of an anticipation, action, and reflection; and if it proves to be satisfying, it will be repeated in other circumstances to create a "scheme," a more automatic response to a situation. Schemes have duration and repetition, and therefore are more easily examined than are isolated actions (see Figure 8.8).

We believe that the construction of schemes by the student is a fundamental component of mathematics learning. We sought to *model and describe students' development and modification of splitting schemes.* Using the triad of the problematic, action, and reflection cycle, we (Confrey, 1989b) provided students with rich problem contexts that invited them to consider ideas of multiplication and exponential functions. We conducted interviews for the teaching experiment. In the interviews, we focused on eliciting students'

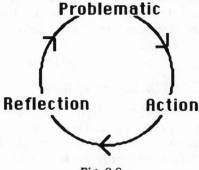

Fig. 8.8

ideas about the problem; we sought to describe their problematic, or call to action. We provided them the resources for diverse forms of action through the use of paper and pencil, calculators, and so on, and then we probed their responses and reasons. By asking them to verbalize their interpretations, actions, and reasons, we promoted their expression and modification of their schemes of internalized actions; hence we accept the label of *teaching experiments* for this work. We attempted to document and describe students' schemes through careful documentation and interpretation immediately after each interview, between interviews (toward revision of the plan for a new interview), and after the duration of the experiment. Interpretation of the interview products, transcription, and video recordings were done through repeated individual and team review. A case was prepared for each student in the teaching experiment from a daily log of their interview actions. Syntheses of these cases led to the development, revision, or dismissal of models for student schemes.

Examples of two episodes will be offered to illustrate how the splitting conjecture resulted from student work.

*Case One.* Dan, a freshman in the an agriculture college was the student who first led us to make the splitting conjecture. His difficulties with negative exponents were not overcome by simply learning an algorithm for moving to or from the top or bottom of a ratio. He first built the idea of an exponent as an index for multiplication and then built the negative exponent as an undoing of the multiplicative action of the exponent. This led us to focus on the operational character of the construction of the exponents and structure that underlies a system of multiplicative relations. Later, Dan persistently argued that repeated multiplication was unlike repeated addition. He drew figures to show that in repeated addition, the parts are still visible (in $9 + 9 + 9$ visualized as 27 dots, one can still see 9 and 18 as partial sums.) In $9 \times 9 \times 9$, one loses the partial multiplicative product. This distinction between repeated addition and repeated multiplication was profoundly important in helping us to articulate the splitting conjecture. Also, this emphasis on the meaning behind the multiplicative operation pushed Dan and us, as interviewers, to consider the operational character of exponential functions. As we watched Dan struggle to figure out an easy way to predict, after ten years, the value of a computer that holds its value by 90 percent per year, we saw how Dan conceptualized the interval as a "jump" and then worked to build jumps of greater and greater size. This exploration encouraged us to pursue a covariation approach to the exponential function. (Confrey and Smith, in press b.)

*Case Two.* Suzanne was also a freshman in the college of agriculture and life sciences. She took part in the second round of teaching experiments and was asked to place a series of dates and events given in a tabular form in scientific notation (for the most part) on a number line. She undertook her task by constructing a number line which eventually had the intervals marked: $1 \times 10^{10}$, $1 \times 10^9$, $1 \times 10^8$, $1 \times 10^7$ . . . After a few revisions, she placed ten spaces between each one. When she then needed to insert points between these values, she was perplexed. She chose one particular task, the placement of values between $1 \times 10^8$ and $1 \times 10^7$ and concentrated on it, but found herself with two competing frameworks. She first considered splitting the interval in half and labeling the middle 1.5, . . . but then said, "No, I'm not going to do that." She counted off her spacing, starting with 1 and reaching 10 on the ninth mark, and said in dismay, "No, if I do, $10 \times 10$ is . . . " and trailed off. She aided herself by writing out the numbers in standard form below the scientific notation. On the interviewer's request, she restated her problem, this time using the standard form of the numbers. "I don't know how to go from 100 million to 10 million. It is just 10—, 10 million times 10 is 100,000. The scientific notation is messing me up. This would be 10 million and 500 thousand, right?" [She pointed halfway between 100 million and 10 million.] "It would be halfway in between, right? So, it would be 1.5 right here?" The interviewer repeated her claim, asking if 10,500,000 is halfway between 10 million and 100 million, and she quickly responded that it was halfway between 10 million and 11 million. She said, "All right, then I have too many spaces. . . . Since this is 10 million, and this is 100 million, I have to go by 10 millions. So it would be, if this is 10 million, 20 million, 30 million, 40 million, . . . , 90 million. Then I'd have an extra box."

What this student has done is to create a log-linear scale. In the language of this proposal, she has combined splitting and counting acts, and one can see how the two compete as viable approaches to putting points in between markings. She has also created a viable model for her task. That is, if she were to try to create a linear number line, even using a scale of 10,000 years to the inch, her number line would be 23 miles long. In later interviews, she also demonstrates her flexibility with this version of the number line, learning to measure between intervals by a distributive law separating the interval sizes. Interpreting this transcript also drew our attention to a fresh view of the structure of scientific notation. It separates the exponential part from the interval between 1 and 10, and, in doing so, allows us to separate counting and splitting worlds. Thus, as one student put it, scientific notation allows us first to determine if numbers are in the same "realm" and then to compare them. (See Confrey, 1989b, for a detailed discussion.)

## USING SPLITTING AS A FOUNDATION FOR EXPONENTIAL FUNCTIONS

From our theoretical analysis of teaching interviews on exponential functions, we concluded that

1. Examining students understanding of operations is a key issue in understanding functions.

2. Focusing on the mapping of real numbers over addition onto the positive real numbers over multiplication as the basis for understanding exponents and logarithms (as is assumed in most texts in the rules of exponents and logarithms) is suspect in that students first need to engage with the relationships within each side of this isomorphism.[2]

3. Discovering that the formal isomorphism between the real numbers over addition and the positive real numbers over multiplication hides the differences in the cognitive schemes used to understand repeated multiplication and those evoked with repeated addition. These results led us to examine multiplication independent from addition, to seek to establish an experiential base in a multiplicative setting,

and to later decide to anchor this experiential base in similarity.

4. Given appropriate contexts, students are able to construct and coordinate both multiplicative and additive concepts of successor and provide meaningful interpretations of their use.

Finally, we see the construction of exponential and logarithmic functions as an action of juxtaposing a counting and a splitting structure or, in contextual situations, identifying some elements we construct through a counting structure and other elements we construct through a splitting structure. In most cases, when we model with exponential functions, we build our counting structure as identifying equal (additive) increments of time, and our splitting structure as involving some measure we create in the physical world (i.e., the amount of accumulated money under compound interest, the amount remaining of a radioactive substance, the population of some organism in certain settings).

## CONCLUSIONS

In this chapter, we examine possible conceptual origins of exponential functions. We locate these origins in an approach to multiplication and division we label *splitting*. Splitting is proposed as a primitive action, rooted in both mental and physical experience, that can lead to the construction of numbers. This number concept that evolves is argued to be independent from counting in its successor action. That is, one can be certain that a split has been accomplished correctly using one-to-one correspondence or symmetric folding. The splitting action creates a formal isomorphism between the splitting world (as a commutative group on $R^+$ for multiplication) and the counting world (as a commutative group on $R$ for addition). The *inter*dependence of these groups is, however, created by naming the outcome of the splitting action consistent with its equivalence in the counting numbers, hence the common identification of multiplication with repeated addition. Therefore, we claim that repeated addition is a sufficient interpretation for a limited subset of the multiplicative splitting world. Finally, we argue that the distinctiveness of splitting and counting becomes most apparent by appealing to geometric perspective, where splitting is intimately linked with similarity, while counting is linked to translation.

Our proposal for splitting could have potentially dramatic implications for the K–12 curriculum. Based on our splitting structure,

1. We suggest that rather than waiting to teach multiplication-division and ratio and proportion until these concepts can be logically derived from the counting numbers, there should be a parallel development for the splitting construct starting in the primary grades.

2. We argue that a geometric approach to numbers should be included immediately with the introduction of arithmetic and should include tree diagrams and similar figures.

3. We hypothesize that young children develop intuitions and insights into splitting that are either ignored or inhibited by our current curriculum.

4. We suggest that the operations of addition, subtraction, multiplication, and division have their roots in primitive experiential actions that are coordinated and integrated to form formal arithmetical operations and that these primitive actions need investigation.

5. We argue that numbers are not uniform entities, rather their behavior is dependent on how they are made within various contextual settings, so that when studying students' number concepts of fractions, ratios and decimals, researchers would be well-advised to pay close attention to the interactions between numbers as objects and the operations on them.

## NOTES

This research was funded under grants from the National Science Foundation, (MDR 8652160), and the External Research Division and the Apple Classrooms of Tomorrow[SM] of the Apple Computer Corporation. The author wishes to recognize the significant contribution of Erick Smith and David Henderson, and Kenneth Carlstrom, Cornell University, for their considerable help on the ideas expressed in this chapter. With support from the National Center for Research in Mathematical Science Education and our NSF project, the Multiplicative Conceptual Field Sub-Group (co-chaired by G. Harel and J. Confrey) has been meeting to discuss issues concerned with multiplication and its representation. A number of the ideas in this chapter were refined after a small group meeting of L. Steffe, M. Behr, P. Thompson, J. Kaput, G. Harel, E. Smith, and J. Confrey. Much appreciation and credit goes to the careful analysis offered by this group.

1. Even though we have such "irrational numbers" (in a counting sense) as $\sqrt{2}$, the splitting world created through multiplication is only as dense as the rationals in a counting world. One can see this by considering that the integer 3 will not be included in a two-split world regardless of its density. Therefore, we will need advanced techniques to allow us to interweave splitting structures together and construct the transcendental.

2. This is analogous to what Vergnaud (1983) refers to as the *scalar relationship* in his isomorphism of measures, although in his case the mapping describes a linear function.

## REFERENCES

Abelson, Harold, and Andrea diSessa. 1982. *Turtle geometry.* Cambridge, MA: MIT Press.

Behr, M., R. Lesh, T. Post, and E. Silver. 1983. Rational-number concepts. In *Acquisition of Mathematics Concepts and Processes,* ed. R. Lesh and M. Landau, 91–126. New York: Academic Press.

Bell, A. 1983. Diagnostic teaching of addition and multiplicative problems. In *Proceedings of the seventh international conference for the psychology of mathematics education,* ed. R. Hershkowitz. Rehovot: Weizman Institute of Science.

Brousseau, G. 1986. Basic theory and methods in the didactics of mathematics. *Proceedings of the second conference on Systematic Cooperation between Theory and Practice in Mathematics Education,* Enschede, The Netherlands.

Carlstrom, K. 1992. Units, ratios, and a "splitting" dimension: Students' construction of multiplicative worlds in a computer environment. Unpublished thesis. Cornell University, Ithaca, NY.

Carpenter, T., and J. Moser. 1983. The acquisition of addition and subtraction concepts. In *Acquisition of mathematical concepts and processes,* ed. R. Lesh and M. Landau, 7–40. New York: Academic Press.

Carraher, T. N. 1986. From drawings to buildings; working with mathematical scales. *International Journal of Behavior Development* 9: 527–544.

Confrey, Jere. 1980. Conceptual change, number concepts and the introduction to calculus. Unpublished doctoral dissertation, Cornell University, Ithaca, New York.

Confrey, Jere. 1988. Multiplication and splitting: Their role in understanding exponential functions. *Proceedings of the tenth annual meeting of the North American chapter of the international group for the psychology of mathematics education.* DeKalb: Northern Illinois University Press.

————. 1989a. Splitting and counting: A conjecture. Paper presented at the annual meeting of the American Educational Research Association, San Francisco.

————. 1989b. The concept of exponential functions: A student's perspective. In *Epistemological Foundations of Mathematical Experience,* ed. Les Steffe. New York: Springer-Verlag.

Confrey, Jere and Erick Smith. 1989. Alternative representations of ratio: The Greek concept of anthyphairesis and modern decimal notation. Unpublished manuscript.

———— and Erick Smith. In press a. Exponential functions, rates of change, and the multiplicative unit. *Educational Studies in Mathematics.*

———— and Erick Smith. In press b. Splitting, covariation and their role in the development of exponential functions. *Journal for Research in Mathematics Education.*

Cook, Theodore Andrea. 1914, reprinted in 1979. *The curves of life.* New York: Dover Publications.

Dawkins, Richard. 1987. *The blind watchmaker.* New York: W. W. Norton.

De Corte, Erik, Lieven Verschaffel, and Veronique Van Coillie. 1988. Influence of number size, problem structure and response modes on children's solutions of multiplication work problems. *Journal of Mathematical Behavior* 7 (3): 197–216.

Dienes, Z. P. 1967. *Fractions: An operational approach.* New York: Herder and Herder.

———— and E. W. Golding. 1966. *Sets, numbers and powers.* New York: Herder and Herder.

Fischbein, Efraim, Maria Deri, Maria Sainati Nello, and Maria Sciolis Marino. 1985. The role of implicit models in solving verbal problems in multiplication and division. *Journal for Research in Mathematics Education* 16 (1): 3–17.

Gerdes, Paulus. 1989. *On possible uses of traditional Angolan sand drawings in the mathematics classroom.* Unpublished manuscript.

Gould, Stephen Jay. 1987. *Time's Arrow—Time's Cycle.* Cambridge, MA: Harvard University Press.

Graeber, Anna O., and Dina Tirosh. 1988. Multiplication and division involving decimals: preservice elementary teachers' performance and beliefs. *Journal of Mathematical Behavior* 7 (3): 263–280.

Greer, Brian. 1988. Nonconservation of multiplication and division: Analysis of a symptom. *Journal of Mathematical Behavior* 7 (3): 281–298.

—— and C. Mangan. 1984. Understanding multiplication and division. In *Proceedings of the sixth annual meeting of the North American chapter of the international group for the psychology of mathematics education*, 27–32. Madison: University of Wisconsin Press.

Hawkins, David. 1974. *The informed vision*. New York: Agathon Press.

Hiebert, J., and M. Behr. 1988. Introduction: Capturing the major themes. In *Number concepts and operations in the middle grades*, ed. M. Behr and J. Hiebert Reston, VA: National Council of Teachers of Mathematics.

Holton, Gerald. 1973. *Thematic origins of scientific thought*. Cambridge, MA: Harvard University Press.

Kaput, James. 1988. Applying technology in mathematics classrooms: It's time to get serious, time to define our own technological destiny. AERA symposium: Classroom Learning and the Challenge of Technology, New Orleans.

Kieren, T. 1976. On the mathematical, cognitive, and instructional foundations of rational numbers. In *Number and measurement*, ed. R. Lesh, 101–144. Columbus, OH: ERIC/SMEAC.

—— and D. Nelson. 1978. The operator construct of rational numbers in childhood and adolescence—An exploratory study. *Alberta Journal of Educational Research* 24 (1): 22–30.

Luke, Clifton. 1988. The repeated addition model of multiplication and children's performance on mathematical word problems. *Journal of Mathematical Behavior* 7 (3): 217–226.

McKibben, Bill. 1989. *The End of Nature*. New York: Random House.

Meadows, Donella, Dennis Meadows, Jorgen Randers, and William Behrens. 1974. *The limits to growth*. New York: Universe Books.

Mechmandarov, I. 1987. The role of dimensional analysis in teaching multiplicative word problems. Unpublished manuscript, Center for Educational Technology, Tel-Aviv.

Millroy, Wendy L. 1990. *An ethnographic study of the mathematical ideas of a group of carpenters*. Unpublished manuscript, Cornell University.

Morrison, Philip, Phylis Morrison, and the Office of Charles Eames and Ray Eames. 1982. *Powers of ten.* New York: Scientific American Library.

National Council of Teachers of Mathematics. 1989. *Curriculum and evaluation standards for school mathematics.* Reston, VA: Author.

Nesher, Pearla. 1988. Multiplicative school word problems: Theoretical approaches and empirical findings. In *Number Concepts and Operations in the Middle Grades,* 19–40. Reston, VA: National Council of Teachers of Mathematics.

Otte, M. 1990. Arithmetic and geometry. *Studies in Philosophy and Education* 1 (1): 41.

Peled, L., and P. Nesher. 1988. What children tell us about multiplication word problems. *Journal of Mathematical Behavior* 7 (3): 239–262.

Piaget, J. 1970. *Genetic epistemology.* New York: Norton and Norton.

Piaget, J., I. Berthoud-Papandropoulou, and H. Kilcher. 1987. Multiplication and multiplicative associativity. In *Possibility and necessity,* ed. J. Piaget, 78–97. Minneapolis: University of Minnesota Press.

Pothier, Yvonne, and Daiyo Sawada. 1983. Partitioning: The emergence of rational number ideas in young children. *Journal for Research in Mathematics Education* 14 (4): 307–317.

Resnick, L., P. Nesher, F. Leonard, M. Magone, S. Omanson, and I. Peled. 1989. Conceptual bases of arithmetic errors: The case of decimal fractions. *Journal for Research in Mathematics Education* 20 (1): 8–27.

Reys, Robert E., Marilyn Suydam, and Mary Lindquist. 1989. *Helping children learn mathematics.* Englewood Cliffs, NJ: Prentice-Hall.

Schwartz, Judah. 1988. Intensive quantity and referent transforming arithmetic operations. In *Number concepts and operations in the middle grades,* ed. M. Behr and J. Hiebert, 41–52. Reston, VA: National Council of Teachers of Mathematics.

Smith, Erick, and Jere Confrey. 1989. Greek concepts of ratio and proportion. Paper presented at AERA.

Sowder, Larry. 1988. Children's solutions of story problems. *Journal of Mathematical Behavior* 7 (3): 227–238.

Spengler, Oswald. 1925. The decline of the West. *Form and Actuality,* 51–90. New York: Alfred A. Knopf.

Steffe, L. 1987. Units and their constitutive operations in "multiplicative contexts". A report prepared under a grant from the National Science Foundation.

———. 1988. Children's construction of number sequences and

multiplying schemes. In *Number concepts and operations in the middle grades*, ed. M. Behr and J. Hiebert. Reston, VA: National Council of Teachers of Mathematics.

————, ed. 1989. *Epistemological Foundations of Mathematical Experience*. New York: Springer-Verlag.

Steffe, L. and P. Cobb. 1983. Multiplicative and divisional schemes. *Proceedings of PME-NA*, 11–27. Montreal.

Steffe, L., E. von Glasersfeld, J. Richards, and P. Cobb. 1983. *Children's counting types: Philosophy, theory, and application*. New York: Praeger Books.

Thompson, D'Arcy. 1917, reprinted in 1961. *On growth and form*, ed. J. T. Bonner. Cambridge: Cambridge University Press.

Vergnaud, Gerard. 1983. Multiplicative structures. In *Acquisition of mathematics concepts and processes*, ed. R. Lesh and M. Landau, 127–173. New York: Academic Press.

————. 1988. Multiplicative structures. In *Number concepts and operations in the middle grades*, ed. M. Behr and J. Hibert, Reston, VA: National Council of Teachers of Mathematics.

# 9 Multiplicative Structures and the Development of Logarithms: What Was Lost by the Invention of Function?

*Erick Smith*
*Jere Confrey*

## I. INTRODUCTION

An increasing concern in recent years about what and how much students are learning in mathematics classes has led to in-depth investigations of specific subject-matter areas. One such area has come to be called *multiplicative structures* (Vergnaud, 1983), which includes topics ranging broadly from multiplication as repeated addition and ratio and proportion at the lower grade levels, leading to concepts of linear, exponential, and logarithmic functions at the secondary level. Broadly viewed, research in this area has taken one of three approaches: (1) analysis of student errors (Fischbein, Deri, Nello, and Marino, 1985; Noelting, 1980a, 1980b; Karplus, Pulos and Stage, 1983; Greer, 1988; Hart, 1988); (2) analysis of the mathematical and semantic structures (Behr, Post, Lesh, and Silver, 1983; Schwartz, 1988; Nesher, 1988; Vergnaud, 1983, 1988; Harel and Behr, 1990; Behr, Harel, Post, and Lesh, this volume); or (3) synthesis of how people construct an understanding of mathematical concepts (Steffe, 1988, this volume; Thompson, this volume; Confrey, 1991a, 1991b).

Although understanding student errors and understanding formal structures are essential and complementary to our work, we believe that learning how to teach more effectively depends fundamentally upon developing an understanding of how people come to know mathematical ideas or, more specifically, being able "to specify the operations involved in constructing a mathematical reality" (Steffe, 1988,

331

p. 119). Therefore the theoretical construct of "splitting" presented by Confrey in this volume can be seen as developing out of a synthesis of ideas that developed from individual teaching interviews with students (Confrey, 1991a; Rizzuti and Confrey, 1988), whole class work (Confrey and Smith, 1989), and analysis of the historical development of these concepts in Western mathematics (Smith and Confrey, 1989a, 1989b).

In undertaking historical analysis, we are not advocating the naive view that individual learning should follow historical development, but rather looking for ways in which a "rational reconstruction" (Lakatos, 1976) of the historical genesis of mathematical concepts may complement our work with students, helping us see students' work from a different perspective. At the same time, listening to students work on problems has helped us appreciate problems in the evolution of historical ideas that would not have been apparent from a solely historical analysis.

This complementary relationship between our historical analysis and work with students was particularly strong in the development of logarithms. In particular, the insistence of one student that repeated multiplication was different from repeated addition (Confrey, 1991a) and the distinctly bioperational number line created by another student (Confrey, 1991b) helped us appreciate the significance of the multiplicative world composed of ratios created by Thomas of Bradwardine and Nicole Oresme. The distinctions made by these two mathematicians in turn helped us define the issues that the students needed to resolve in creating their own understanding and appreciation of the conceptual complexity masked in our conventional definition of the real number line. Likewise the intuitive strength evidenced in students of the notion of function as covariation, rather than our more common notion of function as a correspondence, provided reciprocal insight into the issues Napier needed to resolve to derive a satisfactory model of logarithms (Rizzuti, 1991).

The interweaving of these two themes, the construction of number (or of a number line) through the primitive action of multiplication and a concept of function as the covariation of two quantities, was essential in the historical development of logarithms and provides the thematic structure for the balance of this chapter.

## II. NUMBER CONCEPTS AND THE REAL NUMBER LINE

It is possible to view the elementary and secondary mathematics curriculum in the United States as a twelve-year project to pack many concepts of number and many kinds of actions that people take on these various concepts of number into one entity, the real number line, with its four operations. The crowning touch of this project can be seen as the development of the function concept, allowing the formalization of a relationship between two (or more) subsets of the number line that seems to be independent of the operational origins of the elements of either set. Although this process results in a wonderfully efficient construction that appears to be flexible enough to solve most quantifiable problems that humans create, there is evidence that the project itself is a failure. Many students who go through the process and become relatively facile in performing formal operations on the entity are incapable of using that facility to solve particular problems in specific contexts. In effect, they appear unable to "unpack" the specific concept of number and the particular actions embodied in formal operations that could be useful for them in problem-solving situations.

The problem, one could argue, is that in treating the subject matter of mathematics in this rather uniform and formalized manner, we have confused means with ends. That is, instead of viewing the number line (with its operations) as a flexible tool that allows us to carry out specific actions on specific concepts of number in generalized ways, we neglect the specific for the general; we view the construction of the real numbers as the goal itself, forgetting how we got there. We treat members as uniform entities, points on a line, and operations as formal procedures for manipulating those points. We use functions as algorithms for converting real numbers into real numbers, forgetting their role as a coordinator between number and action in one setting and number and action in another.

Historically, we find evidence for two separate concepts,

number and ratio. In the early Greek work, number was based on counting. Because the early Greeks had considered all magnitudes to be built from a common unit, counting was also seen as sufficient for indicating the size of a magnitude (Smith and Confrey, 1989b). Eventually, as magnitude came to be seen as continuous, the concept of number included integers, rationals, and irrationals, but was still conceived of as a count or as an indicator of the size of a magnitude. Because both counting and measuring create a successor action by the action of addition, addition was the basic operation on numbers, and multiplication was primarily an abbreviation for repeated addition. In Book VII, Euclid defines multiplication as "when that which is multiplied is added to itself as many times as there are units in the other" (Heath, 1956, p. 277). Thus we will talk about "additive worlds" as number structures (lines) built with addition as the successor action.

The concept of number was distinct from the concept of ratio, described by Euclid as "a relation in respect to size between two magnitudes" (Heath, 1956, p. 114), and what Behr et al., called a *comparative index* (1983, p. 95). Section V examines the work of Thomas of Bradwardine and Nicole Oresme, who claim that, whereas the primitive action taken on magnitudes is addition, the primitive action taken on ratios is multiplication. Using this "primitive action," Oresme builds a world whose elements are ratios, but that is, essentially, isomorphic to the real numbers. We will describe multiplicative worlds as number structures (lines) built with multiplication as the successor of action.

## III. CONCEPTS OF FUNCTION

In a review of the historical development of concepts of function, Rizzuti (1991) distinguishes between the classical definition, often associated with Euler, and the modern definition, usually referred to as the *Dirichlet-Bourbaki* definition. An important distinction, according to Rizzuti, is in the importance placed on the notion of covariation between two variables. "The 'classical' definition expresses a relationship between varying quantities, a covarying relationship. The focus of the 'modern' definition, on the other hand, is on the correspondence between two sets . . ." (1991, p. 24). Our own work has led to at least three distinctions in the generation of covarying variables. From the work of Alhazen (Dennis and Confrey, 1991) and our own work with students (Afamasaga-Fuata'i, 1991; Confrey, 1991a) we have seen covariation of two variables constructed from combinations of accumulation and additive rates of change; from Descartes, we see covariation built from geometric construction (Smith, Dennis, and Confrey, 1992); and finally from the time of Aristotle through the work of Thomas of Bradwardine, we see functions arising through the cogeneration of two variables.

In 1837, Dirichlet provided a "correspondence" definition for function: "Y is a function of $x$ if for any value of $x$ there is a rule which gives a unique value of $y$ corresponding to $x$," (attributed to P. L. Dirichlet, 1837; quoted by M. A. Malik, 1980, p. 491). An example, provided by Rizzuti (1991, p. 20), that satisfies this definition yet does not have a sense of covariation of variables is the Dirichlet Function. One form of this function would be: $f(x) = \{1$ if $x$ is rational; $0$ if $x$ is irrational$\}$.

The modern curriculum, with its origins in formal systems and set theory and its heavy reliance on algebraic manipulations, tends to rely on a correspondence notion of function. For example, Dolciani provides the following definition: "A function consists of two sets, the domain and the range, together with a pairing that assigns to each member of the domain exactly one member of the range" (Dolciani et al., 1984, p. 19). Rizzuti argues that introducing functions using definitions based on the Dirichlet-Bourbaki definition plays a major role in the difficulties students have in learning functions: "[It] requires students to learn a definition which is separated from the functional thinking they do outside of mathematics class. It does not build on experiences they have had with functional relationships in their world in which one quantity varies with or depends upon another" (1991, p. 25). In our own work, we have seen the powerful role that covariation can play for students as they often build functional relationships through covarying actions, for example, building down two columns of a table. Because of the importance of these two approaches to functions, we make a distinction between a "correspondence" and a "covariation" notion of function (Confrey and Smith, 1991). We believe that this distinction is central both to understanding the historical development of function concepts and to understanding the work of students in solving problems in functional situations.

The covariation model was central to the development of logarithmic functions and will form the basis for the discussion in the balance of this chapter. In accordance with the preceding discussion of the distinction between number lines generated by additive actions and those build on multiplicative actions, we will see logarithmic functions rising from the cogeneration of an additive and a multiplicative number line or what we might call the juxtaposition of additive and multiplicative worlds.

## Logarithmic Functions in the Curriculum

As noted by Rizzuti, much of the modern curriculum is dominated by the "modern" correspondence concept of function. We would argue that this is the case despite the fact that analytical expressions are almost always used to provide the rule connecting elements of the domain with elements of the range and despite Euler's definition of a covariation model of function based on analytical expressions (see earlier). Although the argument for this claim could be developed at length, it is based on the supposition that students must build an understanding of an analytic expression to use it. In our current algebra curriculum, students build their understanding of analytical expressions primarily as algorithmic procedures that are decoded through a set of rules. Thus, at best, they are taught how to put a value into an expression and determine the output, very much in accord with a correspondence model of function. An alternative approach to algebra would be to see it as a means of coding the actions that one takes when expressing the variation of quantities. Although we believe that much work needs to be done in developing such a curricular approach, it offers the potential of allowing students to build their sense of covariation into their analytic expressions and finally allowing them to see this action in the expressions they and others create.

Like other curricular topics, the development of exponential functions in our secondary curriculum is based on a correspondence model of function. Students first learn to evaluate exponential expressions with positive integer exponents, where the values are generated through repeated multiplication. From this intuitive notion of exponents as counters for repeated multiplication, rules are extracted: $a^x a^y = a^{x+y}$, $(a^x)^y = a^{xy}$, and so forth. These rules are then declared to be universal across the real numbers, conceptually allowing the evaluation of exponential expressions for any real valued exponent and thus allowing continuous exponential functions. Once students have learned to manipulate and evaluate exponential expressions, exponential functions can be defined. This development of exponential function is heavily dependent on the process of starting with an input value and *creating* an output value through a set of algorithmic rules. With this developmental model of exponential functions, it is not surprising to find students having difficulties in their conceptual development of exponential inverses, logarithmic functions. How does one account for or begin to understand the domain of a logarithmic function independent of the exponentials that "created" it? In formal mathematics, we simply say that the domain is the positive real numbers, as if there were no difference between it and the range, the real numbers, except that the domain is a subset of the range. Recently Confrey (1988, this volume) has challenged this idea, positing that a more productive way to look at worlds created through the actions of repeated multiplication is through the development of independent multiplicative structures, which she has called *splitting structures*. Confrey's work would, in effect, allow the development of exponential functions through a cogeneration model, where the elements and structure of the domain and range are cogenerated through simultaneous but independent actions, creating a covariation model of function. With this model, logarithmic functions arise more naturally, being only artifactually distinguished from exponentials.

## Multiplicative and Additive Worlds

The historical development of logarithms offers support for this approach. We find two almost independent developments occurring with their juxtaposition and correspondence culminating in the work of Burgi and Napier. In this chapter, we separate these two worlds as follows:

- *Additive world:* Number worlds developed through the counting numbers and arithmetic sequences, where

successive elements have a constant (additive) difference.
- *Multiplicative world:* Number worlds developed through
geometric sequences, where successive elements have
a constant (multiplicative) ratio.

Historically, both worlds were originally seen as discrete sets
of elements. Over time, both evolved to have first the density
of the rational numbers, then the density of the real numbers; that is both worlds became what we would now call
*continuous.* Part of the argument in this chapter is that as
long as each world is conceived of with only the density of the
rational numbers, the operations that structure each world
are inseparable from the structure of the set of numbers itself, and the individual identities of the two worlds are maintained.[1] However, when the density of the real numbers is
constructed, the operations that created the separate worlds
are subsumed in formal representations, such as axiomatic
systems or the number line, and the multiplicative world becomes seen as a subset of the additive world.

The idea of cogenerating additive and multiplicative
worlds was essential to Napier's development of logarithms.
As opposed to modern function concepts, which would tend
to see an input on one side determining an output on the
other, Napier constructed two independent worlds, a particle
moving arithmetically and one moving geometrically, and by
using time as a basis to visualize their cogeneration, created
a relationship that we now call a (stretched) *log function.* In
the process of making his geometric world continuous,
Napier's model turns out to have what we would now call a
*base* of $1/e$, an intriguing result of his need to be able to
determine the time at which his geometrically moving point
was at any designated position.

## IV. THE DEVELOPMENT OF LOGARITHMS

Looking at the historical development of logarithms, we find
four areas of interest:

1. *The development of arithmetic and geometric sequences.* At the time of Euclid, number meant specifically a
counting number; that is, the set of numbers was specifically

the arithmetic sequence {1, 2, 3, . . . }. Within his discussion of number, Euclid also presented a book on geometric series. Because the elements of his geometric series were (counting) numbers, a major part of his world dealt with geometric series with a finite number of members. For example {81, 54, 36, 24, 16} is a geometric series that cannot be extended within the world of integers. As the concept of number expanded to include what we now call *rational and real numbers*, the variety in these sequences was increased. However, in the Euclidean tradition, number and ratio were primarily viewed as different entities for the next 1500 years. This distinction provided the impetus for the work of Thomas of Bradwardine and Nicole Oresme, which will be discussed later.

2. *The juxtaposition of arithmetic and geometric series.* Archimedes provides one of the earliest examples of placing arithmetic and geometric series side by side. He noted that in such an arrangement, multiplication in one series corresponds to addition in the other (Table 9.1). For example, to multiply 9 times 81, one finds 9 and 81 on the geometric side, finds the corresponding numbers, 4 and 8 on the arithmetic side, adds to get 12, then looks to see what 12 corresponds to on the geometric side.[2] For the next 1700 years, this was seen to be an interesting, but not particularly useful property of this cogenerated form of a table.

With the rise of mercantilism in the fifteenth and sixteenth centuries, interest in finding easier methods of calculation was increasing. Burgi, in the sixteenth century, recognized the potential for simplifying multiplication inherent in

TABLE 9.1. JUXTAPOSITION OF ARITHMETIC AND GEOMETRIC SERIES

| Geometric | Arithmetic |
|---|---|
| 1 | 0 |
| 3 | 2 |
| 9 | 4 |
| 27 | 6 |
| 81 | 8 |
| 243 | 10 |
| 729 | 12 |

tables of geometric and arithmetic series. The problem in tables such as the one in Table 9.1 is, of course, that there are substantial holes between entries on each side. For example, one cannot use that table to multiply 2 times 3. Burgi's solution was to minimize the problem by creating a geometric series with entries very close together. He calculated and constructed extensive tables showing powers of 1.0001 (Table 9.2), which allowed one to find a reasonable approximation of any moderately sized number on the geometric side (Cajori, 1915).

Because it is not typical to think of a logarithmic function as the juxtaposition of a geometric and arithmetic series, it is worthwhile to explicitly define what is meant by a logarithm in a table such as that in Figure 9.2. The definition is *for any number in the geometric series, its logarithm is defined as the corresponding number in the arithmetic series*. Even though Burgi's tables were practical, in that it is possible to find a reasonable approximation for any number on the multiplicative side, there is a conceptual hole in Burgi's work. No matter how small one makes the multiplicative ratio in generating such a table, there are numbers that simply will not appear. Therefore, in the end, this very practical work loses out aesthetically to the work of John Napier who, as we shall see, solved this "missing values" problem ingeniously.

TABLE 9.2. INCREASED DENSITY
OF GEOMETRIC SERIES

| Geometric | Arithmetic |
| --- | --- |
| 1.000000000000 | 0.00 |
| 1.000100000000 | 1.00 |
| 1.000200010000 | 2.00 |
| 1.000300030001 | 3.00 |
| 1.000400060004 | 4.00 |
| 1.000500100010 | 5.00 |
| 1.000600150020 | 6.00 |
| 1.000700210035 | 7.00 |
| 1.000800280056 | 8.00 |
| 1.000900360084 | 9.00 |
| 1.001000450120 | 10.00 |

3. *The development of continuous geometric worlds.* To develop a logarithmic function that both practically and conceptually allows one to find *any* numerical value in the multiplicative world, one must, in effect, find a way to make a geometric series continuous. As is often the case in the historical development of a discipline, this problem was explored, long before the creation of logarithms, by two mathematician's working on an unrelated problem. Although there is no direct historical tie between their work and the model used by Napier, their work makes a substantial contribution to our understanding of multiplicative worlds and thus of logarithms. Therefore we will explore the work of these two men, Thomas of Bradwardine and Nicole Oresme, before turning to that of John Napier.

4. *The cogeneration of continuous additive and continuous geometric worlds.* Once Napier had constructed a model that allowed him to cogenerate continuous additive and multiplicative worlds, he faced another problem. The kinds of actions (i.e., multiplicative) he used to create the multiplicative world were somewhat at odds with the generation of values he needed to create a log table. In particular, he needed to be able to space the entries in his multiplicative world at additive intervals. For example, if his multiplicative world were generated through a constant ratio of 2, entries might be $\{1, 2, 4, 8, 16, \ldots\}$. To make his table useful, however, he needed entries in that world to be spaced additively, $\{1, 2, 3, 4, 5, \ldots\}$. That this was no trivial problem can be seen by noting that *no* (rational) ratio can be successively applied to 1 that can generate both 2 and 3.[3] The major problem facing Napier was to find a way to associate any point in the continuous multiplicative world with a point in the additive world such that the isomorphism remained intact; that is, multiplication of any two numbers in the former would be equivalent to addition of the matching two numbers in the latter. Looking at his solution to this problem will constitute the last section of this chapter.

## V. THE DEVELOPMENT OF A CONTINUOUS MULTIPLICATIVE WORLD

Some of our strongest intuitions deal with the continuity of time and space. In fourteenth century Europe, one learned about space, time, and motion by studying and interpreting what was considered the ultimate authority for such knowledge, the work of Aristotle. Ironically, an interpretation of Aristotle that saw the velocity of an object changing additively when the force on the object changed multiplicatively led Thomas Bradwardine, followed by Nicole Oresme, to create the mathematics of a continuous multiplicative world.

### Thomas of Bradwardine: Ratios as Inherently Multiplicative

Since the time of Aristotle, it had been assumed that the movement of any object was the result of two opposing forces: first, the motive force, $F$, which acted in the direction of the movement; and second, the resistance, $R$, which acted against the motive force. Although Aristotle had claimed that velocity varied with the proportion of these forces, $F/R$, neither he nor subsequent mathematicians had rigorously developed the mathematics of this relationship. This is the problem to which Bradwardine devoted his efforts in his Tractus (Crosby, 1961).

The most literal translation of Aristotle's observation, which had been proposed by others prior to Bradwardine, was

$$(1) \qquad V_2 : V_1 :: (F_2/R_2) : (F_1/R_1)$$

Bradwardine rejected this model for several reasons, the most important being that as velocity becomes smaller, approaching 0, a point would be reached where $F < R$, which violated a common assumption, first stated by Aristotle, that there can be no velocity unless $F > R$.[4] Therefore Bradwardine sought a solution that allowed $V_2$ to approach 0 as $F_2/R_2$ approaches 1 while maintaining Aristotle's claim that $V$ varied in proportion with $F/R$.

Bradwardine states his solution to this problem as, "The proportion (ratio)[5] of the proportions of motive to resistive powers is equal to the proportion of their respective speeds of motion, and conversely"[6] (Crosby, 1961, p. 113). At first glance, this seems almost identical to equation (1). To understand Bradwardine, however, it will be useful to rewrite (1) in a more modern form:

(2)     $V_2/V_1 = (F_2/R_2)/(F_1/R_1)$ or

(3)     $F_2/R_2 = (V_2/V_1)*(F_1/R_1)$

If one imagines having initial conditions $V_1$, $F_1$, and $R_1$, and wants a new velocity, $V_2$, then the question would be, What ratio of forces, $F_2/R_2$ would produce this velocity? Bradwardine answers this by saying, as in (3), that the ratio, $V_2/V_1$, must *act upon the ratio*, $F_1/R_1$, to produce the desired ratio, $F_2/R_2$. Bradwardine makes an important distinction: *Actions on ratios are different from actions on numbers or magnitudes.* It is at this point that our modern notation fails us, for the * operator in (3) no longer means multiplication in Bradwardine's world.

*Applying a Ratio to a Number of Magnitude.* To help see this, let us first review the kind of action that Bradwardine would have seen in a ratio acting on a magnitude or number. In Book VII of the Elements, Euclid applies the ratio $m:n$ to the number $P$ by first separating $P$ into $n$ arithmetic parts, or $n$ arithmetic means. Because the primitive action on numbers is addition, the requirement is that $n$ of these arithmetic parts, added together, equals $P$. Therefore, in modern notation, each of these parts would be equivalent to $(1/n)*P$. The second part of the action would be to add $m$ of these parts together to create a number $Q$, hence $Q$ is the result of applying the ratio $m:n$ to $P$; $Q = (m/n)*P$. (Smith & Confrey, 1989b)

*Applying a Ratio to a Ratio.* For Bradwardine, actions in the world of ratios are inherently multiplicative. Therefore he applies the ratio $m:n$ to the ratio $A/B$ by first separating $A/B$ into $n$ geometric parts, or $n$ geometric means. Because the primitive action on ratios is multiplication, the requirement is that $n$ of these geometric parts, multiplied together, equals $A/B$. Therefore, in modern notation, each of these parts would be equivalent to $(A/B)^{1/n}$. The second part of the action would be to multiply m of these parts together to create a new ratio $C/D$, hence $C/D$ is the result of applying the ratio $m:n$ to the ratio $A/B$; $C/D = (A/B)^{m/n}$.

For Bradwardine, it would be perfectly consistent to write this relationship (in terms of velocities and forces) as

$$(4) \qquad V_2 : V_1 :: (F_2 : R_2) : (F_1 : R_1)$$

for it would be assumed that on the right side of (4), the ratio of ratios would be interpreted as indicating actions appropriate to the world of ratios. In modern symbolism, we have no easy way to write this relationship as a proportion. However if we go back to equation (3) and rewrite it as an action, it would be[7]

$$(5) \qquad\qquad F_2/R_2 = (F_1/R_1)^{(V_2/V_1)}, \text{ or}$$

$$(6) \qquad\qquad \frac{\log(F_2/R_2)}{\log(F_1/R_1)} = \frac{V_2}{V_1}$$

Thus whereas equal increments in $V_2$ for fixed $V_1$ produce an arithmetic series of ratios of forces, $F/R$, in (3), they produce a geometric series of ratios of forces in (5). Thus $V_2$ can approach 0, while maintaining the Aristotelian requirement that $F_2 > R_2$.

Although Bradwardine does not develop the mathematics of this system fully, there are two further points of interest in his model: First is the claim that if $A > B$, then $A:B$ is *not comparable* (not greater than, not less than, and not equal to) to either $A:A$ or $B:A$; and second, $A:A$ is not even a ratio. The basis for this claim is that comparisons are based on actions and, to claim, for instance, that $A:B > A:A$, there would have to be an action that could be performed on $A:A$ that would make it as large or larger than $A:B$, however there is none (there is no number, $n$, such that $(A/A)^n \geq A/B$). Likewise, there is also no possible action on $B:A$ that could create $A:B$. Thus from this action-oriented concept of comparison, Bradwardine excludes all ratios $F:R$ that are less than or equal to unity from his Aristotelian world of motion. Likewise, going back to the Euclidean definition of a magnitude as that which can be increased or decreased, and substituting *ratio* for *magnitude*, effectively disallows unity $(A:A)$ as being considered a ratio in any context.[8]

Thus Bradwardine has posited a world (or mathematical structure) of ratios in which certain types of actions are appropriate (multiplicative) and certain types of actions are not appropriate (additive). His primary interest was the development of a mathematical model consistent with the physics of motion of Aristotle, so he does not develop the mathematics of this world beyond that needed to justify his claim. For this development, we turn to his supporter, Nicole Oresme.

### Nicole Oresme: The Construction of a World of Ratios

Oresme claims that once we accept ratios as the primary entity upon which we will act and operations on ratios as inherently multiplicative, we can return to the classifications of numbers, described by Euclid in Book VII, and recast those classifications within the world of ratios (Grant, 1966). For Euclid, the primary elements were the counting numbers and a part, $P$, of a number, $N$, is what we would now call a unit fraction multiple of the number: $P = N*1/m$, where $m$ is any integer. We have previously proposed (Smith and Confrey, 1989b) that one interpretation of part for Euclid would be in terms of an action: $P$ is an $m$-part of $N$, if $P$ can be added together $m$ times to construct $N$, or $mP = N$, where the multiplication, as defined by Euclid, is repeated addition. Oresme keeps the same notion of part, but in line with Bradwardine, changes the action from repeated addition to repeated multiplication.[9] Thus, the ratio,[10] $A$, is an $m$-part of the ratio $B$, if $A$ can be multiplied together $m$ times to obtain $B: A^m = B$ or $A = b^{1/m}$. For Euclid, whose only elements were the counting numbers, there could be no $m$-part of $N$ unless $N*1/m$ was an integer, otherwise it had no meaning. The Greeks had, however, constructed ratios of incommensurable magnitudes, which, by Oresme's time, were associated with numeric values, (i.e., $\sqrt{2}:1$). If one could posit an action that would create a ratio, then a numeric value was assumed. Therefore, for Oresme, ratio "is divisible, just like continuous quantity"[11] (Grant, 1966, p. 27). Positing an $m$-part of a ratio, $B$, was to posit the action of constructing $m$-1 geometric means between unity and $B$. The action of constructing the geometric means, not the values themselves, created the notion of part.

*Classifications of Ratios.* In parallel to Euclid, Oresme also uses *parts:* $A$ is part of $B$ if $A = B^{m/n}$, where $n > m > 1$ and *commensurable:* $A$ is commensurable with $B$ if there is another ratio $C$ such that $C$ is a part of $A$ ($C = A^{1/n}$), and $C$ is also a part of $B$ ($C = B^{1/m}$). Thus the ratios $2:1$ and $3:1$ are incommensurable, because there is no common part, whereas $4\sqrt{2}:1$ and $2:1$ are commensurable since both are *measured* by $\sqrt{2}:1$.

Oresme's final classification of ratios starts with the major distinction between commensurable and incommensurable. For commensurable ratios, there are three possible types:

1. Two rational ratios that are commensurable: $3:2$ and $27:8$, for example.
2. Two irrational ratios that are commensurable together and are each commensurable with a rational ratio: $(4:1)^{1/3}$ and $(2:1)^{12}$.
3. Two irrational ratios that are commensurable to each other, but incommensurable with any rational ratio: $(8:1)^{\sqrt{2}}$ and $(2:1)^{\sqrt{2}}$.

For incommensurable ratios, there are three parallel types:

1. Two rational ratios that are incommensurable: $3:1$ and $2:1$
2. Two irrational ratios, both of which are commensurable with some rational ratio, but that are incommensurable with each other: $(3:1)^{1/4}$ and $(2:1)^{1/8}$.
3. Two irrational ratios that are each incommensurable with all rational ratios and are incommensurable with each other: $(2:1)^{\sqrt{2}}$ and $(3:1)^{\sqrt{2}}$.

*Constructing the Density of the Rationals in a Multiplicative World.* From a modern perspective, one can view this elaborate classification scheme in relation to Oresme's goals. In effect, he had placed all ratios in a continuum and described a system for classifying them. Any single ratio was classified according to two criteria:

1. It was called *rational* or *irrational* in *relation to the magnitudes* that it compared. Thus, if it was a comparison of two commensurable magnitudes, it was called *rational*. If the two magnitudes were incommensurable, it was called *irrational*.

2. It was called *commensurable* or *incommensurable* in *relation to another ratio*, according to whether the two ratios were multiplicatively commensurable.

From the second criteria we build our concept of a multiplicative world with the density of the rationals, which we will call $W_{Ra}$. Starting with any ratio, $A$, one can apply it twice to get $A^2$, three times to get $A^3, \ldots, n$ times to get $A^n$. Likewise, for any integer, $m$, one can find an $m$-part of $A$ or $A^{1/m}$ as that ratio, $B$, which applied $m$ times is equal to $A$, that is $B^m = A$. Staying completely within this world, that is, limiting ourselves to multiplicative actions, we can construct all ratios of the form $A^{n/m}$, where $n$ and $m$ are any positive integers. Thus from $A$ we construct $W_{Ra}$, a structure with the density of the rationals. However, it will *not* be the set of rational ratios. For example, if we start with the ratio $3:1$, $W_{Ra}$ will be all ratios of the form $3^{n/m}:1$ and will include irrationals such as $\sqrt{3}$, but will not include a rational ratio, $a:1$, unless $a$ is an integer power of 3. Therefore such worlds, although having the density of the rationals, will have a distinctly multiplicative structure.

From this, it is easy to see that, whenever we construct a quantitative world (number or ratio) with the density of the rationals, its operational structure will be evident. Only when we move to create the density of the real numbers are we able to mask this operational structure and lose the distinctions that form the operational basis of numbers.

## VI. JOHN NAPIER AND THE INVENTION OF LOGARITHMS

In many ways Napier's work seems like a marriage between Oresme's world of continuous ratios and the developing work on cogeneration of arithmetic and geometric series, such as that carried out by Burgi. Napier's great invention, which allowed this union, was the creation of a model based on the motion of two particles, each traveling in a straight line, in such a way that additive change in the position of one particle would correspond with proportional change in the position of the second (MacDonald, 1966).

Napier was originally interested in constructing a table of logarithmic values of sines, rather than of numbers. At that time, the sine of an angle was not expressed as a ratio, but as the length of the opposite leg in a right triangle with specified hypotenuse (or radius). Because it was considered desirable to avoid noninteger numbers, it was typical to pick a triangle with a large radius allowing accurate values for the sine to be expressed as whole numbers. Therefore Napier chose a radius of length $10^7$.

In the spirit of Oresme, Napier constructed two quite independent worlds. In the additive world, a point, which we will designate as $A$, is posited to start at point 0, and "in equal times, to be augmented by a quantity always the same" (p. 17) or, in more modern terms, to travel at constant velocity.[12] In the multiplicative world, a point, which we will designate as $M$, starts on the circumference of the circle (position $= 10^7$) and travels toward the center (position $= 0$) such that "in equal times, first the whole quantity then each of its successive remainders is diminished, always by a like proportional part" (p. 17). Therefore, $M$ will never reach the center of the circle.

In his initial table (Table 9.3), Napier plotted the position of $A$ opposite the position of $M$ for the same elapsed time. In this case, we assume that $A$ is moving at 1 length unit per time unit and that, when $A$ is at 0, $M$ is at $10^7$. In this case, determining the position of $M$ starts with a construction that appears very much like those of Burgi; that is, with the construction of a geometric series. Napier proposes that the dis-

TABLE 9.3.   NAPIER'S 'ADDITIVE' AND 'MULTIPLICATIVE' POINTS

| Position of A | Position of M |
|---|---|
| 0.00 | 10,000,000.0000000 |
| 1.00 | 9,999,999.0000000 |
| 2.00 | 9,999,998.0000001 |
| 3.00 | 9,999,997.0000003 |
| 4.00 | 9,999,996.0000006 |
| 5.00 | 9,999,995.0000010 |
| 6.00 | 9,999,994.0000015 |
| 7.00 | 9,999,993.0000021 |
| 8.00 | 9,999,992.0000028 |
| 9.00 | 9,999,991.0000036 |
| 10.00 | 9,999,990.0000045 |

tance to the center be decreased by a ratio of $1/10^7$ in each equal increment of time. The second value is thus $10^7 - 10^7(1/10^7) = 9,999,999$. The third value is $9,999,999 - 9,999,999(1/10^7) = 9,999,998.0000001$.[13]

So far, we have simply juxtaposed a geometric series and an arithmetic series, with the only difference from the typical configuration being that the arithmetic increases while the geometric decreases. This process can be continued to make the set of additively spaced points in the arithmetic sequence or the set of multiplicatively spaced points in the geometric sequence as dense as desired. In fact one could imagine carrying this out to create any rational value for $A$ and at the same time creating a value in $W_{Ra}$ for $M$. In Figure 9.1, for example, if the point $(1, a)$ is given, the next equally spaced arithmetic point, $(2, b)$ can be found, as well as the point $(0.5, c)$, because $c$ will be the geometric mean between $a$ and $R$. Thus it would be possible to extend Table 9.3 and call it a log table, where the number in the left column represents the log of the number in the right column. This would, in fact be compatible with Napier's definition of a *logarithm:* "The logarithm of a given number is that number which has increased arithmetically with the same velocity throughout as that with which radius began to decrease geometrically, and in the same time as radius has decreased to the given

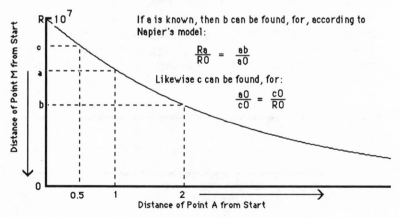

Fig. 9.1
*Using Ratios to Interpolate*

sine" (Napier, p. 19). It is *almost* what Napier does, but not quite.

The problem lies in the fact that no matter how dense one makes the set of positions for $M$ in the model, one will never be able to find the position of $A$ for any arbitrary position of $M$. For example, the preceding methods will not lead to a value for $A$ when $M$ is exactly 9,999,9998. Yet we feel intuitively quite sure that $M$ will pass through 9,999,998 and that, when it does, there will be a unique position for $A$. To see how Napier resolved this dilemma, we will first look at a geometric version of his model. In Figure 9.2, the top line represents the motion of $A$, which has started at $S$ and moves from $S$ to $a'$, from $a'$ to $b'$, from $b'$ to $c'$, and so forth in equal increments of time. The bottom line represents the motion of the geometric point, which has started at $R$ and moves from $R$ to $a$, from $a$ to $b$, and so forth in the same equal increments of time. Thus $R0 = 10^7$ and the ratio of movement to distance from 0 (in a fixed time) is constant, so $Ra:R0 = ab:a0 = bc:b0 = \ldots$, and, according to Napier's definition: $\log(OR) = 0$; $\log(Oa) = Sa'$; $\log(Ob) = Sb'$, and so on. It would seem that one could arbitrarily set $Sa' = 1$ and $Ra = 1$, in which case the logs from this model would be exactly the values in Table 9.3. However, as we have seen, this does not lead to a satisfactory table. For the same reason that no $W_{Ra}$ (a multiplicative world with the density of the rational values) can include all rational values, the specification of *any* initial ratio of decrease for $M$ will automatically exclude the possibility of finding the position of $A$ when $M$ is at certain rational-valued positions, at least with the calculation methods available to Napier.

At this point Napier made an important observation: "A geometrically moving point approaching a fixed one has its

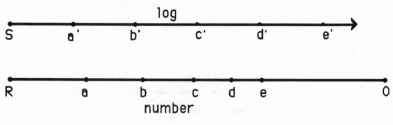

Fig. 9.2
*Napier's Logarithms*

velocities proportionate to its distance from the fixed one" (Napier, p. 18); that is, the velocity of the geometric point in the model will always be proportional to its distance from 0, which, in modern terms, is conceptually equivalent to the statement: $dy/dx = ky$! Although Bradwardine had, rather vaguely, come to the conclusion that, for his model to be consistent, the velocities in the exponent must be what we would now call *instantaneous* (Crosby, 1961), there was no well-described concept of instantaneous velocity and certainly no mathematics of instantaneous rate of change in Napier's time. His "proof," in fact, is simply an argument that because, by hypothesis, in any finite time period, the point moves a distance proportional to its distance from 0, the velocity in any finite time period must then be proportional to the distance from 0. This is the first of two remarkable insights, for it, in effect, allowed the calculation of the log for any arbitrarily chosen value, exactly the problem that seemed unsolvable in the world of geometric series.

Instead of specifying the initial ratio by which the position of *M* decreases, one can specify the initial (instantaneous) velocity (relative to the velocity of *A*) of *M*. Therefore, Napier specified that the initial velocities of the two points be the same. For example, suppose we wish to find the log of 9,999,999 (see Figure 9.3). It does not matter what the initial

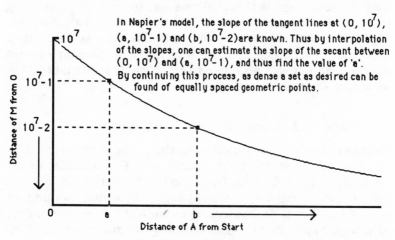

In Napier's model, the slope of the tangent lines at $(0, 10^7)$, $(a, 10^7-1)$ and $(b, 10^7-2)$ are known. Thus by interpolation of the slopes, one can estimate the slope of the secant between $(0, 10^7)$ and $(a, 10^7-1)$, and thus find the value of 'a'. By continuing this process, as dense a set as desired can be found of equally spaced geometric points.

Distance of M from 0

$10^7$

$10^7-1$

$10^7-2$

0      a      b

Distance of A from Start

Fig. 9.3
*Using Tangent Lines to Interpolate*

velocities of the two points are, as long as they are the same.[14] For simplicity, we will assume that the initial velocity of both points is 1 unit of distance per unit of time, making the distance that the arithmetic point moves exactly equal to the elapsed time. If we can calculate the amount of time it takes the geometric point to get to 9,999,999, this will allow us to calculate the distance that the arithmetic point has traveled, the log of 9,999,999. When the geometric point reaches 9,999,999, its velocity will be $9,999,999/10^7$ (velocity is proportional to the distance from 0). Therefore its average velocity in reaching this point is between 1 and $9,999,999/10^7$. If we choose $9,999,999.5/10^7$, then we are correct to six significant figures. At this average velocity, it will require 1.00000005 units of time to reach 9,999,999. In this same amount of time the arithmetic point will travel 1.00000005 units, therefore log(9,999,999) = 1.00000005. Likewise, the average velocity between 9,999,999 and 9,999,998 is about $9,999,998.5/10^7$. It will require the geometric point 1.00000015 units of time to travel from 9,999,999 to 9,999,998, in which time the arithmetic point will travel 1.00000015 units. Therefore the log of 9,999,998 is approximately 1.0000005 + 1.00000015 = 2.0000002. By similar calculation, log 9,999,997 = 2.0000002 + 1.00000025 = 3.00000045, and so forth. Although this method certainly depends on interpolation and approximation, its advantage is that it allows one to begin by specifying the position of the geometric point and then calculate the position of the arithmetic point, rather than vice versa. With a combination of efficient operations and a significant amount of effort, Napier was now able to calculate true log tables[15] to whatever accuracy was desired.

## Napier's Logarithms and Calculations

This is still not a log function in the modern sense; that is, one that obeys the "laws of logarithms." For example, we assume that log $(A/B)$ = log $(A)$ − log $(B)$. In particular, if $B = 1$, then log$(A)$ = log$(A/1)$ = log $(A)$ − log $(1)$, which can be true only if log(1) = 0. In Napier's system, log(1) is well over $10^7$. Although Napier and Briggs eventually reconceptualized logs such that log(1) = 0 and numbers increased as their logs increased, rather than vice versa, this was not necessary to make the original table of logs useful for performing multiplication and division. To use it for this purpose, however, it is helpful to remember that the two worlds (arithmetic and geometric) proposed by Napier operated independently and that it was more common to think of relating actions within worlds (relationships of proportion), than to think of our modern functional concept that relates values between worlds. This is where Napier had a second wonderful insight: "The logarithms of similarly proportioned sines are equidifferent." With this claim, Napier has fully adopted (a specialized version of) Oresme's multiplicative world, for he is claiming that the fundamental relationship in his geometric world, that of multiplication, is continuous, and, thus, holds between any two points one should happen to choose, not just those that arose out of the original geometric series. In this sense his choice of logarithm as "number of ratios" (Moulton, 1915) makes sense, for it is completely consistent with Oresme's sense that an exponent, whether integer, rational, or irrational value, is a measure of the "number" of times a "ratio action" has occurred.

Now we may return to the question of why Napier's logs do not allow us to calculate the way we do with normal log functions. Napier's logs force us to think of multiplication as a "ratio action" rather than a binary operation. A symbolic interpretation of Napier's preceding statement would be $A:B::C:D \Leftrightarrow [\log(A) - \log(B)] = [\log(C) - \log(D)]$. Therefore to multiply 256 times 3978 is to solve the ratio problem: $256:1::A:3978$, and Napier has told us that the difference between $\log(256)$ and $\log(1)$ is the same as the difference between $\log(A)$ and $\log(3978)$. Therefore $\log(A) = \log(256) - \log(1) + \log(3978)$. Likewise to divide 4288 by 333 is to solve the ratio problem: $4288:333::A:1$, from which we find that $\log(A) = \log(4288) - \log(333) + \log(1)$.

### Napier's Logarithms and Modern Mathematics

To model Napier's system in modern functional notation, we need to first model each system separately.

- *Arithmetic point, A:* (1) $A = kt$, where $A$ is the distance traveled, $k$ is the constant velocity, and $t$ is time.

• *Geometric point, G:* The (instantaneous) velocity is proportional to $G$ and decreasing with $t$, where $G$ is distance from the center of the circle.

Therefore,

$$(2) \qquad dG/dt = -rG \text{ or } G = ae^{-rt}$$

When $t = 0$, $G = 10^7$, therefore $a = 10^7$. Likewise, when $t = 0$, $dG/dt = -k$, therefore $k = r(10^7)$, or $r = k/10^7$. Hence (2) can be rewritten as

$$(3) \qquad G = 10^7 e^{(-kt/10^7)}$$

Likewise (1) can be written as $t = A/k$. Substituting into (3) gives

$$(4) \quad G = 10^7 e^{(-A/10^7)} \Rightarrow G/10^7 = (1/e)^{(A/10^7)}, \qquad \text{or}$$

$$(5) \qquad \log_{1/e}(G/10^7) = A/10^7, \qquad \text{or } A = 10^7 \log_{1/e}(G/10^7)$$

From (5), we see that Napier's log function can be interpreted as having a base of $1/e$ and as being both vertically and horizontally stretched by a factor of $10^7$. Also, this depends only on the initial velocities of the two points being the same, not on the particular value of the initial velocity. If the initial velocities were different, it would change the vertical stretch factor in (5). However, another way to write (4) is

$$(6) \qquad G/10^7 = [(1/e)^{(1/10^7)}]^A$$

Letting $z = (1/e)^{(1/10^7)}$, (6) becomes

$$(7) \qquad\qquad G/10^7 = (z)^A \qquad \text{or}$$

$$(8) \qquad\qquad A = \log_z(G/10^7)$$

From (8), another interpretation of Napier logs is that the base is $z$, with a horizontal stretch of $10^7$. With this interpretation, unequal initial velocities would have the effect of changing the base.[16]

## VII. CONCLUSIONS

This historical work not only complements our work with students but offers strong support for Confrey's splitting structures (this volume). In particular, it suggests that multiplication will be a primitive action that in many situations cannot be based on addition, that multiplicative (or splitting) worlds are based on an operational concept of number significantly different from number concepts giving rise to counting structures, and that constructing an understanding of these multiplicative worlds prior to the study of functions may facilitate students' understanding of exponential and logarithmic functions. As we saw, Napier's intuitive understanding of the relationship between position and rate of change within a multiplicative structure allowed him to construct the relationship between an additive and a multiplicative world that gave rise to logarithms.

In addition, we believe that this work offers insights for further work we intend to undertake with students involving rate of change and the development of e. It complements our hypothesis that the relationship between rate of change and position can form a strong intuitive basis for understanding how exponentials are used to model contextual situations prior to any formal introduction to calculus concepts. Also, beginning to look at e as a concept arising out of the basic incommensurability between additive and multiplicative worlds may offer a strong intuitive entrance to natural logs and exponentials.

## NOTES

This research was funded under a grant from the National Science Foundation (MDR 8652160)

1. For example, if we look at the multiplicative world created by all rational powers of two, call it $\{2^r\}$, it has a direct $1-1$ correspondence with the set of additive rational numbers (i.e., what we normally call simply the *rational numbers*) because one can map $2^{a/b} \rightarrow a/b$. Likewise it has the familiar density property of having an element between any two elements. However, this world will not contain many numbers that we normally consider rational, such as $3, 5, 6$, and so on and will contain normally irrational numbers such as $\sqrt{2}$. Its elements are unique to the operation that created it, and

it is a closed set in the following sense: If one takes any element, $z$, that is a member of $\{2^r\}$, and creates the multiplicative world, $\{z^r\}$, the members of $\{z^r\}$ and $\{2^r\}$ will be identical. This will never be true if $z$ is not a member of $\{2^r\}$. In this sense, the claim is made that the operations that structure this world are inherent in the structure of the set of numbers generated by the world.

2. This procedure depends on the particular positioning of the two series. In particular the two identity elements, 1 and 0, must correspond. If they do not correspond, it is necessary to return to the more fundamental relationship: $(g_n/g_m) = (g_p/g_q) \Leftrightarrow (a_n - a_m) = (a_p - a_q)$, which is true independent of the relative positioning of a geometric series, $g_i$, with an arithmetic series, $a_j$. See Section VI for further discussion of this relationship.

3. Suppose for example that for some ratio, $r$, there are integers, $n$ and $m$, such that $r^n = 2$ and $r^m = 3$, then $(r^n)^{m/n} = 2^{m/n} = 3$, but there is no rational power of 2 that yields 3.

4. If $F_1$, $R_1$, $V_1$, and $R_2$ are fixed and $F_2$ decreases, then $V_2$ will decrease. However as long as $F_2 > 0$, $V_2$ will be $> 0$, even when $0 < F_2 < R_2$.

5. Although Bradwardine uses the word *proportion* rather than *ratio*, we will use *ratio* in a sense more consistent with modern usage. A relationship between two magnitudes is a ratio, and equality of ratios gives proportion.

6. As was the custom of his day, Bradwardine does not use mathematical symbolism. Instead, he uses descriptive language to articulate his mathematics. Thus questions of translation and interpretation always plague the modern reader.

7. If we assume $F_1$, $R_1$, and $V_1$ to be initial conditions, and therefore fixed, we might typically, in modern notation, write (5) as $kV_2 = \log(F_2/R_2)$, where $k = [\log(F_1/R_1)]/V_1$.

8. This claim also places a new perspective on the Greeks exclusion of 0 from their number system. Because numbers are related to the size of magnitudes and there is no way to act on a magnitude of size 0 such as to increase or decrease it, one would not consider 0 to be a number.

9. In fact Oresme describes the algorithm for *adding* ratios as multiplication and the algorithm for *subtracting* ratios as division (Grant, 1966, p. 145).

10. In a manner similar to Grant, we will use capital letters to designate the ratios that are the "primary entities," and lower case letters to designate the integers that indicate (multiplicative) actions on these ratios. Oresme, like Bradwardine, wrote his mathematics primarily in descriptive style and did not use exponents as we use them today. Therefore the use of exponents in this chapter is not in the style of the period.

11. Oresme is consistent in his posited actions, thus "division" of a ratio $A$ by $m$ is *not* $A/m$, but $A^{1/m}$.

12. Thus if the velocity were 1 length unit per time unit, the position of this particle would have the same numeric value as the amount of time elapsed, causing one to wonder why Napier bothered to create this additive world, rather than simply using elapsed time. This emphasizes his conception of a logarithm arising from two independent worlds that differ only in their operative structure.

13. Although in his tables, Napier rounded values off to the nearest integer, for calculation purposes, he carried values to seven decimal places.

14. Napier required the same initial velocities to make the model easier to understand. Although he did not think in terms of "bases," having unequal initial velocities would simply change the "base" of his logarithmic table.

15. In the sense that one can create a table with equal additive increments on the geometric side and be able to calculate the required values on the arithmetic side.

16. For any logarithm, a vertical stretch is equivalent to a change of base.

## BIBLIOGRAPHY

Afamasaga-Fuata'i, Karoline. 1991. Students' strategies for solving contextual problems on quadratic functions. Unpublished doctoral dissertation, Cornell University, Ithaca, NY.

Behr, M., Lesh, R., Post, T., and Silver, E. 1983. Rational-number concepts. In R. Lesh and M. Landau (Eds.), *Acquisition of mathematics concepts and processes.* (pp. 91–126). New York, NY: Academic Press, Inc.

Boyer, Carl. 1946. Proportion, equation, function: Three steps in the development of a concept. *Scripta Mathematica* 12: 5–13.

Cajori, Florian. 1915. Algebra in Napier's day and alleged prior inventions of logarithms. In *Napier tercentenary memorial volume*, ed. Cargill Knott, 93–109. London: Longmans, Green and Company.

Confrey, Jere. 1988. Multiplication and splitting: Their role in understanding exponential functions. In *Proceedings of the tenth annual meeting of the North American chapter of the international group for the psychology of mathematics education*, ed. Merlyn Behr, C. LaCompagne, and Montague Wheeler, 250–259. DeKalb: Northern Illinois University Press.

———. 1991a. The concept of exponential functions: A student's perspective. In *Epistemological foundations of mathematical experience*, ed. Les Steffe, 124–159. New York: Springer-Verlag.

——. 1991b. Learning to listen: A student's understanding of powers of ten. In *Radical constructivism in mathematics education*, ed. Ernst von Glasersfeld, 111–138. Dordrecht: Kluwer Academic Publishers.

——. 1991c. Using computers to promote students' invention on the function concept. In *This year in school science 1991*, ed. Shirley Malcom, Linda Roberts, and Karen Sheingold. Washington, DC: American Association for the Advancement of Science.

Confrey, Jere, and Erick Smith. 1988. Student centered design for educational software. In *Proceedings of the first annual conference on technology in collegiate mathematics*, ed. Bert Waite and Frank Demanna. Reading, MA: Addison-Wesley.

—— and Erick Smith. 1989. Alternative representations of ratio: The Greek concept of anthyphairesis and modern decimal notation. Paper presented at the first annual conference of the history and philosophy of science in science teaching. Tallahassee, FL.

—— and Erick Smith. 1991. A framework for functions: Prototypes, multiple representations, and transformations. In *Proceedings of the thirteenth annual meeting of Psychology of Mathematics Education-NA*, ed. Robert Underhill and Catherine Brown, 57–63. Blacksburg: Virginia Polytechnic Institute Press.

Crosby, Lamar, ed. and trans. 1961. *Thomas of Bradwardine: His tractus de proportionibus*. Madison: University of Wisconsin Press.

Dennis, David, and Jere Confrey. 1991. Alhazen's formula for the summations of powers. Project paper, Cornell University.

Fischbein, Efraim, Maria Deri, Maria Sainati Nello,and Maria Sciolis Marino. 1985. The role of implicit models in solving verbal problems in multiplication and division. *Journal for Research in Mathematics Education* 16(1): 3–17.

Grant, Edward, ed. and trans. 1966. De proportionibus proportionum by Nicole Oresme (1375). *De proportionibus proportionum and ad pauca respicientes*. Madison: University of Wisconsin Press.

Greer, Brian. 1988. Nonconservation of multiplication and division: Analysis of a symptom. *Journal of Mathematical Behavior* 7(3): 281–298.

Gridgeman, N. T. 1973. John Napier and the history of logarithms. *Scripta Mathematica* 24: 49–65.

Harel, G., and M. Behr. 1990. Structure and hierarchy of missing value proportion problems and their representations. *Journal of Mathematical Behavior.*

Hart, K. 1988. Ratio and proportion. In *Number concepts and operations in the middle grades,* ed. M. Behr and J. Hiebert, 198–219. Reston, VA: National Council of Teachers and Mathematics.

Heath, T. (ed. and trans.) 1956. *Euclid: The Thirteen Books of the Elements,* V.1–3. NY: Dover Publications.

Karplus, R., S. Pulos, and K. Stage. 1983. Proportional reasoning of early adolescents. *Acquisition of Mathematical Concepts and Processes,* 45–89. New York: Academic Press.

Lakatos, Imre. 1976. *Proofs and refutations: The logic of mathematical discovery.* ed. J. Worrall and E. Zahar. Cambridge: Cambridge University Press.

MacDonald, William, trans. 1966. Mirifici logarithmorum canonis constructio by John Napier (1615). *The construction of logarithms.* London: Dawson of Pall Mall.

Malik, M. A. 1980. Historical and pedagogical aspects of the definition of function. *International Journal of Mathematics Education, Science and Technology* 11(4): 489–492.

Moulton, Lord. 1915. The invention of logarithms, its genesis and growth. In *Napier tercentenary memorial volume,* ed. Cargill Knott, 1–32. London: Longmans, Green and Company.

Napier, John. 1966. *The Construction of Logarithms.* William MacDonald (trans.), London: Dawson of Pall Mall.

Nesher, Pearla. 1988. Multiplicative school word problems: Theoretical approaches and empirical findings. In *Number concepts and operations in the middle grades,* ed. Merlyn Behr and James Hiebert, 19–40. Reston, Virginia: National Council of Teachers of Mathematics.

Noelting, G. 1980a. The development of proportional reasoning and the ratio concept. Part 1—Differentiation of stages. *Educational Studies in Mathematics* 11: 217–253.

———. 1980b. The development of proportional reasoning and the ratio concept. Part 2—Problem-structure at successive stages; problem-solving strategies and the mechanism of adaptive restructuring. *Educational Studies in Mathematics* 11: 331–363.

Rizzuti, Jan. 1991. High school students' uses of multiple representations in the conceptualization of linear and exponential functions. Unpublished doctoral dissertation, Cornell University, Ithaca, NY.

——— and Jere Confrey. 1988. A construction of the concept of exponential functions. In *Proceedings of the tenth annual meeting of the North American chapter of the international*

## 360    SMITH and CONFREY

*group for the psychology of mathematics education (PME-NA)*, ed. Merlyn Behr, C. La Compagne, and Montague Wheeler, 259–268. Dekalb: Northern Illinois University Press.

Schwartz, Judah. 1988. Intensive quantity and referent transforming arithmetic operations. In *Number concepts and operations in the middle grades*, ed. M. Behr and J. Hiebert, 41–52. Reston, VA: National Council of Teachers of Mathematics.

Smith, Erick, and Jere Confrey. 1989a. Ratio as construction: ratio and proportion in the mathematics of ancient Greece. Paper presented at the American Educational Research Association, San Francisco.

—— and Jere Confrey. 1989b. Ratio in a world without fractions: A study of the ancient Greek world of number. Paper presented at the thirteenth international conference for the psychology of mathematics education, Paris.

Smith, Erick, David Dennis, and Jere Confrey. 1992. Rethinking functions: Cartesian constructions. Paper presented at the second international conference on the history and philosophy of science in science teaching, Kingston, Ontario.

Smith, Eugene. 1915. The law of exponents in the works of the sixteenth century. In *Napier tercentenary memorial volume*, ed. Cargill Knott, 81–91. London: Longmans, Green and Company.

Steffe, L. 1988. Children's construction of number sequences and multiplying schemes. In *Number concepts and operations in the middle grades*, ed. M. Behr and J. Hiebert. Reston, VA: National Council of Teachers of Mathematics.

Vergnaud, G. 1983. Multiplicative structures. In *Acquisition of mathematics concepts and processes*, ed. R. Lesh and M. Landau, 127–173. New York: Academic Press.

——. 1988. Multiplicative structures. In *Number concepts and operations in the middle grades*, ed. M. Behr and J. Hiebert. Reston, VA: National Council of Teachers of Mathematics.

# V
# INTUITIVE MODELS

# 10 The Impact of the Number Type on the Solution of Multiplication and Division Problems: Further Considerations

*Guershon Harel*
*Merlyn Behr*
*Thomas Post*
*Richard Lesh*

## INTRODUCTION

Fischbein, Deri, Nello, and Marino (1985) argue that students' conceptions of and performance on multiplication and division application word problems (hereafter multiplicative problems) are unconsciously derived from primitive intuitive models that "correspond to features of human mental behavior that are primary, natural and basic" (p. 15). They suggest their theory to account for conceptions, such as multiplication makes bigger and division makes smaller, identified in previous studies (e.g., Bell, Fischbein, and Greer, 1984; Vergnaud, 1983; Bell, Swan, and Taylor, 1981). These conceptions are in accord with the operations of multiplication and division in the whole number domain but incongruent to these operations in the rational number domain; thus, they block the way to correctly solve many multiplicative problems whose quantities are decimals or fractions.

In our recent effort to better understand the multiplicative conceptual field, and in particular the transition phase from the additive structure to the multiplicative structure, we probed into this question of incongruity. We found that many questions and concerns are still open and need further investigation. Among these we addressed the following: (1) the impact of the number type on the solution of multiplicative problems; (2) the impact of the textual structure on problem interpretation; and (3) solution models used by teachers to solve multiplicative problems. To address some of these questions empirically, we developed an instrument that controls a wide range of confounding variables known to be influential on subjects' performance on multiplicative problems and used it with inservice and preservice teachers. In this chapter we report on an investigation of aspect (1); the investigations on the other two aspects will be reported separately.

## THEORETICAL BACKGROUND

We start with a summary of Fischbein et al.'s models for multiplication and division and the evidence reported by some studies for their impact. Then we address, in this order, questions concerning the instruments used to investigate these models, a specific constraint these models impose on the numerical aspect of the quantities representing the multipliers and divisors, and the relative robustness of the intuitive rules associated with these intuitive models.

According to Fischbein et al. (1985), the model associated with multiplication problems is *repeated addition*. Under this model the roles played by the quantities multiplied are asymmetrical (Greer, 1985). One of the factor quantities, called the *multiplier*, is conceived of as the number of equivalent collections, while the other quantity, called the *multiplicand*, is conceived of as the size of each collection. These conceptions apparently lead subjects to intuit the rule that multipliers must be whole numbers, which, in turn, results in another rule: the product must be larger than the multiplicand, or multiplication makes bigger (Bell et al., 1984; Hart, 1984; Bell et al., 1981). For division, Fischbein et al. suggested two intuitive models, one is associated with equal sharing, or *partitive division* problems, the other with measurement, or *quotitive division* problems. In the partitive division model, an object or collection of objects is partitioned into a number of equivalent fragments or subcollections. The size of the object or the number of objects is represented by the dividend, the number of the equivalent fragments or subcollections is represented by the divisor, and the size of each equivalent fragment or subcollection is represented by the quotient. As a result, certain rules are intuited and become associated with this model: (1) the divisor must be a whole number, (2) the divisor must be smaller than the dividend,

TABLE 10.1. INTUITIVE RULES ASSOCIATED WITH FISCHBEIN'S
THREE MODELS

| Operation | Intuitive rules |
| --- | --- |
| Multiplication | 1. Multiplier must be a whole number |
| | 2. Multiplication makes bigger |
| Partitive division | 1. Divisor must be a whole number |
| | 2. Divisor must be smaller than dividend |
| | 3. Division makes smaller |
| Quotitive division | 1. Divisor must be smaller than dividend |

and (3) the quotient must be smaller than the dividend, or division makes smaller, which is derived from the previous two rules. The quotitive division model is associated with division problems in which it is required to find how many times a given quantity is contained in another quantity. The only constraint imposed by this model is that divisors must be smaller than dividends. Table 10.1 summaries these intuitive rules.

Fischbein et al. (1985) looked for substantiation of their theory in an investigation with fifth, seventh, and ninth graders. Other researchers tested this theory with different populations. For example, Greaber, Tirosh, and Glover (1989) replicated Fischbein et al.'s study with preservice elementary teachers and further investigated the similarity between their conceptions and children's conceptions of multiplication and division. Mangan (1986) addressed Fischbein et al.'s theory in a study that systematically controlled the contextual and the number type variables, with a wide age range of children and adults: primary and secondary school students, continuing education students, university students, and student teachers.

In general, these studies and others (see Greer, 1988) support Fischbein et al.'s theory; that is, they are consistent with the argument that subjects's solution of multiplicative problems is affected by their intuitive models about multiplication and division and by the numerical constraints imposed by them. The following are some of the main results of these studies.

The impact of the repeated addition model was substantiated by the finding that the solution of multiplication word problems is affected by the type of multiplier. It was shown that problems with a whole-number multiplier are significantly easier than problems with a decimal multiplier larger than 1; and the latter were easier than those with a (positive) decimal multiplier smaller than 1 (e.g., Mangan, 1986; De Corte, Verschaffel, and Van Coillie, 1988). Also, consistent with the repeated addition model is the finding that no significant difference in problem difficulty was found whether the multiplicand was represented as a whole number, a decimal greater than 1, or a decimal smaller than one (Luke, 1988).

The impact of the division models also was substantiated. For example, Mangan (1986) showed that problem quantities consistent with the partitive constraints (e.g., $25 \div 8$ and $26.85 \div 9$) yielded the highest level of performance; quantities that violate the constraint that the divisor must be a whole number (e.g., $11.44 \div 4.51$, $32 \div 5.69$, $5.87 \div 0.44$, and $8 \div 0.77$) resulted in a decrease in the percentage of correct responses; and quantities that violate the constraint that the divisor must be smaller than dividend ($7 \div 23$ and $0.38 \div 0.89$) resulted in the lowest level of performance. Similarly, in quotitive division quantities that conform to the quotitive constraint ($25 \div 8$, $26.85 \div 9$, $11.44 \div 4.51$, $32 \div 5.69$, $5.87 \div 0.44$, and $8 \div 0.77$) yielded a higher performance than those that violate it ($7 \div 23$ and $0.38 \div 0.89$).

In an analysis of these and other studies, we made several observations. The first, which has been a concern to other researchers as well (e.g., Nesher, 1988; Goldin, 1986; Lester and Kloosterman, 1985), is that the problems used in these studies controlled only for the number variable, leaving uncontrolled all or some other important variables such as context, text, and syntax, which are known to be important factors in the research on additive word problems. The instrument we developed for this study (described in the next section) was designed to take into consideration this concern.

The second observation from these studies is that it seems that decimal multipliers greater than 1 are treated by subjects like whole numbers: Subjects seem to have little difficulty solving multiplication problems with a non-whole-number multiplier greater than 1 (compared to the difficulty they have with problems with a decimal smaller than 1) despite the fact that such problems violate the same intuitive rule, that the multiplier must be a whole number. Our question was, What is the conceptual basis for this phenomenon? Fischbein et al. noticed this phenomenon and suggested the notion of an *absorption effect* to explain it. They conjectured that a decimal multiplier whose whole part is clearly larger than its fractional part may be treated more like a whole number, as though the whole part "masks" or "absorbs" the fractional part. To support this conjecture, Fischbein et al. (1985) compared performance on several multiplication problems: one with the decimal multiplier 3.25, one with the decimal multiplier 1.25, and four with the decimal multipliers 0.75 or 0.65. They found that compared to decimals like 0.75, 0.65, or 1.25, a decimal like 3.25 has a slighter "counterintuitive" effect when playing the role of multiplier. This explanation raises several questions: What is the conceptual base for the argument that the whole part 3 in the decimal multiplier 3.25 better "masks" or "absorbs" the fractional part 0.25 than 1 does in the decimal multiplier 1.25? Is it merely a matter of the relative size between 3 and 1 or does a more fundamental factor account for the difference? If this is a matter of the relative size, would "large" decimals (e.g., 42.35) be better conceived as multipliers than small decimals (e.g., 3.25)? Would the "absorption effect" apply to decimal multipliers between 2 and 3 (e.g., 2.25)? Does the relative size between the whole part and the decimal part of a decimal multiplier play a role in the "absorption effect"? In this chapter we address some of these questions.

The third observation, related to the previous one, is that the notion of the "absorption effect" has not been investigated with respect to the intuitive partitive division rule, that the divisor must be a whole number, even though the same argument Fischbein et al. (1985) made with the multiplier can be made with the divisor. The argument would be that a non-whole-number divisor whose whole part is clearly greater than its fractional part should be treated like a whole number. That is, a partitive division problem with a divisor such as 2.53 is expected, according to the "absorption effect" notion, to be easier than a partitive division problem with a divisor such as 0.67. Therefore an additional question is, Does the "absorption effect" apply to partitive division as well?

TABLE 10.2.    DISTRIBUTION OF RESPONSES
TO DIVISION PROBLEMS

| Problem | Operation | % Correct (Grade) |
|---------|-----------|-------------------|
| 16 | $5 \div 15$ | 20 (5), 24 (7), 41 (9) |
| 17 | $5 \div 12$ | 14 (5), 30 (7), 40 (9) |
| 20 | $3.25 \div 5$ | 73 (5), 71 (7), 84 (9) |
| 21 | $.75 \div 5$ | 85 (5), 77 (7), 83 (9) |
| 22 | $1.25 \div 5$ | 66 (5), 74 (7), 70 (9) |

The fourth observation is that the intuitive rules do not seem to be equally robust in problem solutions. Consider, for example, Table 10.2, which shows the percentage distribution of responses to problems 16, 17, 20, 21, and 22 from Fischbein et al. (1985, p. 12). All these problems are of partitive division type and violate the same intuitive rule, that the divisor must be smaller than the dividend. Despite this uniformity, the results are strikingly different: The percentages of correct responses on Problems 16 and 17 are much lower than of those on Problems 20, 21, and 22. The explanation to this given by Fischbein et al. (1985) is that in Problems 16 and 17 the students' tendency was to reverse the roles of the numbers as a divisor and dividend. Had they done that in Problems 20 to 22, however, they would have ended up with a decimal divisor! It appears that, faced with having to cope with a violation of the partitive model's rules, the pupils chose instead not to reverse the numbers (p. 13). From this result we concluded that different intuitive rules within the partitive model may not be equally strong in affecting students' solution of partitive division problems: In Problems 20, 21, and 22 the children preferred to cope with the violation of the rule that the divisor must be smaller than dividend than with the violation of the rule that the divisor must be a whole number. Therefore the question of how different rule violations affect differently the choice of operation for solving multiplication and division problems needs to be extended to other intuitive rules.

In the rest of this chapter, we report on the instrument we developed that controls six confounding variables, used in this study with preservice and inservice elementary school teachers. Following this we discuss findings about the constraints imposed by the intuitive rules on the problem operators—multiplier and divisor—and the relative robustness of these constraints. We conclude with a summary and questions for further research.

## METHOD

### Subjects

The subjects who participated in this research were 167 in-service fourth–sixth grade teachers, 148 senior preservice teachers enrolled in a methods course for the teaching of elementary school mathematics (Group S), and 145 junior preservice teachers enrolled in a required sophomore-level content course in mathematics designed for preservice elementary school teachers (Group J). Both groups of students were declared majors in elementary education. The mathematical prerequisite for the methods course is the sophomore content course. For the content course the mathematical prerequisites are one year each of high school algebra and geometry.

### Instrument

As has been indicated earlier, the instrument we developed for this study controls a wide range of confounding variables: number type, text, structure, context, syntax, and rule violation.

*Number Type.* This variable concerns the type of numbers representing the multipliers, multiplicands, divisors, and dividends; these were systematically varied across whole number, decimal greater than 1, and decimal less than 1.

*Text.* Two types of textual problems were included: mapping rule and multiplicative compare (à la Nesher, 1988). Mapping-rule problems are those that involve the phrase *for each*, such as, "For each child there are five bags of candies, there are four children, how many bags of candies are there altogether?"; multiplicative-compare problems are those that involve the phrase *times as many as*, such as, "Tom has five times as many candies as John, John has four candies, how many candies does Tom have?" For more on the conceptual differences between these two types of problems, see Harel, Post, and Behr (1988).

*Structure.* The third variable differentiates multiplicative problems according to an interpretation of their semantic structure: multiplication, partitive division, and quotitive division. Three types of mapping-rule problems were included: the *multiplication mapping-rule* type (e.g., "There are five shelves in Dan's room; Dan put eight books on each shelf; how many books are there in his room?") and the two division mapping-rule problems corresponding to it (e.g., "There are forty books in the room, and five shelves; how many books are there on each shelf if each shelf has the same number of books on it?" and "There are forty books in the room; eight books on each shelf; how many shelves are there?"). Three types of multiplicative-compare problems were also included: the multiplication multiplicative-compare problem (e.g., "Dan has twelve marbles; Ruth has six times as many marbles as Dan has; how many marbles does Ruth have?") and the two division multiplicative compare corresponding to it (e.g., "Ruth has seventy-two marbles; Ruth has six times as many marbles as Dan has; how many marbles does Dan have?" and "Ruth has seventy-two marbles; Dan has twelve marbles; how many times as many as Dan does Ruth have?").

*Context.* All problems used dealt with the familiar context of consumption; examples include the following problems: "Each pound of snow peas costs $3.00. Father buys 2.89 pounds of them. How many dollars does Father spend on snow peas?" and "Each child gets 24 ounces of milk. There are seven children. How many ounces of milk are needed?"

*Syntax.* This variable refers to the wording structure of the problem. More specifically, the syntactical structure of the problems used was controlled with regard to the number of the statements in the problem description, the location of the unknown quantity with respect to the other given quantities, and the coordination of units of measures (for more details see Harel and Behr, 1989; Harel et al., 1988; Conner,

Harel, and Behr, 1988). All problems consisted of three statements: The first two were information statements and the third, a question statement. The coordination of the units of measures of the quantities involved was fixed across each one of problem types used.

*Rule Violation.* We classified multiplicative problems into eleven categories according to all possible rule violations described in Table 10.1. The instrument includes problems from each of these categories. With multiplication, there are three categories of problems:

- M(0) consists of problems that violate *none* of the intuitive rules in Table 10.1.
- M(1.1) consists of problems that violate only the rule that the multiplier must be a whole number. The multiplier in these problems must be a decimal greater than 1.
- M(1.2, 2) consists of problems that violate the two intuitive rules for multiplication: multiplier must be a whole number and multiplication makes bigger (or product must be larger than multiplicand); therefore the multiplier in these problems is necessarily smaller than 1.

With partitive division there are six categories:

- P(0) consists of problems that conform to all the intuitive partitive division rules in Table 10.1.
- P(1.1) consists of problems that violate *only* one rule: divisor must be a whole number. In these problems, therefore, the divisor is necessarily a non-whole number greater than 1.
- P(2) consists of problems that violate *only* the rule that the divisor must be smaller than the dividend.
- P(1.1, 2) consists of problems that violate *exactly* the two rules that the divisior must be a whole number and the divisor must be smaller than the dividend. The divisor in these problems is necessarily a non-whole number greater than 1 (to yet conform to quotient < dividend) and greater than the dividend; that is, this category is the intersection of Categories P(1.1) and P(2).

- P(1.2, 3) consists of problems that violate *only* the two rules that the divisor must be a whole number and that division makes smaller. Consequently, the divisor in these problems is necessarily a number smaller than 1 and smaller than the dividend.
- P(1.2, 2, 3) consists of problems that violate the three partitive rules in Table 10.1. As a result, the divisor in these problems is a number smaller than 1 and bigger than the dividend; this category is the intersection of the Categories P(1.2, 3) and P(2).

With quotitive division there are two categories:

- Q(0) consists of problems that conform to the intuitive quotitive division rule in Table 10.1.
- Q(1) consists of problems that violate the only rule for quotitive division, that the divisor must be greater than the dividend.

A summary of this classification of problems according to rule violations is given in Table 10.3.

## RESULTS AND DISCUSSION

Let us recall the main findings from studies investigating Fischbein et al.'s theory:

1. The relative difficulty of multiplication word problems is affected by the type of the multiplier, and the index for this relative difficulty is a multiplier represented by the number 1: Problems with a whole-number multiplier were easier than problems with a decimal multiplier larger than 1; and the latter were significantly easier than those with a (positive) decimal multiplier smaller than 1.

2. The multiplicand has no impact on problem difficulty.

3. The relative difficulty of partitive division word problems is affected by the type of the divisor and its order relation by magnitude to the dividend: Problems with a whole-number divisor smaller than the dividend yielded the highest level of performance; problems with a divisor greater than the dividend resulted in a decrease in the percentage of correct responses; and problems with a nonintegral decimal divisor

TABLE 10.3. CLASSIFICATION OF PROBLEM QUANTITIES ACCORDING TO RULE VIOLATIONS

| Multiplicative situation | Intuitive rule | Type of problem quantities violating the corresponding intuitive rule | Categories used | Example of of operation |
|---|---|---|---|---|
| Multiplication | 1. Multiplier must be a whole number | 1.1. Multiplier is a non-whole number greater than 1 | M(1.1) | 16.5 × 3 |
| | 2. Multiplication makes bigger | 1.2. Multiplier is a non-whole number smaller than 1 Product is smaller than multiplicand | M(1.2, 2) | 0.34 × 1.25 |
| Partitive division | 1. Divisor must be a whole number greater than 1 | 1.1. Divisor is a non-whole number | P(1.1) | 11 ÷ 2.53 |
| | 2. Divisor must be smaller than dividend | 1.2. Divisor is a non-whole number smaller than 1 | P(2) | 3 ÷ 5 |
| | | Divisor is greater than dividend | P(1.1, 2) | 12 ÷ 24.67 |
| | 3. Division makes smaller | | P(1.2, 3) | 6 ÷ 0.67 |
| | | Quotient is bigger than dividend | P(1.2, 2, 3) | 0.35 ÷ 0.79 |
| Quotitive division | 1. Divisor must be smaller than dividend | 1. Divisor is greater than dividend | Q(1) | 83 ÷ 193 |

smaller than the dividend resulted in an even lower level of performance.

4. The relative difficulty of quotitive division word problems is affected by the order relation between the magnitudes of the divisor and the dividend: Problems with a divisor smaller than the dividend yielded a higher performance than those with a divisor greater than the dividend.

Some of our data are consistent with these findings and some do not fully agree with it; other data refine and even extend the observations about the previously documented cognitive obstacles to the solution of multiplicative problems. These are discussed in the following.

*Multiplication.* Table 10.4 shows the percentage distribution of responses to three categories of multiplication problems by the three samples of subjects who participated in this study: junior preservice teachers (Group J), senior preservice teachers (Group S), and inservice teachers. As can be seen, the percentage of correct responses on problems with a whole number multiplier (thus, conforming to Fischbein's multiplication model) is very high. On problems whose multiplier is a decimal greater than 1 (and therefore violating the rule that the multiplier must be a whole number) it drops slightly by an average of about 13 percent; and on problems whose multiplier is a decimal smaller than 1 (violating the exact same rule and the rule that the multiplication makes bigger) it drops by an average of about 41 percent. These results indicate a moderate effect on the level of difficulty from changing the type of multiplier from a whole number to a decimal greater than 1, but a great impact on the level of difficulty when the multiplier changes from a whole number or a decimal greater than 1 to a decimal smaller than 1. These results are consistent with results obtained in other studies, and therefore they support Fischbein et al.'s intuitive models for multiplication, the repeated addition model.

The percentage distribution of correct responses to the partitive division and quotitive division problems is included in Table 10.5. These results are consistent with the constraints of the models associated with these types of problems as were suggested by Fischbein et al. (1985). This can be seen by comparing the percentage of correct responses on P(0) prob-

TABLE 10.4. DISTRIBUTION OF RESPONSES TO MULTIPLICATION PROBLEMS

| Type of Multiplier | Type of Multiplicand | Example of operation | % Correct Responses (N) | | |
| --- | --- | --- | --- | --- | --- |
| | | | Preservice | | Inservice |
| | | | Group J | Group S | |
| Whole number | Whole number | 14 × 8 | 96.5 (146) | 94.5 (60) | 98 (22) |
| | Decimal > 1 | 24 × 2.28 | 93 (150) | 94.5 (60) | 100 (22) |
| | Decimal < 1 | 15 × 0.38 | 95 (146) | 97 (68) | 99 (132) |
| | | Mean | 94.8 | 95.3 | 99 |
| Decimal > 1 | Whole number | 16.5 × 3 | 82 (146) | 80.5 (67) | 91 (22) |
| | Decimal > 1 | 43.61 × 2.37 | 74 (145) | 75.5 (72) | 87 (132) |
| | Decimal < 1 | 2.5 × 0.75 | 79 (150) | 89 (60) | 96 (22) |
| | | Mean | 78.3 | 81.6 | 91.3 |
| Decimal < 1 | Whole number | 0.45 × 150 | 69 (146) | 78 (67) | 77 (22) |
| | Decimal > 1 | 0.34 × 1.25 | 45 (150) | 50 (60) | 55 (22) |
| | Decimal < 1 | 0.75 × 0.6 | 45 (146) | 41 (61) | 40 (22) |
| | | Mean | 53 | 56.3 | 57.3 |

TABLE 10.5. DISTRIBUTION OF RESPONSES

| Category | Rule(s) violated | Example of operation | Correct responses, % (N) | | |
|---|---|---|---|---|---|
| | | | Preservice | | Inservice |
| | | | Group J | Group S | |
| M(0) | No rule violation | $24 \times 2.28$ | 93.6 (148) | 95 (60) | 99 (44) |
| M(1.1) | Multiplier must be a whole number | $16.5 \times 3$ | 78.3 (137) | 79.2 (60) | 90.4 (55) |
| M(1.1, 2) | Multiplier must be a whole number and Multiplication makes bigger | $0.75 \times 0.62$ | 51 (147) | 55.8 (60) | 58 (22) |
| P(0) | No rule violation | $68 \div 17$ | 86.5 (103) | 91.5 (96) | 94.5 (22) |
| P(1.1) | Divisor must be a whole number | $11 \div 2.53$ | 32.5 (62) | 36 (94) | 49 (77) |
| P(2) | Dividend must be greater than divisor | $3 \div 5$ | 64.3 (88) | 69.3 (107) | 78.5 (88) |
| P(1.1, 2) | Divisor must be a whole number and Dividend must be greater than divisor | $12 \div 24.67$ | 19.3 (88) | 22 (103) | 31 (44) |
| P(1.2, 3) | Divisor must be a whole number and Quotient must be greater than dividend | $6 \div 0.67$ | 44 (145) | 46.3 (67) | 52 (92) |
| P(1.1, 2, 3) | Divisor must be a whole number and Dividend must be greater than divisor and Quotient must be greater than dividend | $0.35 \div 0.79$ | 33 (146) | 37 (67) | 44 (131) |
| g(0) | No rule violation | $175 \div 35$ | 90.6 (145) | 91.5 (83) | 95.1 (88) |
| g(1) | Dividend must be greater than divisor | $83 \div 193$ | 38.4 (145) | 35.2 (89) | 63.7 (39) |

lems (those that conform to the partitive division model) to P(1.1), P(2), P(1.1, 2), P(1.2, 3), and P(1.2, 2, 3) problems (those that violate some or all of the partitive division rules), and the percentage of correct responses on Q(0) problems (those that conform to the quotitive division model) to Q(1) (those that violate the rule that the dividend must be greater than the divisor): The performance on P(0) problems is significantly higher than the performance on each of the other partitive division categories, except P(2); and the performance on Q(0) problems is significantly higher than the performance on Q(1) problems.

These findings are consistent with other studies' findings; therefore they support Fischbein et al.'s argument that intuitive models, which are not necessarily in accord with the operations of multiplication and division, govern subjects' solutions of multiplicative problems. However, as is discussed later, a further look at our data leads to new observations.

*The Impact of the Operators, Multiplier and Divisor.* Recall Fischbein et al.'s "absorption effect" notion that they suggested to explain the conceptual basis for why multiplicative problems with "small" (e.g., 0.65 and 1.25) decimal operators (i.e., multipliers and divisors) are more difficult than those with "big" (e.g., 3.25) decimal operators: "We conjectured that when the whole part of a decimal is clearly larger than the fractional part, the pupil may treat it more like a whole number (as though the whole part 'masks' or 'absorbs' the fractional part). . . . Although the decimal operator still appears as a source of difficulty, one can see that compared to decimals like 0.75, 0.65, or 1.25, a decimal like 3.25 has a slighter counterintuitive effect when playing the role of operator" (p. 11). From this explanation it is expected that an increase in the value of a decimal operator (multiplier or divisor) greater than 1 will increase the effect of masking the decimal part of the operator by its whole number part, which, in turn, will decrease problem difficulty. As can be seen in Table 10.6, our data on subjects' performance on multiplication problems do *not* dovetail with this pattern: An increase in the whole part of the multiplier has not led to a decrease in problem difficulty. Further, the data in Table 10.5 suggest

TABLE 10.6.   DISTRIBUTION OF RESPONSES
TO MULTIPLICATION PROBLEMS WITH DECIMAL
MULTIPLIERS GREATER THAN 1

| | | % Correct responses (N) | | |
| | | Preservice | | |
| Multiplier | Multiplicand | Group J | Group S | Inservice |
|------------|--------------|---------|---------|-----------|
| 1.45 | 2.86 | 71(93) | — | 87(23) |
| 2.5 | 0.75 | 87(150) | 90(60) | 96(22) |
| 2.89 | 3 | 88(146) | 89(62) | 91(22) |
| 10.5 | 18.25 | 94(145) | 93(72) | 100(22) |
| 16.5 | 3 | 76(146) | 79(67) | 91(22) |
| 43.61 | 2.37 | 54(146) | 58(72) | 74(132) |

that the "absorption effect" does not apply to partitive division problems. This can be seen by comparing the performances of problems whose divisor is a decimal greater than 1 (P(1.1)) and problems whose divisor is a decimal smaller than 1 (P(1.2, 3)). In both cases the level of performance is relatively low compared to the level of performance on P(0) problems (those which conform to Fischbein et al.'s partitive model).

Further, note that the performance on the division problem with the divisor 24.67 ((P1.1,2)) is very low (less than 30 percent; see Table 10.5), even though its whole part is relatively large. One might argue that this low result is attributable to the fact that this problem violates another rule, that the divisor must be smaller than dividend. Although it is likely that this rule violation has, to some extent, affected subjects' performance, we do not think that its effect is that strong. This is supported by the fact that the performance on problems that violate only this rule (i.e., P(2)) is relatively high (Table 10.5).

*Levels of Robustness.* In Table 10.5, there are two conspicuous results concerning how the intuitive rules effect differently the solution performance: First, the performance on M(1.1) problems (Table 10.5) that violate only the rule that the multiplier must be a whole number is higher than the performance on M(1.2, 2) problems that violate two rules:

that rule and the rule that multiplication makes bigger. Second, the performance on P(2) problems (those that violate only the rule that the dividend must be greater than the divisors) is higher than the performance on the other categories of partitive division problems that violate one or a combination of intuitive rules. The latter result indicates that, when the divisor is a whole number, the rule that the dividend must be greater than the divisor is the least robust in the solution of partitive division among other combinations of the intuitive rules. This is not surprising, because problems with this rule violation (i.e., a number of objects, $x$, is to be equally divided into a whole number of sets, $y$, where $x <$ $y$) are quite common in everyday situations. On the other hand, problems that violate the rule that the divisor must be greater than the dividend and, in addition, violate the rule that the divisor must be a whole number (i.e., P(1.1, 2)) scored the lowest level of performance.

In looking at the subjects' solutions to problems in Category P(1.1), we found that almost all subjects who did not solve these types of problems correctly offered an inverse expression to the correct division expression (for example, if the solution of the problem was $11 \div 2.53$, the incorrect solution was $2.53 \div 11$). Many more subjects chose the inverse expression (mean = 54.75) than solved the problem correctly (mean = 32.5). Note that, although the correct expression violates the rule that the divisor must be a whole number, the inverse expression violates the rule that the divisor must be smaller than the dividend. We interpret this result to indicate that the former rule is more robust that the latter one.

*The Impact of the Multiplicand.* Looking back at Table 10.4, one can see that our data do not fully dovetail with other studies that have observed no impact of the multiplicand. Indeed, consistent with other studies, Table 10.4 shows that, as long as the multiplier is a whole number or a decimal greater than 1, the type of the multiplicand as a whole number, decimal greater than 1, or decimal smaller than 1, has almost no effect on the problem's difficulty. On the other hand, when the multiplier is smaller than 1, our data show that the type of the multiplicand seems to have some effect: The percentage of correct responses on problems with a whole-number multiplicand is much higher than that on problems with decimal multiplicand.

This observation could, of course, be a result of a perturbation or noise in our data. However, assuming this is not so, we speculate the following explanation: The presence of a whole-number quantity in the problem helps subjects to sort out the role of the quantities involved, which, in turn, enables them to choose a correct operation for solving the problem.

## SUMMARY

In this chapter, we dealt with several aspects of the impact of the number type on the relative difficulty of multiplicative problems. We reexamined the findings from other studies concerning this impact, investigated the "absorption effect" notion suggested by Fischbein et al. to account for differences in subjects' performance on multiplicative problems with different non-whole-number operators and probed into the level of robustness of the intuitive rules derived from Fischbein et al.'s models. The observations and findings reported in this chapter are summarized as follows:

1. The instruments used in Fischbein et al. and the studies that followed it do not control for many variables known to be influential in problem solution. We offered a framework for an instrument that controls for a wide range of confounding variables: number type, text, structure, context, syntax, and rule violation.

2. Our data is consistent with the finding that subjects' model for multiplication is the repeated addition model, and for division, subjects' models are partitive division and quotitive division.

3. Our data do *not* support Fischbein et al.'s notion of the "absorption effect": No significant difference in performance was found between multiplication problems with multipliers whose whole part is relatively large and those with multipliers whose whole part is relatively small. Moreover, the absorption effect does not apply to division problems.

4. Evidence was shown for differential robustness of the intuitive rules associated with Fischbein et al.'s models.

5. The type of multiplicand seems to have an impact on problem solution when the multiplier is smaller than 1.

## FURTHER RESEARCH QUESTIONS

This research has raised several questions for further investigations. First, this and other studies focused on one type of problem quantities: decimal numbers. The question of whether subjects encounter similar difficulties with multiplication and division problems that involve fractions has never been directly addressed. There is a reason to believe, however, that fractions and decimals do not have the same effect on the solution of multiplication and division problems. More specifically, it seems easier to solve multiplication and division problems in which the operator (i.e., the multiplier or divisor) is a fraction than in those in which the operator is a decimal. A rationale for this is based on the fact the naming rule of fractions is different from the naming rule of decimals: Under these naming rules, it is easier to identify the role of a problem quantity as an operator or operand if the quantity is a fraction than if it is a decimal; therefore it is easier to recognize its relation to other problem quantities. For example, the two propositions in the statement, "*John had 5 ounces of ice cream* and *he ate x of the amount he had*" are easier to connect if (the operator) $x$ is a fraction, say 2/5, than if $x$ is a decimal, say 0.40.

Second, a further distinction among the intuitive rules derived from Fischbein et al.'s model is that some of the rules are associated with the *problem information*, others with the *problem solution*. In multiplication, the rule that the multiplier must be a whole number imposes a constraint on the type of multiplier provided in the problem information; in contrast, the rule that the multiplication makes bigger restricts the problem solution to be a number greater than the multiplicand. Similarly, in partitive division, the rules that the divisor must be a whole number and the divisor must be smaller than dividend are problem information rules, whereas the rule that the quotient must be greater than dividend is a problem solution rule. Finally, in quotitive division, the rule that the divisor must be smaller than dividend is a problem information rule; no problem solution rule is involved. This raises the question of whether problem information rules are equally robust as the problem solution rules.

Finally, when we looked at the other studies' data on multiplication and division, we raised the question, Why are problems with a multiplier greater than 1 relatively easy for the subjects despite the fact that they are in conflict with the model of multiplication as a repeated addition? If indeed this model governs subjects solution of multiplication problems, it is not at all clear why the intuitive rule derived from it—that the multiplier must be a whole number—is substantially less robust in the case of a non-whole-number multiplier greater than 1 than in the case of a multiplier smaller than 1. Further, it is not all clear what is the conceptual basis for the multiplier 1 being an index for the relative difficulty of multiplication problems. In fact, Fischbein et al. in their explanation to the observation that the intuitive rule that the multiplier must be a whole number does not equally affect multiplication problems with decimal multipliers, did not differentiate between multipliers greater than 1 and those less than 1. Rather, they suggested the "absorption effect" notion, which differentiates between multiplicative problems according to the size relationship between the whole number part and the fractional part of their multipliers. In this chapter we reported data that are *not* consistent with this explanation; therefore, further theoretical and empirical investigations are needed to answer these questions.

## NOTE

This research was funded in part by grant #CRG.890977 from NATO.

## REFERENCES

Bell, A., E. Fischbein and B. Greer. 1984. Choice of operation in verbal arithmetic problems: The effect of number size, problem structure and context. *Educational Studies in Mathematics* 15: 129–147.

Bell, A., M. Swan, and G. Taylor. 1981. Choice of operations in verbal problems with decimal numbers. *Educational Studies in Mathematics* 12: 399–420.

Conner, G., G. Harel, and M. Behr. 1988. The effect of structural variables on the level of difficulty of missing value proportion problems. Proceedings of the tenth PME-NA annual meeting, DeKalb, IL.

Fischbein, E., M. Deri, M. Nello, and M. Marino. 1985. The role of implicit models in solving verbal problems in multiplication and division. *Journal of Research in Mathematics Education* 16: 3–17.

De Corte, E., L. Verschaffel, and V. Van Coillie. 1988. Influence of number size, problem structure and response mode on children's solutions of multiplication word problems. *Journal of Mathematical Behavior*, 7: 197–216.

Goldin, G. A. 1986. Review of A. Bell, E. Fischbein, and B. Greer: Choice of operation in verbal arithmetic problems: The effects of number size, problem structure and context. *Investigations in Mathematics Education* 19 (1): 1–7.

Greaber, A., D. Tirosh, and R. Glover. 1989. Preservice teachers' misconception in solving verbal problems in multiplication and division. *Journal of Research in Mathematics Education* 20: 95–102.

Greer, B. 1985. Understanding of arithmetical operations as models of situations. In *Cognitive processes in mathematics*, ed. J. Sloboda and D. Rogers. London, Oxford University Press.

———. 1988. Nonconservation of multiplication and division: Analysis of a symptom. *Journal of Mathematical Behavior* 7: 281–298.

Greer, B., and C. Mangan. 1984. Understanding multiplication and division. In *Proceedings of the Wisconsin meeting of the PME-NA*, ed. T. Carpenter and J. Moser, 27–32. Madison: University of Wisconsin Press.

Harel, G., and M. Behr. 1989. Structure and hierarchy of missing value proportion problems and their representations. *Journal of Mathematical Behavior*.

Harel, G., T. Post, and M. Behr. 1988. On the textual and semantic structure of mapping rule and multiplicative compare problems. Proceedings of the tenth international conference of PME, Budapest.

Harel, G., M. Behr, T. Post, and R. Lesh. in press. Teachers' solution strategies of multiplicative problems. *Journal of Mathematical Behavior*.

Hart, K. M. 1984. *Ratio: Children's strategies and errors in secondary mathematics project*. Windsor, Berkshire: NFER-NELSON.

Lester, F. K., and P. Kloosterman. 1985. Review of E. Fischbein, M. Deri, M. Nello, and M. Marino: The role of implicit models in solving verbal problems in multiplication and division. *Investigations in Mathematics Education* 18: 10–13.

Luke, C. 1988. The repeated addition model of multiplication and children's performance on mathematical word problems. *Journal of Mathematical Behavior* 7: 217–226.

Mangan, C. 1986. Choice of operation in multiplication and division word problems. Unpublished doctoral dissertation, Queen's University, Belfast.

Nesher, P. 1988. Multiplicative school word problems: Theoretical approaches and empirical findings. In *Number concepts and operations in the middle grades*, ed. M. Behr and J. Hiebert. Reston, VA: National Council of Teachers of Mathematics.

Vergnaud, G. 1983. Multiplicative structures. In *Acquisition of Mathematics Concepts and Processes*, ed. R. Lesh and M. Landau. New York: Academic Press.

# VI
## SUMMARY

# 11 Multiple Views of Multiplicative Structure

*Thomas E. Kieren*

In a recent paper Bereiter (1990) exhorted the educational research community to select conceptual objects of analysis that were large and complex enough to provide an adequate framework for assisting learning and teaching as it occurs in schools. He argued that research of psychologists who focused on more microscopic objects, although helpful to psychological or neuropsychological theories, did not provide an adequate foundation for complex learning of complex fields such as potentially occurs in school math. As examples of such large-scale objects, Bereiter suggested studying the "school learning modules" of children. He maintained that most children quickly build well-developed schemes for functioning in school and that learning under such schemes is about school functioning rather than constituting the important conceptualizing of the real world as suggested by Vergnaud in this volume (although I suppose one might argue that school functioning is highly practical to children in our society). Bereiter suggested that a few students develop an "intentional learning module" that allows them to learn ideas at a sophisticated level for the sake of organizing their own worlds. As important as this module would seem to be for researchers, observers, or teachers, Bereiter unfortunately is vague about its qualities and ways in which we might effectively study its growth and its sponsorship in children.

But we need not wait for a clearer definition of the intentional learner module. The growth of multiplicative structures, a study of which is the topic of the chapters in this volume, is critical for a person's conceptualizing or bringing forth the world in which he or she lives. We are fortunate that the study of multiplicative structures (and other multifaceted conceptual fields or schemes of children's mathematics) has

387

been under way in recent years, following the earlier work of Vergnaud (e.g., 1983). In his 1983 essay, Vergnaud describes the conceptual field of multiplicative structures as a network of different but interconnected concepts such as multiplication, division, fractions, ratios, rational numbers, and linear and nonlinear functions. In this volume Vergnaud sees the main problems in building multiplicative structures as consisting of building dimensional relationships in simple and multiple proportional situations and in the extension of concepts and operations of ratios and rates to ever more complex situations.

The studies of persons building multiplicative structures in this volume have been done in an attempt to provide a grounding for educational practice with respect to the complex school mathematics related to these structures in children. A major question for a reader of the chapters in this volume to ask is "What do we now know about multiplicative structures?" Beyond the problems raised by Vergnaud, what are the ways by which children build such structures? What cognitive operations underlie such structures? In what way do such multiplicative structures manifest "their own intrinsic organization which is not reducible to additive aspects" (Vergnaud, 1983, p. 128)?

The chapters in this volume look at multiplicative structures from two perspectives. One perspective looks at the formal mathematics itself, which is related to multiplicative structures. Although the content, leading to abstract algebra and analysis, can be considered abstract and austere, the studies here try to consider how such mathematics is humanly knowable; that is, what mental structures and organizations of structures are needed to build such mathematics? This approach is perhaps most vividly seen in Smith and Confrey's tracing of the historical development of mathematical operations on ratios (rather than quantities), which showed the exponential and logarithmic nature of multiplicative structures. But many of the other chapters make explicit use of and reference to mathematical analysis as well.

The second perspective is human in focus. What are humans like—what kinds of operations do they manifest, build, and organize that lead to multiplicative thinking and eventually to the interconnected conceptual network suggested by Vergnaud? To me, this approach suggests that such mental structures are not disembodied. The studies here seem to be consonant with Lakoff's (1987) argument that the mind (or here multiplicative structures) is embodied and that individual personal experiences form the basis and essence of concepts. This perspective raises the questions as to what actions, action schemes, and mental actions or operations lead to the concepts and understandings that an observer sees as a person's multiplicative structures.

Because the very concept, multiplicative structures as a conceptual field, implies that individuals build and organize their knowledge structures, the chapters in this volume take a constructivist stance, at least implicitly. Nonetheless, this central philosophical agreement does not imply that there are no differences among the authors. The most vivid difference occurs between the exponential "splitting" approach taken by Confrey and supported by the historical analysis of Smith and Confrey, and the growth to multiplicative acts based on the number sequence structure as studied by Steffe. The former approach observes multiplicative actions as independent of addition ideas, whereas the latter appears to consider at least early multiplicative acts as making natural (or provoked) use of counting-based mental structures. Although no conclusive answer is offered in this volume, it is educationally important to raise the question of how, independent from a child's additive structures, one might think of, observe, and teach for that child's multiplicative structures. If one followed Lakoff's (1987) or Johnson's (1987) argument that (mathematical) forms or ideas emerge from bodily functioning in and kinaesthetic images of the world, one might ask about the nature of the images that would be the basis for the concepts of multiplicative structures. Are there actions and images that go beyond those that support counting (e.g., entities, sequences, pluralities), which are a necessary base for multiplicative structures? Is it clear that at least some of these chapters would say "yes."

Several chapters (Thompson, Lamon, Kaput and West) rather directly address the problems associated with building multiplicative structures described by Vergnaud. Before turning to those issues it seems worthwhile to pursue the notion of *splitting* described by Confrey. Her work suggests actions and images that might be foundational to the various structures suggested by Vergnaud, but also enlarges the scope of multiplicative structures to include exponential functions. Confrey defines a *split* as "an action of creating equal parts or copies of an original . . . , an operation . . . that requires only a recognition of the type of split and of the requirement that the parts are equal." She goes on to argue that this basic action scheme provides a noncounting basis for multiplicative structures and would provide a basis for multiplication and division that would not lead to some of the intuitive rules (products of a child's school module?) such as "multiplication makes bigger" that have been noted to be held by children and adults as well (e.g., Greer, and Harel et al. in this volume).

There has been considerable research supporting one form of splitting as a basis for fractional and rational number learning. In fact, Pothier and Sawada (1983) observe the phenomenon of splits being generated from other splits in a multiplicative manner in a sharing-partitioning situation.

Is there any indication of splitting as a generator of multiplicative intuitions in a more general way? In studies of two regular classes ($n$ = 24, 26) of 8-year-olds, I have been able to observe effects of one of the splitting actions—folding. To facilitate the experience we started with newsprint-sized sheets as our units, which allowed for a number of successive folds. It was found that all students could predict the number of pieces resulting from successive "half-folds"— e.g., after four folds there were sixteen subparts and all children could predict the number of parts after five folds even without carrying out the action. Further, all but three students in one class and four in the other could accurately describe the relationship between the sizes of parts after various numbers of foldings; that is they could compare the size of fourths (the product of two folds) with the result of two more folds. It appears that the act of folding (in two equal parts) and the resulting folded unit provide kinaesthetic and visual images that can serve as a basis for multiplicative intuition.

A later activity with these students had them folding a meter-long strip of paper using "half" folds. They were to record the results of their actions in a table with the following four headings: folds, bends, chunks, and chunksize. All but two children in one class and three in the other completed the table for one to five folds; many did so without bothering with the actions. One child, Joe, accurately found the number of chunks for fifteen folds. Another child, Kris, asked him, "How did you know how to do that?" Joe replied, "It's easy. The number of folds just tells you how many times you have to multiply two over and over." A gifted child, Sandy, generated the number of parts or chunks and actually related these numbers to exponential notation. He was able to associate no folds with $2^0$ as well as with a result of one piece. Therefore, Sandy could easily see that exponential notation (a symbol system he picked up outside of class) could characterize a splitting environment. Further, he was able to construct "exponential" properties in a splitting environment.

A third activity with one of the classes led to folding into three equal parts. Being able to successfully do this fold physically led all children to create what Confrey would call six splits, twelve splits, and other $3*2^n$ splits. Twelve of the twenty-six generated at least two more powers of three splits, and one child, Sandy, imaginatively generated fractions involving three and ten splits. "I've noticed an interesting fraction—$1/270$."

The purpose of recounting these episodes is not to report research in detail. These observations were meant only to support Confrey's hypothesis that even young children can use splitting actions as a basis for intuitions of multiplicative structures—in this case, fractions, fractional multiplication, and exponential functions (or at least geometric series).

In posing "splitting" as an action scheme basic to the building of solid multiplicative structures, Confrey argues against typical teaching approaches as relying too exclusively on repeated addition and its underlying basis in counting. What of this claim? Is her claim a challenge to the basis of, or extent of, some other work reported in this volume? To resolve the conflict it is useful to look at the historical analysis done by Smith and Confrey. In looking at Oresme's notion of ratio (and other historical analysis as well), Smith and Confrey focus on the *action as creating the mathematical object* and, in fact, focus on *actions on actions*. In looking at the early number and part-part whole addition research, one might think of additive structures as generative of a calculus of the organization of *objects* whereas multiplicative structures generate a calculus of organization of actions. In fact,

Confrey's splitting analysis seems to point to actions on actions as opposed to number scheme research (e.g., Steffe, this volume), which points to units of units.

Confrey points to actions (e.g., joining, annexing) that she claims are used to support additive approaches to multiplication and would obviously be related to a "repeated addition" approach. As we look at other research, do we see "actions" that could be related to counting but that point beyond "repeated addition"? One such action scheme is "iterating." Steffe distinguishes iterative multiplying or iteration from the usual interpretation of 6 * 4 as six groups of four things. This latter interpretation can be seen as a basis for repeated addition and can support the product, twenty-four, as simply a count. In iterative multiplying, the 6 indicates six iterations of an iterable unit, 4 (as a chunk of identified quantity). As an example, consider the following (real) situation. A father was watching a horse jumping show with his two daughters. One was 6-years-old and the other was 8-years-old. The older daughter had studied multiplication in school. It was observed that if a horse knocked down a fence, it was assessed 4 penalty points. The father asked, "Suppose a horse knocked down six fences, how many penalty points?" Both girls giggled and said that the horse and rider would just stop, but soon after that each child independently said, "24." When asked how they knew, the younger child said she counted with fours. The older one did not model the situation with 6 * 4 (a fact she knew), but said she added fours. Although both girls understood the "reality" of the situation and could envision (but not see) six fences, this product was clearly not six sets of 4 for them. Both girls appeared to iterate a unit (4 penalties per knocked-down fence) over six fences. Steffe would see this action as beyond counting and simple adding, although those might have been used in the actual computation of the result. Iterating is a primitive form of carrying or *distributing* one quantity across another. The 4-units here are not six different sets of 4. The child in effect mentally created replicates of a 4 points per 1 fence unit.

Of course, this action scheme does not lead to a complete mature scheme of multiplication in and of itself. But such a simple action scheme can also underlie the solution to simple proportion and simple linear function problems as seen in Vergnaud's examples. It is also related to the build-up and build-down schemes described by Kaput and West in which children create and *replicate* a two-part correspondence or unit. The iterable unit noted by Steffe and the one just illustrated are particularly simple examples of such a two-part unit. In these units, the number associated with the second part is 1. Of course, both Vergnaud and Kaput and West would argue that such particular iterable units or ratios are not sufficient for proportional reasoning because they are too situation bound. That is, if the numbers in a problem do not "match" replicas of such a particular unit, then the child using such a simple action scheme would be confused. Yet Kaput and West argue and show in the manipulation of computer-generated linked quantity mats, that one can grow from using such iterating actions on particular ratios to generate and use the multiplicatively more general rate scheme.

The iterating and build-up schemes just described have several features that are key to their nature. The first is the seeing of a set or a pair as a unit. The second is the distribution of this quantity across another. And finally these schemes have a recursive nature. The replicating act is applied to the result of a previous replicating act. Thus, the 4 in the 6 * 4 example, becomes "4 more than" applied to results of itself. As with the splitting scheme, it can be argued that iterating and building-up schemes have bases that are embodied even more in such everyday acts as linking and replicating (Lakoff, 1987).

Both the splitting and especially the iterating and building-up schemes focus on the notion of units. Thompson makes a convincing argument that units such as those just described are the results of individual persons' actions in reconstituting objects of experience so that they have qualities an observer sees as mathematical. He claims that the distinctions made in various recent "definitions" of *ratio* and *rate* come from analyzing situations and their features, rather than from observing the mental operations of people who constitute or construct the rate and ratio situations. Although it is conventional in schools to think of speed as a quotient of distance and time, Thompson shows that a child's unit of speed can be very different from that situation based definition and arises out of the child's mental operations.

Thompson observed (and created situations calculated

to trigger changes in) a child's different and growing understanding of speed in a very clever computer-simulation situation. This child at first constructed speed of motion as distance happening over time with an explicit emphasis on the distance. With experience in this situation with pairs of objects traveling at varying rates, this child reconstructed this limited idea of speed until it took on the quality of a multiplicative quantitative relationship between distance and time. The critical point that Thompson makes, which is similar to that made less explicitly by Kaput and West, Steffe and Confrey, is that the child's proportional problem solving strategy was based on her construct of rate or ratio and not on the situation or question facing her. More important, the child can change and recapitulate this construction in the face of appropriate experience. It would be a general conclusion from research in this volume that one does not determine the nature of a child's operations by looking at whether or not a child successfully completes a sequence of questions. The observed nature of the operations of a child comes from analyzing what the child does and how this changes in a situation.

## CONCLUDING REMARKS

### Theoretical Conclusions

A major thrust of the work in this book has been to identify action schemes and mental operations that underlie and constitute multiplicative structures in children. Some of these, the linked units developed by Kaput and West, the unitizing of Lamon, and the analysis of the change in the nature of a child's concept of speed by Thompson elaborate Vergnaud's idea of dimensional relationships. Other action schemes such as iterating, building up, distributing one quantity across another, or norming (Lamon) to allow for computation with the quantities associated with multiplicative structures, such as rational numbers, appear to point to the relationship between number sequence structures, additive structures, and multiplicative structures. Confrey, in talking about splitting, seems to challenge the role of the number-related schemes in building up multiplicative structures. In the larger scheme of mathematical thinking, multiplicative acts and related mathematical objects can be seen in analytic-functional, as well as algebraic, contexts. In resolving the apparent conflict posed by Confrey it might be thought that the splitting scheme contributes to the "analytic-functional" side of the conceptual field whereas the others more fundamentally relate to the "algebraic" aspect of multiplicative structures.

One could see the conceptual units analysis of Behr et al. as a way of addressing at least in part the connection between the splitting-analytical and the distributive-algebraic aspects of multiplicative structures. To do this, the units analysis, well-illustrated in the chapter by Behr et al. to theoretically illuminate the distributive aspect of multiplicative structure, would have to be shown as pertinent as well to the splitting circumstances, especially those such as fractions when a unit is split by action rather than measurement.

Of course, one significance of the units conceptual analysis is to point out the number of different kinds of units that manifest themselves in multiplicative situations. Consider the following activities in which 8-year-olds engaged to build fraction ideas:

1. The children folded units in half successively to generate halves, fourths, eighths, and so on.
2. The children were given a "half-fractions" kit that contained two units worth each of halves, fourths, eighths, and sixteenths.

In the folding setting, children came to use fraction language to reflect amounts related to action; thus "five-eighths" referred to the actions of folding a unit in half three times and indicating five of these eight pieces. Five-eighths here was not a quantity independent of actions. It was observed that a large majority of the class who first functioned capably in the folding setting were not able to automatically use language such as "five-eighths" to signal actions for themselves in the "fraction kits setting." If one asks why, the units conceptual analysis provides one answer. In the folding case, an eighth referred to a product of their own action on a unit: one-eighth of a one unit. Five-eighths was $5/8$ of a one unit. In the second case, although the parts looked the same to an adult, they were different for the child. Here five-eighths referred to five one-eighth units. Thus, the conceptual analysis would predict that conceptual transformation and enlargement would be required for a child to understand the relationship between units in these environments; that is, that five-eighths of a 1-unit could become five one-eighths units. Although by no means offered as a validation, this evidence points to the possible role of units of quantity and their transformation as a key and unifying activity in a child building a multiplicative structure. It also points to the need for curriculum builders and teachers to be aware of units and their transformation as they build activities for and observe children.

## Practical Conclusions

I would like to conclude by saying that the chapters in this volume have contributed to our understanding of multiplicative structures by elaborating some of the mental operations theorized to underlie their building, and proposing some new ones, as well as by showing how children build such structures and how such structures can change with experience. Part of the evidence showing the practical need for such new understanding comes from the chapter by Harel et al. They show that even under rigorous test conditions, adults training to be teachers can be shown to have intuitions related to multiplicative structures that hinder the growth and use of such structures. Unless one thinks that such intuitions as "multiplication makes bigger" or "one can only divide by a whole number" are wired in, one must conclude that these intuitions or schemes are based on these persons' experiences (and even may have been a result of their successful functioning in school). Therefore, it seems that the nature of such experiences must change.

It has often been the criticism of educational research that it points out the faulty products of educational activity, but gives no clear advice for improving the situation. Such is not the case with the research described here. It carries with it numerous suggestions for large-scale curricular change, with greater integration of the topics related to multiplicative structures, for more emphasis on the observation of and development of students' structures and the deemphasis of algorithmic approaches, formulas, or syntactic schemes (e.g., arithmetic of units). Specific examples are offered for helping children with the vertical expansion of their multiplicative thinking, some very well illustrating constructive uses of computer technology (Greer, Thompson, Kaput and West). These suggestions emphasize means by which children can develop more sophisticated reasoning capabilities. Greer also suggests that horizontal expansion of multiplicative structures is needed as well. That is, students must have explicit and interrelated experience with the variety of canonical situations in which multiplicative structures would be useful.

Therefore, although the research in this book is not the result of tightly coordinated efforts, it does make theoretical and practical contributions to our understanding of the conceptual field of multiplicative structures.

## ACKNOWLEDGMENT

Research for this chapter was supported in part by Social Sciences and Humanities Research Council of Canada Grant 410 900735.

## REFERENCES

Bereiter, Carl. 1990. Aspects of an educational learning theory. *Review of Educational Research* 60(4): 603–624.

Johnson, Mark. 1987. *The body in the mind: The bodily basis of meaning, imagination, and reason*. Chicago: University of Chicago Press.

Lakoff, George. 1987. *Women, fire, and dangerous things*. Chicago: University of Chicago Press.

Pothier, Yvonne, and Daiyo Sawada. 1983. Partitioning: The emergence of rational number ideas in young children. *Journal for Research in Mathematics Education* 14(4): 307–317.

Vergnaud, Gerard. 1983. Multiplicative structures. In *Acquisition of Mathematics concepts and processes*, ed. R. Lesh and M. Landau, 127–174. Orlando, FL: Academic Press.

# NAME INDEX

# SUBJECT INDEX